BLACKBODY RADIATION

A HISTORY OF THERMAL RADIATION COMPUTATIONAL AIDS AND NUMERICAL METHODS

Optical Sciences and Applications of Light

Series Editor
James C. Wyant
University of Arizona

BLACKBODY RADIATION

A HISTORY OF THERMAL RADIATION COMPUTATIONAL AIDS AND NUMERICAL METHODS

Seán M. Stewart • R. Barry Johnson

CRC Press
Taylor & Francis Group
Boca Raton London New York

CRC Press is an imprint of the
Taylor & Francis Group, an **informa** business

Dedication

To those scientists and engineers, immortalized in this work, whose tireless investigations into blackbody radiation created the beginnings of quantum physics;
and
to Donald and Cathy Stewart, loving parents,
and Marianne F. Johnson, loving wife,
for their unceasing support and encouragement.

*A computer who must make many difficult calculations usually
has a book of tables... or a slide rule, close at hand.*
Maurice L. Hartung writing in 1960 in the preface
to "How to Use Log Log Slide Rules."

The spectral density of black body radiation ... represents
something absolute, and since the search for the absolutes
has always appeared to me to be the highest form of research,
I applied myself vigorously to its solution.
-Max Planck
In Michael Dudley Sturge, *Statistical and Thermal Physics* (2003), 201.

Contents

Preface

It is often said mathematics is the queen of all the sciences. Computation, then, is its most loyal and humble servant. Computation has always been an integral part of science and until quite recently was the sole domain of the human computer. As anyone who has ever been required to do so can confirm, computation by hand is a slow and error prone task. Not surprisingly, much effort went into the creation of a number of computational strategies and mechanical aids designed to relieve the burden imposed by such work. Before the advent of the digital computer the speed and ability with which calculations could be made often placed very real human limits on the types of problems one could realistically hope to solve. Nowadays, the swiftness and ease with which calculations can be made makes it all too easy to forget a time, not that long ago, when even relatively simple calculations were a minor undertaking. Who today pays much attention to how the calculations we seek are made, as long as they can be made? Fast is all that matters.

Until computers became affordable and widely available in the mid-1970s, escape from the computational mire for many scientists and engineers came in the form of a number of surprisingly simple and unassuming aids. The oldest and perhaps most well-known examples are those of tables. One conservative estimate puts tables as having been with us for some 4500 years, and having been used as the main aid to calculation for at least the last two millennia. They were, however, not the only aids developed and used. In addition to tables, by the late nineteenth century graphical aids such as nomograms and graphs, and mechanical aids such as slide rules and calculating machines, had become available and were beginning to play an important role in aiding computation. Collectively their importance lay in the substantial time savings that were to be had from their use. No longer burdened by having to perform many tedious and time consuming calculations, it allowed scientists and engineers to focus their time and attention on the real problem at hand.

At its heart, this book is about computation. It focuses on those methods used for one very specific, yet historically important, problem — that of *blackbody* radiation. In doing so, we examine the often overlooked and largely forgotten computational aspects relating to an important area of science and engineering before and after the coming of the digital age. We develop and discuss many of the mathematical techniques and methods that have been developed and continue to be applied to evaluating quantities associated with blackbody radiation. The book also concerns itself with the many aids, such as nomograms and graphs, slide rules, and tables, that were produced, together with their more recent digital arrivals that continue to be developed to aid calculations encountered in this field. The story of these aids, from their initial inception until what for many was the time of their final demise at the hands of the impending digital age, forms a central part of the narrative we attempt to tell.

The genesis of this book lies in an article the first author wrote in 2011 entitled "Blackbody radiation functions and polylogarithms." The paper concerned itself with a way of writing many of the functions associated with blackbody radiation in terms of a special function known as the polylogarithm function. In setting the stage for such work, its short introduction attempted to place the importance of these functions within their proper historical context. In so doing he was to discover that in the not too distant past, many slide rules and extensive sets of tables had been developed and used for finding values to these and other closely related functions. Intrigued, he resolved to further explore these "analogue" aids of old once time permitted. His searches with requests for information led him to the second author who, back in the day, had been a user of such aids. Still in possession of many examples, a fruitful correspondence ensued with the end result being the book you now have before you.

The second author began using radiation slide rules in the late 1960s for the development of infrared systems for medical, industrial, and military applications. In the early 1970s he was part of a group studying the dynamic behavior of materials of interest in the ballistic missile defence field. As this required untold millions of evaluations of Planck's equation, it led him to investigate the possibility of using Gauss–Laguerre quadrature to significantly reduce the amount of computational time required in order to integrate Planck's equation over any specified spectral band. It worked, and resulted in significant savings in time per each run. Four decades later, he still keeps a radiation slide rule nearby that he continues to use for quick estimates when a computer is not immediately available but an answer is needed.

Historically, thermal radiation, the radiation emitted from an object due to its temperature alone, not only played a pivotal role in the development of modern physics, particularly in what later became known as quantum mechanics, but technologically its importance over time grew as more and more applications begun to emerge. Reflecting the growing paramountcy of the technology, many mathematical attempts were made to find efficient means for the evaluation of the most common quantities associated with thermal radiation, while in time, a large array of aids would eventually become associated with the field. The latter were designed to help assist in the growing computational demands being placed on the working scientist and engineer. In their day, they were recognized as an important labor saving device and were widely used. While they may only represent a small aspect of the scientific enterprise, in most accounts until now, they are either relegated to a mere footnote, or entirely ignored altogether.

By the last decade of the nineteenth century, driven largely by growth in industries such as electrical lighting and photography, the study of radiation emitted from an object due to its temperature alone had become a subject of intense scientific investigation. The occurrence of very hot objects changing color as their temperature increases is familiar to most. For example, a solid iron rod glows a dull red color at a much lower temperature than when it is glowing a bright, yellowish-white color. The technological problem faced by the electric lighting industry at the turn of the twentieth century was one of efficiency. How was the maximum amount of visible light emitted from an object to be achieved using the least amount of energy? It was a question whose answer ultimately depended on a sound theoretical understanding of the laws of thermal radiation. Knowing the body that emits the greatest amount of radiation compared to all other bodies at the same temperature is of considerable importance. Knowledge of such a body's characteristics allows it to serve as the ideal theoretical standard against which all real radiating bodies can be compared. Known as a blackbody, it was first introduced as a theoretical concept by the German physicist Gustav Robert Kirchhoff in 1860. A further forty years would pass before a correct mathematical description of the radiation emitted by a blackbody was given final form by Max Planck in 1900. Embodied in his now famous law for thermal radiation, calculations made using Planck's equation is a slow process owing to its particularly cumbersome mathematical form. Planck's equation, and those functions closely related to it, thus readily lend themselves to being given a graphical, mechanical, or numerically tabulated form for their evaluation.

This book attempts to give a comprehensive account of the development of Planck's equation, and other closely related functions, together with the many methods used in their evaluation. Due importance is paid to computational techniques and the various aids developed to facilitate such calculations. To help achieve this goal, the book has been divided into three parts. In the first, which consists of Chapter 1, thermal radiation and the blackbody problem is introduced and discussed. It examines early developments made by the experimentalists and the theoreticians as they strove to understand the problem of a blackbody. Introduced here are all the well-known early "classical" laws discovered before 1900 which one today associates with thermal radiation such as the Stefan–Boltzmann law, Wien's homologous law, and Wien's displacement law. Though each law antedates Planck's law, each is a consequence of the latter's law. In attempting to demonstrate their essentially classical character, techniques from dimensional analysis are introduced and used to show how these laws follow directly from properties of their basic physical dimensions.

The second part of the book concerns itself with a number of theoretical developments that stem directly from Planck's law, and the various computational matters that arise when numerical evaluation is required. Some basic elements of radiometry that tie together and use many of the theoretical and computational ideas developed are also considered. These matters are taken up and developed in Chapters 2 through 4.

The different spectral representations that can be used to represent Planck's law are often not fully appreciated. The selection of a particular spectral representation has an important affect not only on the shape of the spectral distribution curve, but also on its corresponding peak location. Chapter 2 begins with a general discussion on spectral representations for continuous radiation sources such as a blackbody and the consequences this has on the form of Planck's law. The development of Planck's equation is also considered in this chapters as are the all important fractional functions of the first and second kinds which result when Planck's equation is integrated between finite limits. For the latter functions, they are expressed in closed form in terms of a special function known as the polylogarithm function and is done using the recently introduced polylogarithmic reformulation for the blackbody problem.

Chapter 3 focuses on the various computational methods developed to obtain numerical values from the various quantities associated with thermal radiation. Here all the well-known approximations, together with many of the not so well-known approximations to Planck's law, are introduced and considered while various methods used to evaluate the fractional functions such as rapidly converging infinite series, Gauss–Laguerre quadrature, integrand approximations, and asymptotic expansions methods are developed and analyzed. Concepts as they pertain to basic radiometry and its measurement methods are introduced in Chapter 4. The latter makes use of many of the ideas developed in the preceding two chapters and is used as an example to show how the forgoing theoretical and computational developments can be used in practice.

Shortly after Planck introduced his law, early users of his equation were quick to discover how overwhelming the task of having to perform many calculations by hand can be. Unsurprisingly, it was not long before the first computational aids designed to alleviate this burden first started to appear. These quickly grew in importance and number as their general utility was realized. Widely used, they would go on to relieve generations of scientists and engineers alike from the onerous task of human computation in an age before digital computers existed, and even after they did, before digital computers became widely available. These aids are the subject of the third part of the book which are explored in Chapters 5 through 8.

These individuals or groups of individuals who embarked on creating the various computational aids used for thermal radiation calculations and what drove them on is often just as interesting as the aids they produced. Much effort, first human then computer, went into the creation of these aids. It was laborious work, and for their respective creators, lacked the prospect of any new physical or technological insight being gained from their production. And yet aids of the highest quality continued to be made for the better part of the twentieth century. An absence of any reputation to be gained as a result of their production usually meant their creation was undertaken on an ad hoc basis rather than as a result of any general desire to produce such aids. Many tables, for example, were created in this way as the by-product of other work, while the first of the radiation slide rules only came about when a need to quickly estimate the integral of Planck's equation arose due to the pressures of World War Two. Occasionally the task of creating specific aids was carefully planned in advance. Most notably, tables appearing in book form were usually produced in this way. To their credit, many saw the creation of aids as important work, and if it had not been for their collective efforts, we would have not been left with as rich a computational legacy as the one we now have.

The greatest need for computational aids designed to assist in calculations involving thermal radiation would not come until the technological importance of the infrared spectrum was fully realized. At first there seems to have been little interest in infrared systems. After Theodore Case developed the first infrared system in 1917 for ship-to-ship communications, his work remained largely ignored and forgotten until the mid-1930s when the drums of war began to distantly play once more. As part of his postgraduate work in Germany, Edgar Kutzscher discovered that several classes of lead salt materials could be used for infrared detection. His discoveries lead to a variety of German military applications. At about the same time, researchers in both England and the United States were pursuing similar developments. The coming of the Second World War, however changed everything. With their obvious military applications, by the war's end the importance of utilizing detection devices capable of operating in the infrared portion of the electromagnetic spectrum had been firmly established. Mirroring these developments, the number of new aids produced in the years during and immediately following the war steadily grew.

While several different types of slide rules and a number of nomograms and graphs appeared, it was in the production of tables where the greatest number of aids for thermal radiation were to be found. Chapter 7 is devoted to a discussion of these aids. They appeared surprisingly quickly, just a few years after Planck first introduced his equation. For the most part, these early tables consisted of direct tabulations of Planck's equation as a function of wavelength and temperature. The production of these early tables was driven largely by the needs of those working in the field of illumination, such as photometrists and colorimetrists and it would be several decades before the first tables produced specifically for those working in the infrared emerged.

During the early part of the twentieth century, values for the fundamental constants of nature were still very much in a state of flux. Early tables quickly became out of date with each new revision in the values for the fundamental constants. To overcome the problem of frequent and continual changes in the values of these constants, tables in dimensionless form started to appear. Since their values no longer depended on the values used for the fundamental constants, tables of this type had a considerable advantage over previously produced tables that were not dimensionless. While the longevity of tables produced in this manner was assured, compared to tables which gave the quantity of interest directly, a number of additional calculations from the dimensionless tabulated value had to be made before the quantity of final interest could be found. Despite the uncertainties in the values for the fundamental constants, by the late 1920s the importance of blackbody radiation had been firmly established. Demand for tabulations of these functions increased and remained high for many decades to come. From the earliest of times, tables for these functions were included in a number of the more widely available handbooks and general texts on tables. Later, tables relating to thermal radiation were included in many discipline-specific texts and they remain one of the few places today where tabulations for these functions continue to be found. Of all the aids, tables dominated the computational aid landscape for thermal radiation in terms of total number produced from the time of their inception at the beginning of the twentieth century until the late 1970s when they were finally displaced by the arrival of affordable digital computers.

The advent of high-speed digital computers in the 1950s changed the task of table making but did not, at least initially, remove the need for tables altogether. While the digital computer may have shifted the computational effort from slow and error prone human computers to fast and accurate digital machines, they remained extraordinarily rare, and access to them was limited to the very few. A decade later, as mainframe computers and minicomputers slowly began their inextricable march through the universities, government laboratories, and scientific agencies of the industrialized world, turnaround time remained long, with a day to three days not being uncommon. Coinciding with this period, infrared technology was rapidly evolving and drove the need for increasingly accurate values for many blackbody radiation functions.

As the breadth of infrared system applications continued to grow throughout the 1960s, a marked increase in the number of new tables produced during this period is found. From the mid-1950s until the late 1970s many tables, and often for very specific purposes, appeared. On account of being computer rather than human generated, tables produced during this period also differed from their predecessors in two important ways. The first was in an increase in the number of significant figures used. Values tabulated between seven to nine significant figures were not uncommon. The second was in their sheer physical size. Large extensions in the range of values used for the independent variables coupled with smaller interval sizes resulted in some tables running to unprecedented numbers of pages. Finding tables produced during this period with anywhere from a hundred to several hundred pages is not unusual. During this time, the vast majority of the tables produced were published as either technical reports or in book form. Unlike many of the tables which appeared as technical reports, those published as books were available to a far wider and more general audience. Books of tables prepared by Marianus Czerny and Alwin Walther, and by Mark Pivovonsky and Max R. Nagel, proved to be incredibly popular. Each had their origin in Germany, appeared in 1961, were highly cited in the literature, and went on to have a lasting influence in those fields where they were put to widest use.

Beginning in the mid-1920s the very earliest of the nomograms start to appear and that is the subject of Chapter 5. The nomogram was a relatively recent addition to the class of graphical methods used to aid calculation, having only been developed in the 1880s. Compared to early tables, where values for Planck's equation were typically given to three or four significant figures, nomograms were only suitable for order-of-magnitude estimates. Their advantage lay in their ability to provide rapid estimates without the need of having to interpolate between adjacent values found in tables. A number were produced, and the inventiveness displayed in some is quite remarkable. Graphs from which values for various quantities relating to thermal radiation could be estimated also start to appear around this time and would go on to be produced for many years to come. Often found in texts, like nomograms, they were again only suitable for order-of-magnitude estimates. Many more graphs appeared, but as we shall see, are far less versatile as a graphical aid compared to those of nomograms.

The final years of the Second World War saw the development of the first of the radiation slide rules and is the subject of Chapter 6. By this time the slide rule was already a very old invention. The English mathematician and clergyman William Oughtred is generally credited as its inventor in 1622, however, it would not be until the 1850s that usage in countries such as England, France, and Germany started to become common. By the turn of the twentieth century, many general and special purpose slide rules were in widespread use across the industrialized world. Motivated by military needs the German experimental physicist Marianus Czerny designed and constructed the first of the radiation slide rules in 1944. In the immediate decades after its introduction, the radiation slide rule was widely adopted by those working in the infrared and recognized as a useful and important tool for engineers and scientists alike. The accuracy of the slide rules varied considerably, but it was not uncommon to achieve estimates of one per cent or better from such devices. More accurate than a nomogram and far quicker to use than tables meant that in the early decades of infrared technology, the radiation slide rule was the definitive "workhorse" for a generation of those involved in infrared systems design and engineering. By the latter half of the 1970s, with the appearance of affordable, hand-held programmable calculators and digital desktop computers, the radiation slide rule went into steep decline and all but disappeared within a few short years.

Collectively, computational aids such as nomograms and graphs, slide rules, and tables made more rapid development possible at a time when work requiring excessive computation could severely limit the rate of progress. They had a profound impact on the field of illumination from the early 1920s onwards, and on the entire infrared industry during its early growth years, beginning in the 1950s. Beyond radiometry, photometry, colorimetry, and optical pyrometry, they also played a more modest role in other related fields where a need for such things occasionally arose. These were mostly in the form of tables produced for and by those working in the fields of astrophysics or meteorology.

Astrophysical work in the early 1930s was one area where a need for suitable computational aids arose. Stemming from his work on planetary nebulae and stellar temperatures, tables for the integral of Planck's equation over a finite spectral band were devised by the Dutch astronomer Herman Zanstra in 1931. Often, early aids produced in one field went for many years unnoticed by those working in other fields as was the case with the tables prepared by Zanstra. Tables for the astrophysical and meteorological communities also tended to differ from many of the tables produced for work in infrared or illumination work. Tables intended for astrophysical use were prepared to very-high values for the temperature while exceptionally narrow terrestrial temperature ranges with far smaller interval sizes were the norm for tables intended for meteorological use.

By the late 1970s the growing computational abilities offered by relatively cheap and widely available digital computers meant that there no longer existed a need for most of the computational aids of old. Calculations formerly performed by the use of graphical, mechanical, or numerically tabulated means were rapidly replaced by computations that could now be performed on demand by the user using their own programmable calculator or personal computer. Their use rapidly declined and within the space of a few years most analogue aids had became antediluvian. A number of digital aid replacements did however emerge. Most were in the form of either online calculators, or programs for particular programming languages or certain electronic devices, with the best of these frequently inspired by their analogue predecessors. It is this recent and continually unfolding story we take up and tell in Chapter 8.

Assembling the many historical documents and artefacts considered in parts of this book would not have been possible without the gracious help of others. Special thanks are extended to Prof. William L. Wolfe, who kindly supplied photographs of the now very rare ARISTO Nr. 10048 slide rule and its prototype, to Ms Nina Senger-Mertens from Arithmeum in Bonn who provided the photograph of the equally rare ARISTO Nr. 922 rule, and to Prof. John E. Greivenkamp from the University of Arizona for providing photographs of the Vahlo slide chart. We are also grateful to Barbara Grant from *Lines and Lights Technology* for providing copies of the EG&G Judson and the Engineering Summer Conferences Infrared Radiation Calculators, to Chari Gallen from Teledyne Judson Technologies and Gary Wilson from BAE Systems for providing copies of their company's latest Infrared Radiation Calculators gratis, to Brent Lindstrom, Sales and Marketing Director at Electro Optical Industries in Santa Barbara, for providing a number of beautiful specimens of their commemorative Blackbody Radiation Slide Rule, and to Steven Peters who not only drew our attention to the recent Infrared Radiation Calculator from BAE Systems but also answered a number of questions regarding his Blackbody Radiation Calculator app for Android.

Locating copies of some of the more obscure documents cited in this work was a particular challenge. Many individuals met requests for copies of tables either they or their institutions had produced. They were, in no particular order: Prof. Stephen M. Robinson from the University of Wisconsin-Madison, Prof. Yalçın A. Göğüş from the Middle East Technical University in Turkey, Perry Abramowitz from RAND Corporation, Julie Niesen from the Nelson Institute Center for Climatic Research at the University of Wisconsin-Madison, Sharon Wilson from the National Physical Laboratory in the UK, Steven D. Vanstone from Redstone Arsenal in the United States, and Sugiura Yoshio from the National Institute of Advanced Industrial Science and Technology in Japan. All are gratefully acknowledged. We thank Shingo Ushida from the Japan Meteorological Agency for providing high-quality copies of two nomograms reprinted in this book. We would also like to thank the following individuals for their recollections of using radiation slide rules: Prof. William L. Wolfe, Prof. Hans J. Queisser, Prof. Raymond J. Chandos, Mr Max J. Riedl, Mr Jack R. White, and Mr Arthur Cussen and Mrs Marjorie Cussen. These have been invaluable. Finally, we are both indebted to the Deputy Editor-in-Chief of the *Journal of Infrared and Millimetre Waves*, Dr Hong Shen, who drew our attention to relevant work published in the Chinese literature.

The first author would like to extend his gratitude to the College of Engineering, Technology and Physical Sciences at Alabama A&M University in Huntsville who hosted him for two weeks during the summer of 2015 as the final draft of the manuscript for the book was being completed. Support from the first author's home institution in the form of a small grant for such an undertaking is also gratefully acknowledged.

A book of this nature readily leads itself to the inclusion of many figures, particularly figures from the original sources. We have made every effort to determine original sources and obtain permissions for the use of these illustrations where necessary. Credit is to be found in the figure captions. Figures and photographs created by the authors are typically not given any particular credit. While all reasonable effort has been made to contact the holders of copyright material reproduced in the book, any omissions will be rectified in future printings if notice is given to the publisher.

Seán M. Stewart
Astana

R. Barry Johnson
Huntsville, Alabama
January, 2016

List of Figures

List of Tables

Authors

Seán M. Stewart is an Associate Professor at Nazarbayev University and previously was at The Petroleum Institute for over a decade. Dr. Stewart has written numerous technical articles in the infrared field and developed the polylogarithmic formulation to the blackbody problem described in this book. He is a Member of the Institute of Physics, the Australian Institute of Physics, the Optical Society of America (OSA), and SPIE – The International Society for Optics and Photonics.

R. Barry Johnson, a Senior Research Professor at Alabama A&M University, has been involved for over 40 years in infrared technology. Dr. Johnson developed the Gauss–Laguerre quadrature method of integrating Planck's equation. He is an SPIE Fellow and Life Member, OSA Fellow, an SPIE past president (1987), has been awarded many patents, published numerous technical articles, and was awarded the 2012 OSA/SPIE Joseph W. Goodman Book Writing Award for *Lens Design Fundamentals, Second Edition*.

Section I

The blackbody problem

1 Thermal radiation and the blackbody problem

All objects emit radiation as a result of their temperature. At room temperature an object does not appear self-luminous, as almost all the radiation emitted is in the infrared portion of the electromagnetic spectrum.[1] As the temperature increases, the emitted radiation becomes visible, first appearing as a faint, almost colorless grey glow[2] before becoming a deep, dark, blood red in color. Continuing to increase the temperature further, the glow gradually changes color from a dull red to a bright red, before appearing yellowish and finally white. At still higher temperatures its color would change from a white glow to one which appears bluish-white in color.

Known as thermal radiation, the body which radiates the greatest amount of energy compared to all other bodies at the same temperature is referred to as a blackbody. The German experimental physicist Gustav Robert Kirchhoff (1824–1887) was the first to recognize the important theoretical role blackbodies played in understanding the emission of radiation from real bodies. An understanding of the nature of blackbody radiation was one of nineteenth-century physics' crowning achievements and paved the way to the quantum revolution that followed during the early part of the twentieth century.

A blackbody is defined as a body which absorbs all thermal radiation incident upon it. The name given to the body is appropriate.[3] Provided a body is not hot enough so as to appear self-luminous, as no radiation is reflected by the body, it appears completely black. The complete absorption of radiation by a blackbody holds true for radiation at all wavelengths and for all angles of incident upon the body. For a blackbody in thermal equilibrium with its surroundings it means that all radiation received through absorption must be emitted if its temperature is to remain constant as there is no other mechanism available to the body to lose energy without a corresponding increase in its temperature. A blackbody therefore radiates more energy per unit time in any given wavelength interval and more total energy per unit time over all wavelengths than any other body at the same temperature and is independent of the nature (its size, shape, or material from which it is made) of the radiating body. The radiation from a blackbody is referred to as *blackbody radiation*. By the late nineteenth century, the problem of the blackbody had become one of finding a mathematical expression that could describe the amount of energy emitted within a given spectral range as a function of both wavelength and temperature.

Not surprisingly it was Kirchhoff, having first proposed the blackbody problem, who made the first serious attempt at its solution. In the winter of

1859–60, following a simple yet brilliant line of physical reasoning based on the second law of thermodynamics, he postulated that the radiation emitted from a blackbody within a given wavelength interval is a universal function depending only on the temperature of the body [354, 355]. It was an important first step, and constitutes what today is known as *Kirchhoff's law of thermal radiation*, but there was still a long way to go. Kirchhoff did not give the mathematical form for the universal function associated with blackbody radiation but understood its importance and suspected it must be relatively simple.[4] Its universal character did however, ensure that the problem remained an important one in need of a solution. Kirchhoff left the job of determining the final form to others. The task they inherited, however, proved far more difficult than even Kirchhoff himself could have anticipated. Its solution, when finally found, shook the very foundations of physics and ultimately led to the development of a completely new branch of physics which today we know as quantum mechanics.

In this chapter we consider a number of events that historically were important in understanding the nature of the blackbody problem. The chapter sets the scene for understanding the wider problem of blackbody radiation by introducing quantities central to its description and the laws it obeys. These, in turn, will be discussed in greater detail in the chapters to follow. The important interplay to be found between theory and experiment resulting from the various attempts made towards arriving at a final solution to the problem are traced out and considered. Along the way many of the well known names associated with the early history of the field are introduced. The chapter closes with an examination of two important laws for thermal radiation, which while initially found using arguments based on classical physics and thermodynamics, will be arrived at using arguments based on a powerful though often overlooked technique known as dimensional analysis.

1.1   TOWARDS A SOLUTION TO THE BLACKBODY PROBLEM

By the close of the nineteenth century the exact mathematical form for the universal function of blackbody radiation continued to allude scientists. But it was not from a want of trying. By the late 1880s a number of theoretical attempts to derive the form for Kirchhoff's universal function had been made. The earliest of these was perhaps that of the German physicist Eugen Cornelius Joseph von Lommel (1837–1899) in 1878, then holder of the chair in experimental physics at the Friedrich-Alexander-Universität Erlangen-Nürnberg [394]. Lommel sought to understand, using mechanical principles, absorption; a process he saw as an interaction between light waves and the atoms of a body; and was convinced Kirchhoff's law of absorption followed from his theory. After Lommel, other attempts quickly followed [173] but advanced little in the search for the ultimate form for Kirchhoff's universal function.

Like Kirchhoff, all who approached the problem theoretically were convinced of the simplicity in form the universal function should take.

Reiterating Kirchhoff's words some thirty years later in 1889, the great nineteenth century English physicist Lord Rayleigh (1842–1919) [545] wrote "...the function ... being independent of the properties of any particular kind of matter, is likely to be simple in form," before quickly adding its expected simplicity had led to "...speculations [that had] naturally not been wanting."[5]

In the opinion of Lord Rayleigh, speculations that had been wanting were two attempts that came in the late 1880s. These were the theoretical attempt made by the Russian physicist Vladimir Aleksandrovich Michelson (1860–1927)[6] in 1887 [444, 445] and the empirical law advanced by Heinrich Friedrich Weber (1843–1912), a Professor of Theoretical and Technical Physics at the Polytechnikum, Zürich, in 1888 [677]. While neither adequately described the solution to the blackbody problem, each made a number of predictions which, at least in the case of Weber, proved correct.[7]

While these early attempts to find a correct form for Kirchhoff's universal function were quickly found to be inadequate, beyond trying to theoretically understand the spectral distribution of radiation from a blackbody with temperature, two important results relating to the radiation emitted by a blackbody would be established before the close of the century. The first of these related to the total amount of energy emitted from the surface of a blackbody at a given temperature and is today embodied in what we call the *Stefan–Boltzmann law*. It states the total energy emitted from the surface of a blackbody at temperature T per unit time per unit area in all directions into the half space above a surface is proportional to the fourth power of the temperature. Mathematically

$$M_e^b(T) = \sigma T^4. \tag{1.1}$$

Here M_e^b is the total radiant exitance and is measured in units of watts per square meter [W·m^{-2}] while σ is the Stefan–Boltzmann constant.[8] The law is named in honor of Jožef Štefan (1835–1893) who deduced it empirically in 1879 [619][9] and Ludwig Boltzmann (1844–1906) who derived the result theoretically some five years later using arguments based on thermodynamics [92]. While the Stefan–Boltzmann law allowed the total amount of energy radiated from the surface of a blackbody at a given temperature to be found, as it gave no information about the energy radiated from a blackbody within a specific wavelength interval, it meant the form of Kirchhoff's universal function remained as elusive as ever.

The second important result to be found in the last decade of the nineteenth century related to the particular functional form Kirchhoff's universal function could take. In addition to knowing the total energy radiated by a blackbody at a given temperature, understanding how this energy is distributed throughout the spectrum as a function of wavelength is equally important. By the late nineteenth century it was experimentally known that the radiation emitted from a blackbody was spread continuously over a singly peaked spectrum consisting of all wavelengths. However, finding the mathematical form that described this spectral distribution became one of the great unsolved problems facing physics at the close of the nineteenth century.

In 1893 the recently licensed docent at the Universität zu Berlin, Wilhelm Carl Werner Otto Fritz Franz Wien (1864–1928), was able to extend the work of Boltzmann in an important way. The universal function of Kirchhoff's, for a given wavelength λ, can be identified with what is today known as the *spectral radiant exitance* $M_{e,\lambda}^b(\lambda, T)$. The spectral radiant exitance will be discussed in greater detail in Chapter 2, where the sub- and superscripts appearing here will be clearly explained. Physically, the spectral radiant exitance corresponds to the amount of energy emitted by a body into a hemispherical envelope in space per unit time per unit area within the unit wavelength interval λ to $\lambda + d\lambda$. Using thermodynamic arguments together with a principle related to the change in wavelength a wave experiences as it moves relative to a source known as the Doppler effect, Wien was able to deduce theoretically [684] the important result equivalent[10] to

$$M_{e,\lambda}^b(\lambda, T) = \frac{1}{\lambda^5} F(\lambda T). \tag{1.2}$$

Equation (1.2) shows that Kirchhoff's universal function, which depends on both wavelength and temperature, can be reduced to an unknown universal function F of a single variable equal to a product between the wavelength and temperature only. At first sight this may not seem very significant. After all, the final form for the unknown function F remains to be determined and would appear to bring one no closer to a solution of the blackbody problem than before. Superficially at least, it appears one unknown function has simply been replaced by another unknown function. And indeed it has, but to see it only in this light is to completely miss the point.

To understand the true significance of Eq. (1.2), in this result we see that if the spectral radiant exitance is known at one temperature its form at any other temperature can be found. Importantly, any temperature will do. So once the spectral distribution of a blackbody is known at a single temperature, Wien's result ensures the spectral distribution for a blackbody at any other temperature can be found. Graphically, if $\lambda^5 M_{e,\lambda}^b(\lambda, T)$ were to be plotted against the wavelength–temperature product λT, a single spectral curve for the radiation from a blackbody results — regardless of its temperature. The result of this homologous relationship is generally known as *Wien's law* or sometimes as *Wien's general displacement law*. The suitability of either of these terms to describe the law is however, far from ideal. In order to avoid confusion with a number of other laws due to Wien that are to follow, neither of these terms will be used. Instead, the homologous relationship for the law suggests a more appropriate name would be *Wien's homologous law* and that is the name we intend to use.[11]

While Wien's homologous law was a tremendous step forward, it remained to determine the spectral radiant exitance for a blackbody at a single temperature. Of course, it could be obtained experimentally but intellectually is far less satisfying compared to a proper theoretical treatment. Throughout the late 1890s the experimentalists quickly provided many curves for the spectral radiant exitance at various temperatures to great accuracy out to wavelengths extending deep into the infrared. A correct theoretical description however, remained illusive and would have to wait until the close of the nineteenth century and the work of the great German theoretical physicist Max Karl Ernst Ludwig Planck (1858–1947).

Despite the ultimate shortcomings in Wien's homologous law, surprisingly it can still be used to deduce the Stefan–Boltzmann law. The total radiant exitance is found by summing the spectral radiant exitance over all wavelengths. Mathematically this corresponds to integrating $M_{e,\lambda}^b(\lambda, T)$ from zero to infinity, namely

$$M_e^b(T) = \int_0^\infty M_{e,\lambda}^b(\lambda, T) d\lambda. \tag{1.3}$$

Substituting Wien's homologous law into Eq. (1.3) gives

$$M_e^b(T) = \int_0^\infty \frac{F(\lambda T)}{\lambda^5} d\lambda = T^4 \int_0^\infty \frac{F(x)}{x^5} dx, \tag{1.4}$$

after the change of variable $x = \lambda T$ is made. The second of the improper integrals, provided it exists, is just a number since it is dimensionless and the Stefan–Boltzmann law follows with the proportionally to the fourth power of the temperature.

The special case of Wien's homologous law is often cited. If λ_{max} is the wavelength at which $M_{e,\lambda}^b(\lambda, T)$ attains its maximum value, the peak in the spectral curve will be determined by a fixed value of λT, that is, by

$$\lambda_{max} T = b. \tag{1.5}$$

Here b is a constant referred to as Wien's displacement law constant[12] and it is in this special form the law is known simply as *Wien's displacement law*.[13] The law derives its name from the fact that as the temperature increases, the peak in the curve for the spectral radiant exitance, when plotted as a function of wavelength, becomes "displaced" towards shorter wavelengths.

Wien himself in his original publication of 1893 [684] and subsequent publications of 1893 [685] and 1894 [686] on the same topic did not give either of the two laws we have associated with his name in the form given by Eqs (1.2) and (1.5). For his displacement law Wien wrote

$$\lambda \vartheta = \lambda_0 \vartheta_0, \tag{1.6}$$

where λ_0 and ϑ_0 were some fixed wavelength and temperature (here Wien used the symbol ϑ for absolute temperature in place of the modern day T) respectively and wrote:

> In the normal emission spectrum of a blackbody each wavelength shifts as the temperature changes in such a way that the product of temperature and wavelength remains constant.[14]

Wien did not come to any conclusion about the displacement of the maximum in the spectral curve which today his law has become most widely associated with, instead preferring to give a wider interpretation to the form he presents. It is likely he knew of the more limited form of the spectral maximum to which his displacement law could be applied, since at the end of his first paper of 1893 he wrote that the result found agreed with the shift in the spectral maximum correctly deduced by Weber five years earlier. Weber had used a form for the spectral distribution function of a blackbody he had deduced empirically but which, as we have already noted, later proved to be incorrect. The writing of Wien's displacement law chiefly in terms of the maximum is due largely to the extensive experimental work performed by the German experimental physicist Louis Karl Heinrich Friedrich Paschen (1865–1947) between the years 1895 and 1901 [479, 480, 481, 483, 484, 487].

On his homologous law, as presented in Eq. (1.2), Wien again did not give it in the form we have given. Instead the form Wien gives towards the very end of his first 1893 paper on the subject, in terms of the spectral energy density ϕ_λ; a quantity proportional to the spectral radiant exitance; was

$$\phi_\lambda = \phi_{\lambda,0} \frac{\vartheta^5}{\vartheta_0^5}. \tag{1.7}$$

Here $\phi_{\lambda,0}$ and ϑ_0 are fixed values for the spectral energy density and absolute temperature respectively. If Eqs (1.6) and (1.7) are combined, one obtains

$$\phi_\lambda \lambda^5 = \phi_{\lambda,0} \lambda_0^5, \tag{1.8}$$

and is constant. Obviously, as seen in Eq. (1.7), the spectral energy density for a blackbody is a function of the absolute temperature. But as Wien had already shown, the wavelength–temperature product is constant, so one may rewrite Eq. (1.8) as

$$\phi_\lambda = C\lambda^{-5} f(\lambda T), \tag{1.9}$$

for some constant C and some unknown universal function f. Note Eq. (1.9) reduces to Eq. (1.2) on setting $Cf(\lambda T) = F(\lambda T)$.

In terms of priority it was the Northern Irish physicist Sir Joseph Larmor (1857–1942) who, at the Seventieth Meeting of the British Association for the Advancement of Science in 1900, first gave the form for Wien's homologous law we have used here [375, 384, 376]. We will refer to it as the Larmor form for the homologous law. However, unbeknownst to Larmor, two years earlier an equivalent form

$$M_{e,\lambda}^b = T^5 \psi(\lambda T), \tag{1.10}$$

had been given by Lord Rayleigh [547], but attracted little attention, so much so it was independently given for a second time two years later by the German theoretical physicist Max Ferdinand Thiesen (1849–1936) [644]. In this form we will refer to it as the Rayleigh–Thiesen form for the homologous law. The two forms can be seen to be equivalent if we write

$$M_{e,\lambda}^{b} = \lambda^{-5} F(\lambda T) = T^5 \left[\frac{F(\lambda T)}{(\lambda T)^5} \right] = T^5 \psi(\lambda T). \qquad (1.11)$$

Wien's result, as would finally become apparent only a few years later, represented the limit of what could be determined for a blackbody using classical physics and thermodynamics alone. Wien had managed to bring one tantalizingly close to the solution of the blackbody problem, closer than anyone before him had managed to do, yet its final solution continued to remain as elusive as ever. It appeared the complete solution to the problem of blackbody radiation would only come from a radical new way of thinking about the problem. One would not have to wait long and it was to come from a theoretical physicist who also found himself working at the epicenter of research into the nature of the blackbody problem in the late 1890s. It, of course, was Planck himself.

1.2 PLANCK AND THE BLACKBODY PROBLEM

Despite what had been achieved by the close of the nineteenth century, the final form for the universal function $F(\lambda T)$ deduced by Wien remained unknown. This is not to say explicit forms for it had not been proposed. They had, but under closer experimental scrutiny all were found wanting in one way or another in their description of the radiation emitted from a blackbody over all wavelengths.

In 1896 Wien himself proposed a radiation function which initially appeared to provide a solution to the blackbody problem [687, 688] and for a time it seemed to be supported by the available experimental evidence [481]. Further theoretical support to Wien's proposed form for the distribution law of a blackbody would come from Planck himself. Using arguments quite different from those used by Wien, in a series of several long papers commencing in 1897 and culminating in 1900, Planck also managed to arrive at the same result as Wien [503, 504, 505, 506, 507, 508, 509]. *Wien's displacement law*, as it become known, took the form

$$M_{e,\lambda}^{b}(\lambda, T) = c_1 \lambda^{-5} \exp\left(-\frac{c_2}{\lambda T}\right). \qquad (1.12)$$

Here c_1 and c_2 were two unknown radiation constants to be determined.

Wien had proposed his law in June of 1896. In the same month, and quite independently of Wien, Paschen, based on his own extensive series of measurements, proposed a similar looking empirical law for the spectral distribution function of a blackbody [481]

$$M_{e,\lambda}^{b}(\lambda, T) = c_1^* \lambda^{-\alpha} \exp\left(-\frac{c_2^*}{\lambda T}\right). \qquad (1.13)$$

Here c_1^* and c_2^* were again two radiation constants to be determined while α was an adjustable parameter. Paschen found the best fit to his data was obtained when $\alpha = 5.660$.

The closeness between the two forms was convincing evidence in itself that Wien's law was correct while the later theoretical work of Planck only added to the conviction of correctness. Only with newly improved experiments for radiation at wavelengths out to an incredible 51 µm, which put them deep in the infrared, were experimentalists able to show for the first time inconsistencies between those predictions based on Wien's law and experiment [410, 411, 572, 574]. The final solution to the blackbody problem, which for a time was thought to be solved, once more found itself at the beginning of the new century the object of intense theoretical and experimental investigation.

As inconsistencies between predictions based on Wien's law and the work of the experimentalists emerged, others quickly turned their attention back to one of the nineteenth century's greatest unsolved problems. The first to do so was the German theoretical physicist Max Thiesen, who working in Berlin, was a colleague of Planck's. He made his attack on the problem in March of 1900 [644]. On the basis that any distribution function proposed must satisfy Wien's homologous law,[15] Thiesen gave the following family of solutions for the distribution law

$$\phi[x] = \phi_m \left[\frac{x_m}{x} \exp\left(1 - \frac{x_m}{x}\right) \right]^\alpha. \tag{1.14}$$

Here ϕ_m and x_m were two universal constants while the index α was a number between two and five. Setting $\alpha = 5$ Thiesen's law immediately reduces to Wien's law with the two constants appearing in Wien's formula being equal to

$$c_1 = \phi_m x_m^5 e^5 \quad \text{and} \quad c_2 = 5 x_m. \tag{1.15}$$

But in order to provide a better fit with the most recent experimental data, Thiesen found he had to set the value for the index α equal to 4.5. For this value his distribution law takes the form

$$M_{e,\lambda}^b(\lambda, T) = c_1' \sqrt{\lambda T} \, \lambda^{-5} \exp\left(-\frac{c_2'}{\lambda T}\right), \tag{1.16}$$

where c_1' and c_2' are once again two radiation constants to be determined.

Next into the fray was Lord Rayleigh. In June of 1900, using reasoning which appeared to him to be more probable a priori, being based, as it was, on an analogy between the behavior of radiation whose wavelengths are large with the theory of sound, suggested Wien's law could be modified as follows

$$M_{e,\lambda}^b(\lambda, T) = \bar{c}_1 \lambda^{-4} T \exp\left(-\frac{\bar{c}_2}{\lambda T}\right). \tag{1.17}$$

Once more $\bar{c}_1$ and $\bar{c}_2$ were two radiation constants to be determined.

Based on his analogy with sound, Lord Rayleigh had in fact arrived at the following form for the distribution law

$$M_{e,\lambda}^b(\lambda, T) = \bar{c}_1 \lambda^{-4} T, \tag{1.18}$$

which he stated was valid in the long wavelength, high temperature limit. The additional exponential factor was added afterwards to ensure the final distribution function tended to zero in the short wavelength, low temperature regime.

Whether his proposed distribution law more closely represented the experimental facts Lord Rayleigh was not in a position to say but hoped "... the question may soon receive an answer at the hands of the distinguished experimenters who have been occupied with this subject" [548].[16] The final solution however, when it came in late 1900, would require a complete conceptual break in how physics up until that time had conceived energy to be. And it was to come from the man who would go on to become one of the greatest theoretical physicists of all time — Max Planck.

Planck, as we saw, first turned his attention to the problem of blackbody radiation in 1897. He had initially been drawn to the problem by the universal character of the law required by Kirchhoff's law. At the time, Planck was almost forty years old and it would have been tempting to think his best work lay behind him. Up until this point he had spent his entire scientific career investigating the second law of thermodynamics and its consequences. The prospect of him solving one of the late nineteenth century's great unsolved problems seemed remote. After three years of solid effort working on the problem, by the end of 1900, by considering the energy emitted from a blackbody is not continuous but rather discrete and only comes in amounts made up of multiples of some fundamental "quanta," Planck was finally able to give what later proved to be the correct mathematical description for the spectral radiant exitance of a blackbody valid for all wavelengths.

Planck initially achieved his correct form for Kirchhoff's universal function using what has undoubtedly become one of the most famous and productive interpolations ever made in the history of theoretical physics. By taking Wien's radiation law, a result known to be valid in the short wavelength, low temperature limit, together with Lord Rayleigh's classical result from June 1900 and known to be valid in the long wavelength, high temperature limit, and interpolating between the two, by October 1900 [510] Planck was able to arrive at, at least mathematically to begin with, a form for the radiation law that now bears his name [511, 518]. While Planck had an equation which seemed to fit all known experimental data better than any of the previously proposed distributions laws, particularly at the high temperature, long wavelength limit, it lacked physical justification. After what he later described as the most intense period of work in his life, by December 1900 he had a physically plausible theory to back his earlier result obtained through interpolation [512, 519]. Finessing his ideas the following year [513, 514, 515] it is now known as *Planck's law* for thermal radiation and has held up to all subsequent experimental scrutiny [58, 570, 672, 575, 576, 621, 151]. In modern form his law takes the form

$$M^{\mathrm{b}}_{\mathrm{e},\lambda}(\lambda, T) = \frac{2\pi hc^2}{\lambda^5 \left[\exp\left(\dfrac{hc}{k_B \lambda T} \right) - 1 \right]}. \tag{1.19}$$

Here the fundamental constants c, h, and k_B are the speed of light in vacuo, Planck's constant, and Boltzmann's constant respectively.[17] Developing his law required Planck to introduce the latter two new fundamental constants of nature.[18]

While the Stefan–Boltzmann and Wien displacement laws antedate Planck's formulation, as we will show in Chapter 2, both are direct consequences of the latter's law. The application of classical physics together with the general principles of thermodynamics had carried one only so far along the road to theoretically understanding the spectral distribution of radiation emitted from a blackbody. It had enabled one to determine the general form the spectral distribution function must take but could not deliver its ultimate final form. Only by departing from classical physics could a solution to the problem of blackbody radiation be found. As Planck would recall years later, it was only through "an act of sheer desperation" that a solution could finally be found. But found it was, and like all good revolutions paved the way for what in the twentieth century was to become the quantum revolution in physics.

1.3   THE WORK OF THE EXPERIMENTALISTS

The 1890s was a period when work on the nature of the blackbody problem was greatly extended by a small but brilliant group of experimentalists working in Germany. Between them they performed what has been described as not only some of the most important experiments undertaken into the nature of blackbody radiation, but in the history of early modern physics [239]. Despite this, their story is often overshadowed by the work of the theorists whose contributions are often seen as having had the greatest influence on shaping and advancing our understanding of the nature of the blackbody problem. As we will see, the contributions of the experimentalists during this time were no less important than those of the theoreticans.

Today the work of the experimentalists is often relegated to a minor footnote in the history of blackbody radiation, with little of their work being known by the modern reader. Often their names are cast aside, or if recalled at all, are but peripheral characters compared to the names associated with the great theoreticians. Other than Wien, who of us today remembers the quintuple of contemporary German experimental physicists Friedrich Paschen, Otto Richard Lummer (1860–1925), Ernst Pringsheim (1859–1917), Heinrich Rubens (1865–1922), and Ferdinand Kurlbaum (1857–1927),[19] and the landmark work they performed in the years leading up to 1900?[20]

Advances in the ability to detect ever smaller amounts of radiation out to wavelengths deep in the infrared coupled with the final realization of a usable blackbody during this period allowed the laws of thermal radiation to be probed to an unprecedented degree of accuracy. At the same time as Wien, Lord Rayleigh, Thiesen, and Planck were working theoretically on the problem, the work of the German experimentalists was not only important in its own right but led to a particularly significant symbiosis between theory and experiment. Here advances made by one side acted as the impetus for further advancement by the other side.

After the work of Kirchhoff the rate of experimental progress was initially slow, being largely a reflection of the very poor state measurement that existed for detecting thermal radiation. Not until the introduction of the bolometer by the American astrophysicist Samuel Pierpont Langley (1834–1906) in 1880, and its subsequent refinement and improvement, could radiation measurements capable of discerning the spectral distribution of a sufficently hot thermal source be made with any degree of accuracy and confidence. Improvements in the detection capabilities of the bolometer was an important early step, first taken by Lummer and Kurlbaum in 1892 [407], and modified and further improved a few years later by Lummer and Pringsheim [408, 414].

The next important step came with the experimental realization of a usable blackbody source. It was taken by Wien and Lummer in 1895 [690] and was based on the proposal Kirchhoff had suggested some thirty-five years earlier when he first introduced the theoretical notion of a perfect radiator. In the realization of a blackbody source Kirchhoff proposed a cavity with a small opening be used. The radiation emanating from the opening would now be "black" and it was in this configuration Kirchhoff referred to the radiation produced in this manner as *Hohlraumstrahlung* or *cavity radiation*. What Lummer and Wien constructed thirty-five years later in 1895 was, surprisingly, exactly this — an enclosure with blackened walls capable of being uniformly heated with a small opening. The radiation emitted from the small opening, to a very good approximation, was then black.[21]

With the modifications made in Langley's bolometer, large sensitivity gains in the measurement of radiation extending out to wavelengths deep into the infrared were now possible. With this, together with a usable blackbody source, the work of the German experimentalists centered around the newly created *Physikalisch-Technische Reichsanstalt* in Berlin-Charlottenburg proceeded at breakneck speed.

First to be tested was the Stefan–Boltzmann law and the dependency of the total radiant exitance on the fourth power of temperature for a blackbody. It was initially confirmed by August Ludwig Eduard Friedrich Schleiermacher (1857–1953) [590] and others [690, 695] for bodies which were not quite perfectly black, and more fully for a blackbody a few years later by Lummer and Pringsheim [409, 413], Lummer and Kurlbaum [408], and Kurlbaum [365].

Starting around 1895 Paschen, a Privatdozent working at the *Technischen Hochschule* in Hannover at the time, and later with the assistance of Heinrich Wanner, began to focus their research efforts on the painstaking search

Paschen hoped would lead him empirically to the law governing Kirchhoff's universal function. Working with wavelengths in the near to mid-infrared ranging from 1 to 8 μm over a temperature range from 400 to 1400 K, Paschen attempted to measure and understand how radiant energy was distributed throughout its spectrum. In the process of doing so he was able to confirm Wien's displacement law and found an experimental value for the constant that appears in this law to great accuracy [479, 480, 481, 482, 489, 490].

Not only did Paschen's experimental work confirm Wien's displacement law, as we have already seen, in 1896 he proposed an empirical formula of his own to describe the distribution law of a blackbody that turned out to be almost identical with the independently proposed theoretical law of Wien's [481]. Although Wien had obtained his distribution function using some rather arbitrary assumptions on how radiation is emitted from vibrating molecules; something Lord Rayleigh had heavily criticized him for, describing the assumptions he made as little more than conjecture [548]; its fit was surprisingly close to the experimental data Paschen, and later Wanner, were to obtain in the course of the next few years, particularly at low temperatures and over the short wavelengths they considered [481, 482, 483, 484, 489, 490, 485, 486, 671].

Paschen had good reason to believe that the spectral distribution function for a blackbody in the form of Wien's law had been discovered. Besides his own experimental work, further confirmation of Wien's law was to shortly follow in the theoretical work of Planck. As already mentioned, Planck had deduced Wien's law in an entirely different manner to how Wien had arrived at his result using arguments based on the electromagnetic theory of light [503, 504, 505, 506, 507, 508, 509]. However, like Wien, Planck had also been compelled to make assumptions that were somewhat arbitrary and opened his entire approach up to similar criticisms Wien had faced only a few years earlier. So despite the hope of Kirchhoff's universal function having been found, such hope quickly faded as it turned out to be premature.

As we now know, the reason for the closeness in fit between Paschen's measured experimental data and Wien's proposed distribution function at relatively low temperatures resulted from one being in the region $\lambda T < 3200$ μm·K for the wavelengths and temperatures considered by Paschen. In this region the difference between Wien's and Planck's formula is less than one per cent, an amount well within the limit of experimental error obtainable at the time.

With Paschen focusing his attention primarily within a band of wavelengths about the maximum of the spectrum distribution curve, commencing in 1896, Lummer and Pringsheim started to explore the emission spectrum of a blackbody at wavelengths deeper into the infrared [402, 403, 404]. Using a cavity radiator which Lummer together with Wien had realized a year earlier for their blackbody source, like Paschen, their work initially focused on wavelengths from 1 to 6 μm within a temperature range of 600 to 1600 K. The success of their intended program of research was immediate, and it was not long before the Stefan–Boltzmann law [409, 408, 413] and Wien's displacement law [410, 411] had been quickly confirmed. Further success with more profound consequences on the state of the blackbody problem would soon follow.

As the work of Lummer and Pringsheim progressed at the upper limit of temperatures obtainable with their blackbody, systematic differences between their experimentally measured data and the distribution function predicated by Wien started to emerge [410, 411]. Undeterred, they pressed on. Working with Jahnke first, Lummer proposed empirically a distribution function slightly different in form from Wien's which more closely fitted his data [406]. And to test both his and Jahnke's new distribution function against Wien's, Lummer, back now working with Pringsheim, started anew their experimental program of research. This time they extended their measurements still deeper into the infrared to wavelengths ranging from 12 to 18 μm. After this second phase of their work was complete they arrived at a definitive and profound conclusion, Wien's law was not valid in the long wavelength, high temperature limit [412, 405].

The experimental work of Lummer and Pringsheim was performed at the *Physikalisch Technische Reichsanstalt* located in Berlin-Charlottenburg. A second group, working independently of them, was also to be found within the walls of this august institution and their work would have a direct influence on Planck and the discovery of his law for thermal radiation. These two men were Rubens and Kurlbaum.

After the initial success Rubens achieved in producing "residual rays" (*Reststrahlen*) — the infrared rays that remain in a beam of thermal radiation after a series of multiple reflections from the surface of a solid crystal – with his colleague Emil Aschkinass (1873–1909) in 1898 [571] out to a staggering 51.2 μm, convinced Rubens of the enormous potential impact such rays could have if put to the study of the blackbody problem. By 1900, working now with his collaborator Kurlbaum, found them measuring the spectral distribution for a blackbody at unprecedented long wavelengths of 24, 31.6, and 51.2 μm for temperatures ranging from 85 to 1773 K [572, 573, 574].

Like Lummer and Pringsheim before them, they too observed Wien's distribution law was no longer valid in the long wavelength, high temperature limit. As important as this observation was, their work is however, noteworthy for another far greater reason. It was their initial results that were to have a direct and profound influence on the direction Planck's own theoretical investigations of the blackbody problem were to subsequently take.

Rubens and Kurlbaum's work was first reported by the latter before the *Deutschen Physikalischen Gesellschaft* (German Physical Society) on October 19, 1900 [366]. Ostensibly, their work re-confirmed what Lummer and Pringsheim had found earlier — the failure of Wien's distribution law at the long wavelength, high temperature limit. Had it been for this alone, Rubens and Kurlbaum's work would have represented further experimental evidence towards the long sought-after distribution function for a blackbody.

Two weeks previously, on a visit to Planck, Rubens had shared with him the results of his latest experimental findings. While no doubt a fortuitous occasion in the history of physics, it prompted Planck to abandon his previously held commitment to Wien's distribution law in favor of what, as we have seen, turned out to be a slightly modified form of it.[22] Planck presented this publicly for the first time after Kurlbaum's report before the *Deutschen Physikalischen Gesellschaft* on October 19, 1900 [511].

One week later on 25 October, 1900, before the *Königlich Preussische Akademie der Wissenschaften* (Royal Prussian Academy of Sciences), the results of Rubens and Kurlbaum's work were presented more fully [572]. Here their experimental findings were compared to a number of distributions functions which had by now been proposed. These included those by Thiesen, Wien, Rayleigh, Lummer and Jahnke, along with Planck's latest result. Planck's formula was a last minute addition and was only included after their experimental work had been concluded. Of the five distribution functions presented, Rubens and Kurlbaum concluded that the one which reproduced their observations most closely within the limits of experimental error was that of Planck's.

After some debate concerning the validity of certain experimental aspects between Paschen, who continued to support Wien's result, and Lummer and Pringsheim, who were firmly of the view that Planck's equation fitted their data better [488, 321, 415], by the end of 1901 all agreed Planck's equation fitted the available experimental data better than any of the previously proposed laws [487, 405, 533, 534, 573, 574]. The closeness in the agreement between the predictions made by Planck's equation and experiment moved Pringsheim to write a few years later:

> Planck's equation is in such good agreement with experiment that it can be considered, at least to a very close approximation, as the mathematical expression for Kirchhoff's function.[23]

Kirchhoff's universal function had been finally found. It had only taken the best part of forty years to find since first being postulated and its form turned out to be not as simple as had originally been supposed. In the immediate years after its introduction Planck's equation would be confirmed to a high degree of accuracy [58, 570, 672, 575, 576, 621, 151] and thereafter accepted by all[24] as the universal law describing thermal radiation from a blackbody.

1.4 THERMAL LAWS FROM DIMENSIONAL ANALYSIS

The success in deriving theoretically the Stefan–Boltzmann law and Wien's homologous law had pushed to the very limit what was possible using only the laws of classical physics and thermodynamics. Beyond this, as we have seen, further understanding on the nature of the blackbody problem would only come from the adoption of ideas in physics that represented a radical break from the past. Despite this, only a short time after the fact did it emerge both the Stefan–Boltzmann law and Wien's homologous law could be found using the nascent ideas of a technique that would later become known as dimensional analysis.

For any physical problem, the main objective is in obtaining relationships among the quantities that characterize the phenomenon under consideration. Dimensional analysis is one way relationships between the physical quantities involved in a particular phenomenon can be found. It is a technique which restructures the original dimensional variables of a problem into a smaller subset of dimensionless products using the constraints imposed upon them by their dimensions. In doing so it allows one to identify the form a particular physical law should take even before its final form is known. While dimensional analysis is not able to deliver the exact mathematical form for a particular relationship sought, it nevertheless is an important tool that can be used in helping to better understand and characterize physical phenomena.

Essentially, any physically meaningful problem described by an equation must have the same dimensions on either side of the equality and stems from a requirement of dimensional homogeneity. While the idea behind dimensional analysis is rather simple and was known to Isaac Newton [617], it was not formalised in the literature in a form that could be readily understood by many until the work of Edgar Buckingham (1867–1940) in 1914 and his so-called Π-theorem [111].[25]

Because the Buckingham Π-theorem is central to the discussion which is about to follow, we give a clear statement of it before applying it to the problem of blackbody radiation.

Theorem 1.1: Buckingham Π-theorem

Let $q_1, q_2, \ldots, q_n$ be n dimensional variables that each represent the magnitude of a physical quantity relevant to the given problem and are inter-related by an (unknown) dimensionally homogeneous set of equations which can be expressed by a functional relationship of the form

$$f(q_1, q_2, \ldots, q_n) = 0.$$

If k is the number of different physical dimensions required to describe the n variables, then there are $n - k$ dimensionless groups $\Pi_1, \Pi_2, \ldots, \Pi_{n-k}$ constructed from $q_1, q_2, \ldots, q_n$ by $n - k$ dimensionless equations.

In terms of the Π-groups the functional relationship for f reduces to

$$\Phi(\Pi_1, \Pi_2, \ldots, \Pi_{n-k}) = 0.$$

The Π-groups are all dimensionless monomials which are independent of each other and are of the form

$$\Pi_i = \prod_{j=1}^{n} q_j^{\alpha_{j,i}}, \quad 1 \leqslant i \leqslant n - k.$$

Here the exponents $\alpha_{j,i}$ are rational numbers and are chosen so as to make each Π_i monomial dimensionless. ∎

For a proof of Buckingham's Π-theorem and the many refinements and improvements that have been made to the theorem and its proof since its initial conception, the reader is directed to the relevant literature [102, 147, 312, 85, 220, 154, 64, 440, 522].

The Buckingham Π-theorem tells one how many dimensionless Π monomials are needed from a given list of dimensional variables used to describe the problem. Importantly, it gives the new dimensionless relation in terms of k fewer variables compared to the original dimensional relation. The choice for the set of Π monomials however, is not unique. The Buckingham Π-theorem provides a way to generate such a set, but is not able to choose among all the sets found which is the most physically meaningful. In such cases a reduction in their number can typically be made by an appeal to some aspect of the physical phenomenon under consideration. However, once found the Π monomials form a complete, independent set.

In application, dimensional analysis rests heavily on assumptions made concerning the number and nature of the physical constants involved. In the case of thermal radiation the English physicist Sir James Hopwood Jeans (1877–1946) was the first to apply the method of dimensional analysis in a schematic way to the problem of a blackbody [327]. At the time the work was performed, in 1905, not only was the theory of dimensional analysis not fully developed but the central role the universal constants h, Planck's constant, and k_B, the Boltzmann constant, play in blackbody radiation was yet to be fully appreciated. Unsurprisingly, due to the inherent non-uniqueness of the solution in dimensional analysis Jeans' work was heavily criticized by the Austrian-born Dutch theoretical physicist Paul Ehrenfest (1880–1933) shortly after its publication [205]. In a series of exchanges [329, 206, 330] within the German periodical *Physikalische Zeitschrift*, each rejected the other's criticisms and counterclaimed with new arguments of his own. The nature of the criticisms as far as Ehrenfest was concerned rested entirely on the essentially arbitrary decision made by Jeans in his choice for the set of Π monomials. As each Π monomial is dimensionless, so too is any power, and the problem becomes one of choosing the correct power.

Only later, once the theory of dimensional analysis had been more fully developed, particularly the formal enunciation following the Buckingham Π-theorem, and the acceptance by physicists of the central importance Planck's and Boltzmann's constants both play in statistical phenomena such as thermal radiation, were the initial criticisms levelled against Jeans' use of dimensional

Table 1.1

Physical variables and fundamental constants for the problem of finding the form for the total radiant exitance.

Quantity	Symbol	SI unit	Dimension
Total radiant exitance	M_e^b	kg/s^3	MT^{-3}
Temperature	T	K	Θ
Speed of light	c	m/s	MT^{-1}
Boltzmann's constant	k_B	m^2·kg/(s^2·K)	$ML^2T^{-2}\Theta^{-1}$
Planck's constant	h	m^2·kg/s	ML^2T^{-1}

analysis finally laid to rest. As can be seen in the later work of Carl David Tolmé Runge [578], Ludwig Hopf [297], Fritz Reiche [552], and others [613, 214, 475, 42, 642, 639], the application of dimensional analysis is now well founded and it is from this modern vantage point that we proceed.

Choose M for the dimension of mass, L for the dimension of length, T for the dimension of time, and Θ for the dimension of absolute temperature as a basis for the basic physical dimensions of the problem. To find the form for the total radiant exitance, namely the form for the Stefan–Boltzmann law, we postulate $F(q_1, q_2, q_3, q_4, q_5) = F(M_e^b, T, c, k_B, h)$. Here $q_1 = M_e^b$ is the total radiant exitance, $q_2 = T$ is the absolute temperature, $q_3 = c$ is the speed of light in a vacuum, $q_4 = k_B$ is Boltzmann's constant, while $q_5 = h$ is Planck's constant. The five dimensional quantities (two physical variables and three fundamental constants), their SI units, and associated dimensions are listed in Table 1.1.

As the number of dimensional quantities in this problem is five while the number of different fundamental dimensions is four (mass, length, time, and absolute temperature), there is only one dimensionless Π monomial for this problem since $n - k = 5 - 4 = 1$. Here

$$\Pi_1 = \prod_{j=1}^{5} q_j^{\alpha_j} = (M_e^b)^{\alpha_1} T^{\alpha_2} c^{\alpha_3} k_B^{\alpha_4} h^{\alpha_5}. \tag{1.20}$$

The exponents α_1 to α_5 are unknown. They are determined in such a way that when found, Π_1 as given by Eq. (1.20) will be dimensionless. To do this we first construct a table for the exponents of the fundamental dimensions for the five dimensional quantities. These we list in Table 1.2.

The unknown exponents can now be readily found using matrix methods from linear algebra. Writing a dimensional matrix whose rows are the four dimensions of Table 1.2 and whose columns are the five variables of Table 1.2

Table 1.2

Exponents for the fundamental dimensions corresponding to the five dimensional quantities related to finding the form for the total radiant exitance.

	M_e^b	T	c	k_B	h
L	0	0	1	2	2
M	1	0	0	1	1
T	-3	0	-1	-2	-1
Θ	0	1	0	-1	0

the problem reduces to solving the following homogeneous system of linear equations

$$
\begin{bmatrix}
0 & 0 & 1 & 2 & 2 \\
1 & 0 & 0 & 1 & 1 \\
-3 & 0 & -1 & -2 & -1 \\
0 & 1 & 0 & -1 & 0
\end{bmatrix}
\begin{bmatrix}
\alpha_1 \\ \alpha_2 \\ \alpha_3 \\ \alpha_4 \\ \alpha_5
\end{bmatrix}
=
\begin{bmatrix}
0 \\ 0 \\ 0 \\ 0
\end{bmatrix}.
\tag{1.21}
$$

As there are four equations in five unknowns, the system will not have a unique solution. Instead the solution to the system will be in terms of a non-zero parameter. How the linear system is solved is a matter of personal taste. We prefer to solve it by reducing it first to row echelon form by performing elementary row operations on the argumented dimensional matrix before back substituting. If we set $\alpha_1 = t$, where the parameter t is a non-zero real number, one finds for the solution to the linear system: $\alpha_1 = t$, $\alpha_2 = -4t$, $\alpha_3 = 2t$, $\alpha_4 = -4t$ and $\alpha_5 = 3t$. Equation (1.20) becomes

$$
\Pi_1 = (M_e^b)^t T^{-4t} c^{2t} k_B^{-4t} h^{3t},
\tag{1.22}
$$

or after rearranging

$$
M_e^b(T) = \frac{\Pi_1' k_B^4}{c^2 h^3} T^4 \propto T^4.
\tag{1.23}
$$

Note here we have written $\Pi_1' = \sqrt[t]{\Pi_1}$, and like Π_1, is just another dimensionless number to be determined.

The proportionaly of the total radiant exitance to the fourth power of temperature is readily apparent and is in accordance with the Stefan–Boltzmann law. In Chapter 2 the dimensionless number Π_1' will be shown to be equal to $2\pi^5/15$, it being obtained from further knowledge of the physics of the problem.

Table 1.3

Physical variables and fundamental constants for the problem of finding the form for the spectral radiant exitance.

Quantity	Symbol	SI unit	Dimension
Spectral radiant exitance	$M_{e,\lambda}^b$	kg/(m·s^3)	$ML^{-1}T^{-3}$
Wavelength	λ	m	L
Temperature	T	K	Θ
Speed of light	c	m/s	MT^{-1}
Boltzmann's constant	k_B	m^2·kg/(s^2·K)	$ML^2T^{-2}\Theta^{-1}$
Planck's constant	h	m^2·kg/s	ML^2T^{-1}

So far, so good. But what else does dimensional analysis have to say about the blackbody problem? For example, what information does dimensional analysis offer one about the form of the radiation emitted from a blackbody at a given temperature, and how it is distributed throughout its spectrum as a function of wavelength? To this end, unlike the previous analysis where one was led to the functional form for the Stefan–Boltzmann law, an additional variable of the wavelength needs to be introduced. A small change from total radiant exitance to spectral radiant exitance also needs to be made.

The problem now consists of six dimensional quantities; three physical variables together with the same three fundamental constants used previously. We therefore postulate the spectral radiant exitance will be of the form $F(q_1, q_2, q_3, q_4, q_5, q_6) = F(M_{e,\lambda}^b, \lambda, T, c, k_B, h)$. Here $q_1 = M_{e,\lambda}^b$ is the spectral radiant exitance, $q_2 = \lambda$ is wavelength, $q_3 = T$ is the absolute temperature, $q_4 = c$ is the speed of light in a vacuum, $q_5 = k_B$ is Boltzmann's constant, while $q_6 = h$ is Planck's constant. Each of these quantities, their associated SI units, and dimensions are listed in Table 1.3.

With the number of dimensional quantities increased to six while the number of different fundamental dimensions remains the same at four (mass, length, time, and absolute temperature), two dimensionless Π monomials enter the problem since $n - k = 6 - 4 = 2$. Each Π monomial takes on the form

$$\Pi_i = \prod_{j=1}^{6} q_j^{\alpha_{j,i}} = (M_e^b)^{\alpha_{1,i}} \lambda^{\alpha_{2,i}} T^{\alpha_{3,i}} c^{\alpha_{4,i}} k_B^{\alpha_{5,i}} h^{\alpha_{6,i}}, \qquad (1.24)$$

where $i = 1, 2$. Table 1.4 presents the exponents of the fundamental dimensions for the six dimensional quantities.

Table 1.4
Exponents for the fundamental dimensions corresponding to the six dimensional quantities related to finding the form for the spectral radiant exitance.

	M_e^b	λ	T	c	k_B	h
L	-1	1	0	1	2	2
M	1	0	0	0	1	1
T	-3	0	0	-1	-2	-1
$\ominus$	0	0	1	0	-1	0

The six unknown exponents can be found from the following homogeneous linear system of equations

$$
\begin{bmatrix}
-1 & 1 & 0 & 1 & 2 & 2 \\
1 & 0 & 0 & 0 & 1 & 1 \\
-3 & 0 & 0 & -1 & -2 & -1 \\
0 & 0 & 1 & 0 & -1 & 0
\end{bmatrix}
\begin{bmatrix}
\alpha_{1,i} \\
\alpha_{2,i} \\
\alpha_{3,i} \\
\alpha_{4,i} \\
\alpha_{5,i} \\
\alpha_{6,i}
\end{bmatrix}
=
\begin{bmatrix}
0 \\
0 \\
0 \\
0
\end{bmatrix}. \tag{1.25}
$$

With six unknowns and only four equations no unique solution exists. Instead the solution this time will be in terms of two parameters. If we set $\alpha_{1,i} = t$ and $\alpha_{6,i} = \tau$ where t and τ are parameters which cannot simultaneously both be equal to zero, the following solution is found: $\alpha_{1,i} = t$, $\alpha_{2,i} = 4t - \tau$, $\alpha_{3,i} = -t - \tau$, $\alpha_{4,i} = -t + \tau$, $\alpha_{5,i} = -t - \tau$, and $\alpha_{6,i} = \tau$. With such a solution the question becomes one of how to proceed from here. A judicious choice in the values for either parameter is needed. In making such a choice, as a guide, we demand the spectral radiant exitance should occur explicitly in the functional relation found. To achieve this, the following choices are made. For $i = 1$ choose $t = 1$ and $\tau = 0$ while for $i = 2$ choose $t = 0$ and $\tau = 1$.

With each of these choices we arrive at the following solutions. For $i = 1$ one has $\alpha_{1,1} = 1$, $\alpha_{2,1} = 4$, $\alpha_{3,1} = -1$, $\alpha_{4,1} = -1$, $\alpha_{5,1} = -1$, and $\alpha_{6,1} = 0$. While for $i = 2$ one finds $\alpha_{1,2} = 0$, $\alpha_{2,2} = -1$, $\alpha_{3,2} = -1$, $\alpha_{4,2} = 1$, $\alpha_{5,2} = -1$, and $\alpha_{6,2} = 1$. The two Π monomials can therefore be formed as follows

$$
\Pi_1 = (M_{e,\lambda}^b)^1 \lambda^4 T^{-1} c^{-1} k_B^{-1} h^0 = \frac{M_{e,\lambda}^b \lambda^4}{k_B c T}, \tag{1.26}
$$

and

$$
\Pi_2 = (M_{e,\lambda}^b)^0 \lambda^{-1} T^{-1} c^1 k_B^{-1} h^1 = \frac{hc}{k_B \lambda T}. \tag{1.27}
$$

From the Buckingham Π-theorem, the functional relationship between the Π monomials can be expressed as

$$\Phi(\Pi_1, \Pi_2) = 0,$$

or equivalently as

$$\Pi_1 = \phi(\Pi_2). \tag{1.28}$$

Substituting Eqs (1.26) and (1.27) into (1.28), after rearranging algebraically, yields

$$M_{e,\lambda}^{b}(\lambda, T) = \frac{k_B T c}{\lambda^4} \phi\left(\frac{hc}{k_B \lambda T}\right) = \frac{hc^2}{\lambda^5}\left(\frac{hc}{k_B \lambda T}\right)^{-1} \phi\left(\frac{hc}{k_B \lambda T}\right), \tag{1.29}$$

or

$$M_{e,\lambda}^{b}(\lambda, T) = \frac{hc^2}{\lambda^5} \varphi\left(\frac{hc}{k_B \lambda T}\right). \tag{1.30}$$

Here $\varphi(hc/(k_B \lambda T)) = [hc/(k_B \lambda T)]^{-1}\phi(hc/(k_B \lambda T))$ is an unknown universal function. Equation (1.30) is the form one expects for the spectral radiant exitance for a blackbody and is nothing more than Wien's homologous law. The unknown universal function necessarily comes from an understanding of the physics of thermal radiation and was something, as we saw, that had to wait until the work of Planck at the beginning of the twentieth century.

The form for the spectral radiant exitance Jeans arrived at in 1905 was

$$M_{e,\lambda}^{b}(\lambda, T) = \lambda^{-4} T f(\lambda T), \tag{1.31}$$

where f was an unknown universal function. In his analysis Jeans did not use either of the fundamental constants of Planck or Boltzmann. Rather, in their place he used the electronic charge and mass of an electron, q_e and m_e respectively, and the ideal gas constant R. Also included was a quantity related to the inductive capacity of the aether K. Jeans' form as given by Eq. (1.31) was only finally arrived at after an assumption about the size of one of the monomials being small was made and accordingly dropped, together with the dropping of the four constants q_e, m_e, R, and K; something he was heavily criticized for doing by Ehrenfest.

Wien too, writing in 1909 in his article "Theorie der Strahlung" (Theory of radiation) for the *Encyklopädie der Mathematischen Wissenschaften*, also did not find Jeans' use of dimensional methods in his derivation of the homologous law entirely convincing [689]. Wien particularly objected to the approximation Jeans used which he thought had been introduced in a rather ad hoc fashion and as such could not be regarded as a proof for his homologous law. The same year of 1909 also saw Albert Einstein (1879–1955) apply a dimensional consideration to the blackbody problem [208]. Unlike Wien, Einstein, who was also aware of Jeans' dimensional work, thought a dimensional consideration of the blackbody problem was extremely important as it allowed him to confirm which quantities played a role in the law of thermal radiation. By slightly modifying some of those points used by Jeans in his analysis, Einstein proceeded to give a brief account of his own development of the problem from a dimensional perpective before arriving at Wien's homologous law.

While Jeans was the first to apply the emerging methods of dimensional analysis in a systematic way to the problem of a blackbody, evidently he was not the first to apply such methods. That honor instead seems to go to the English theoretical physicist Henry Cabourn Pocklington (1870–1952). In a paper read before the Seventieth Meeting of the British Association for the Advancement of Science held in September, 1900, using dimensional analysis Pocklington was able to show the total radiant exitance was proportional to the fourth power of temperature and the spectral radiant exitance could be expressed in the form

$$M_{e,\lambda}^{b}(\lambda, T) = T^4 \lambda^{-1} c q_e^{-6} g\left(\frac{\lambda T}{q_e^2}\right), \tag{1.32}$$

where g is an unknown universal function. Like Jeans, Pocklington had not used either Planck's constant or Boltzmann's constant in his analysis but unlike Jeans had used the electronic charge of an electron as an additional fundamental constant.

Unfortunately, the account of Pocklington's work given in the Association's subsequently published *Report of the Meeting* [523] appears only in adumbrate form. It was further summarised in an account of the Association's meeting which appeared in the October 4 issue of *Nature* of 1900 [384]. Both summaries, lacking details on how he obtained his results, simply list the results of the work. Other than a single reference to Pocklington's work buried deep within a footnote of Heinrich Gustav Johannes Kayser's (1853–1940) magisterial tome *Handbuch der Spectroscopie* of 1902 [345], the work was largely overlooked and quickly forgotten. In all likelihood it is probably correct to infer Jeans' work of 1905 was performed without prior knowledge of Pocklington's earlier results.

The two forms as given by Jeans and Pocklington are equivalent if we note

$$M_{e,\lambda}^{b}(\lambda, T) = \lambda^{-4} T \left[c\left(\frac{\lambda T}{q_e^2}\right)^3 g\left(\frac{\lambda T}{q_e^2}\right)\right] = \lambda^{-4} T f(\lambda T), \tag{1.33}$$

and each in turn is equivalent to the form we found using dimensional analysis since from Eq. (1.29) one can write

$$M_{e,\lambda}^{b}(\lambda, T) = \lambda^{-4} T \left[k_B c \phi\left(\frac{hc}{k_B \lambda T}\right)\right] = \lambda^{-4} T f(\lambda T). \tag{1.34}$$

Interestingly, without knowing any further physical information about the nature of the blackbody problem Wien's displacement law can be shown to follow directly from Eq. (1.30). To see this we seek the value of λ for which

$M_{e,\lambda}^{b}$ is a maximum for a given temperature. Assuming one exists, a fact which in the absence of any physical theory can only be verified experimentally,[26] it can be found from $\partial M_{e,\lambda}^{b}/\partial \lambda = 0$. Letting $z = hc/(k_B \lambda T)$ we find for the partial derivative

$$\frac{\partial M_{e,\lambda}^{b}}{\partial \lambda} = -\frac{hc^2}{\lambda^6}[5\varphi(z) - z\varphi'(z)]. \tag{1.35}$$

Setting the result equal to zero the position of the maximum is given by the equation

$$5\varphi(z) - z\varphi'(z) = 0. \tag{1.36}$$

Denoting the real positive root of this equation as $z = z^*$ corresponding to $\lambda = \lambda_{\text{max}}$, it follows that

$$\lambda_{\text{max}}T = \frac{hc}{k_B z^*}, \tag{1.37}$$

where $hc/(k_B z^*)$ corresponds to Wien's displacement constant b. The value for the constant b can only be determined once the form for the unknown function φ is known. So we have arrived at the extraordinary result of being able to deduce Wien's displacement law using arguments whose roots lay in dimensional analysis alone.

The form found for the spectral radiant exitance allows the total radiant exitance to be readily found. If we integrate Eq. (1.30) over all wavelengths one has

$$M_e^{b} = \int_0^{\infty} M_{e,\lambda}^{b}(\lambda, T)\, d\lambda = hc^2 \int_0^{\infty} \frac{1}{\lambda^5}\varphi\left(\frac{hc}{k_B \lambda T}\right) d\lambda. \tag{1.38}$$

After the change of variable $x = hc/(k_B \lambda T)$ is made one finds

$$M_e^{b} = \left[\frac{k_B^4}{h^3 c^2} \int_0^{\infty} x^3 \varphi(x)\, dx\right] T^4. \tag{1.39}$$

Assuming the improper integral $\int_0^{\infty} x^3 \varphi(x)\, dx$, which is dimensionless, converges, it will be a number and evidently Eq. (1.39) takes on the form of the Stefan–Boltzmann law, namely σT^4, where σ is a constant. The constant itself, which again in the absence of any physical theory, must be determined experimentally and is nothing more than the well known Stefan–Boltzmann constant. The Stefan–Boltzmann law therefore follows directly from Wien's homologous law.

Lastly, the maximum for the spectral radiant exitance $M_{e,\lambda_{\text{max}}}^{b}$ can be found by substituting $\lambda = \lambda_{\text{max}}$ into Eq. (1.30). By making use of Wien's displacement law, namely $\lambda_{\text{max}}T = b$, one finds

$$M_{e,\lambda_{\text{max}}}^{b} = \frac{hc^2}{\lambda_{\text{max}}^5}\varphi\left(\frac{hc}{k_B \lambda_{\text{max}}T}\right) = \left[\frac{hc^2}{b^5}\varphi\left(\frac{hc}{k_B b}\right)\right]T^5 = aT^5. \tag{1.40}$$

Here a is a constant since the argument for the universal function φ is constant. This shows the maximum value in the spectral peak for a blackbody is proportional to the fifth power of the temperature.

1.5   TRANSITION AND NEW BEGINNINGS

The years between 1895 and 1901 were no doubt a period of intense theoretical and experimental investigation into the nature of blackbody radiation. Within the space of a few short years the pace of development in understanding the fundamental laws associated with blackbody radiation increased markedly.

Commenting on the almost frenetic pace of development in understanding made during the latter half of the final decade of the nineteenth century, observing from afar a few years after the event, Arthur L. Day and C. E. Van Orstrand from the US wrote:

Tentative advances and partial solutions of the problem of [blackbody] radiation have been made in considerable number, greatly stimulated by the construction recently of an experimental blackbody capable of such exact manipulation as to outstrip, for the moment, the analytical advances. The steps in this development have almost all been contributed from abroad, and they have followed each other with such rapidity that those of us who have been compelled to follow their progress from a distance have hardly been able to see what has stood and what has fallen in the keen contest between the mathematical and experimental development.[27]

Experimentally, a number of modifications made to the design of Langley's original bolometer produced a precision instrument capable of measuring minute quantities of radiation over a broad range of wavelengths from the visible to the far infrared. Together with the final realization of a usable black body during this period, experimentalists were finally in a position to uncover and carefully test the fundamental laws of radiation. As we saw, these efforts were led by the German experimental physicists Paschen and Wanner, Lummer and Pringsheim, and Rubens and Kurlbaum. Theoretically, a number of worthy spectral distribution functions proposed for a blackbody came from Wien in 1896, Lord Rayleigh in 1900, Max Ferdinand Thiesen in 1900, and finally Planck himself, first in 1900, and more fully in 1901.

While our understanding of the blackbody problem increased tremendously during this period, thereafter, as attention quickly turned from uncovering to verifying the thermal laws of radiation to a greater degree of precision experimentally and in refining and trying to better understand these laws theoretically, the attention and focus of others slowly started to shift as they began to apply these laws to a range of applications [416, 535].

In an age before digital computers the cumbersome mathematical form of Planck's law made it particularly toilsome to work with. In those early days, whenever Planck's law was needed it meant the resulting hand computations that followed were always going to be an especially lengthy and onerous task. Naturally this would have been only a minor inconvenience to a handful of people had it not been for the multitude of applications that would come to be found for thermal radiation in the years that followed. With large numbers of people being inconvenienced by laborious intermediate calculations, it was only a matter of time before ways would be found to avoid what was otherwise painfully slow and burdensome work. To this end, it was not long before the first tables consisting of tabulations based on Planck's equation appeared. The great age of computation related to thermal radiation had begun. These, together with other aids produced shortly after Planck introduced his radiation law up until modern times, will be taken up and discussed at great length in the chapters to follow.

Notes

[1] How much radiation is found in the various parts of the electromagentic spectrum as a function of an object's temperature at its surface is taken up and considered in Section 2.14 of Chapter 2.

[2] The faint glow first seen by the eye in the visible part of the spectrum as the temperature of an object increases is occasionally referred to as "gespenster grau" or "ghost grey."

[3] The term blackbody, *Schwarzer Körper* in German, was first coined by Kirchhoff in his paper 'Ueber das Verhältniss zwischen dem Emissionsvermögen und dem Absorptionsvermögen der Körper für Wrme and Licht' published in 1860 [354]. On page 277 he writes: Ich will solche Körper *vollkommen schwarze*, oder kürzer *schwarze*, nennen. In the same year, when published in translation in the *Philosophical Magazine*, it was rendered into English as: I shall call such bodies *perfectly black*, or, more briefly, *black* bodies (the italics given here appeared in each of the original publications) [355]. The term was, however, not without its detractors. For example, in 1904 Aurthur L. Day and C. E. Van Orstrand writing in *The Astrophysical Journal* commented that given an experimental blackbody in the form of a cavity radiator which only starts to glow visibly at a temperature of about $525°C$, the terms "blackbody" and "black radiation" were somewhat misleading at higher temperatures as the body, at least to one's eye, no longer appeared black [173]. A decade later, during the discussion of a paper by Clifford Copland Paterson and B. P. Dudding on the estimation of the temperature of incandescent sources, Sylvanus P. Thompson, a professor of physics at the City and Guilds Technical College in Finsbury, England, made his now famous remark concerning the inadequacies of a language that spoke of "white" light from a self-luminous object in terms of the amount of radiation radiated from a "black" body [491].

[4] As Kirchhoff had shown, the universal function associated with blackbody radiation was a function of wavelength and temperature only. Since it did not depend on the nature of the body from which the blackbody itself was made he expressed his expectation that such a function should be simple as all functions encountered up to his time which did not depend on the properties of a body were simple in form. How wrong he would turn out to be.

[5] Lord Rayleigh, 1889 'On the character of the complete radiation at a given temperature,' *The London, Edinburgh, and Dublin Philosophical Magazine and Journal of Science* [Fifth Series], **27**(169), p. 460.

[6] In the past the transliteration of Michelson's first name from the Cyrillic to the Latin script was rendered as "Wladimir," and hence in the literature his work appears under the author of "W. Michelson" or "Wladimir Michelson." Today the accepted transliteration is "Vladimir."

[7] Though the form for the spectral distribution of a blackbody advanced by Weber was soon found to be incorrect, he was the first to correctly suggest the wavelength where the maximum in the spectral distribution curve occurred moved towards shorter wavelengths as temperature increased took the form $\lambda_{max}T = \text{constant}$. That this is the case will be shortly seen.

[8] All values quoted for the fundamental constants throughout this book are those of the 2014 adjusted values recommended by the Committee on Data for Science and Technology

(CODATA) for international use [466]. For the Stefan–Boltzmann constant, its value is $5.670\,367 \times 10^{-8}$ W·m^{-2}·K^{-4}

[9]The story of Štefan's empirical deduction of the law that now partly bears his name is an interesting one [588, 477, 171, 189]. Based on earlier experimental work performed by the Irish physicist John Tyndall (1820–1893) in 1865 [651], where Tyndall found the radiation emitted from a spiral of platinum wire heated by an electric current was 11.7 times as great when the wire appeared white-hot compared to when it appeared red-hot. Štefan, estimating each of these temperatures to be 1200°C and 525°C respectively, noticed the ratio of 1473 K to 798 K raised to the fourth power was equal is 11.6, and based on this evidence, stated the law empirically which he thought was valid for all radiating bodies. However, as Boltzmann later showed, the fourth power law was only valid for blackbody radiation, something the radiation from platinum is far from.

[10]As we point out in a few paragraphs' time, Wien did not actually express the result in the form given here. Instead the result Wien gives turns out to be equivalent to the result we give here, and as we shall see, is the far more convenient form in which to express this result. It seems Wien himself did not express his law explicitly in this form until 1909 [689].

[11]Calling this law *Wien's law of homology* was first suggested by the Northern Irish physicist Sir Joseph Larmor (1857–1942) in 1900 [375]. While it was the term Larmor used when referring to the law in his later writings [376, 377], the term did not gain wide acceptance. This is unfortunate as the term homologous is particularly germane. In order to avoid the inexorable confusion a single eponym related to the same person covering several slightly different but interrelated laws inevitably causes, it is the term we intend to use.

[12]Its currently accepted value is $2.897\,7729 \times 10^{-3}$ m·K. In the past it was thought Wien's displacement constant b could be expressed only as a numerical value due to the appearance of a mathematical constant arising from the non-trivial root of a transcendent equation. As we shall see in the following chapter it is possible to express the constant in terms of a number of alternative closed-form expressions.

[13]The name "displacement law" (in German *Verschiebungsgesetz*) first appeared in 1899 in a paper by the two German experimental physicists Otto Lummer and Ernst Pringsheim [410]. In the Russian literature the law is often referred to as the *Golitsyn–Wien law* [268]. Named after the Russian physicist and seismologist Boris Borisovich Golitsyn (1862–1916), it recognizes his independent theoretical discovery of the law. In his master's thesis "Investigations in mathematical physics" presented to the Faculty of Mathematics and Physics of Moscow University in 1893 Golitsyn, in the second part to his thesis, considered theoretically a number of aspects relating to thermal radiation [188, 349]. Presented are several new ideas relating to the problem of thermal radiation. In addition to arriving at Wien's displacement law theoretically, Golitsyn introduced the idea of "thermal temperature." A radical idea for its time, but now considered essential for a proper understanding of the problem, Golitsyn suggested that just as material substances can be assigned a certain temperature, so too can the radiant energy emitted from a blackbody. The idea was ferociously attacked by a number of eminent physicists at Moscow University, causing Golitsyn not only to leave Moscow but to switch fields altogether; he would go on to become one of the founding fathers of seismology.

[14]Im normalen Emissionsspectrum eines schwarzen Körpers verschiebt sich mit veränderter Temperatur jede Wellanlänge so, dass das Product aus Temperaur und Wellenlänge constant bleibt. W. Wien, 1893 'Eine neue Beziehung der Strahlung schwarzer Körper zum zweiten Hauptsatz der Wärmetheorie,' *Sitzungsberichte der Königlich Preussischen Akademie der Wissenschaften zu Berlin* (16 Februar 1893), p. 62.

[15]Actually the Rayleigh–Thiesen form of Wien's homologous law which Thiesen independently proposed.

[16]Lord Rayleigh, 1900 'Remarks upon the law of complete radiation,' *Nature*, **49**(301), p. 540.

[17]The values for these constants have evolved over time as experimental methods and techniques have improved. The latest 2014 adjusted values recommended by the CODATA for these constants are: $2.997\,924\,58 \times 10^8$ m·s^{-1} for the speed of light, a value now taken to be exact, $6.626\,070\,040 \times 10^{-34}$ J·s for Planck's constant, and $1.380\,648\,52 \times 10^{-23}$ J·K^{-1} for

the Boltzmann constant. For an interesting history on the evolution of Planck's constant see Steiner [620].

[18]Interestingly, Boltzmann himself never used the constant now named after him in his own work. In his Nobel Prize lecture of 1920 Planck wrote [520] "This constant is often referred to as Boltzmann's constant, although, to my knowledge, Boltzmann himself never introduced it – a peculiar state of affairs, which can be explained by the fact that Boltzmann, as appears from his occasional utterances, never gave thought to the possibility of carrying out an exact measurement of the constant."

[19]This quintuple, with the exception of Pringsheim, all worked at one time or another in either the optics laboratory of the newly created *Physikalisch-Technische Reichsanstalt* located in Berlin-Charlottenburg or at the nearby *Technischen Hochschule Berlin-Charlottenburg*.

[20]An interesting sentiment on the loss of such figures from the history of science can be found at the beginning of Allan Franklin's delightful little book *The neglect of experiment* [239]. Franklin writes:

> One of the great anticlimaxes in all of literature occurs at the end of Shakespeare's *Hamlet*. On a stage strewn with noble and heroic corpses – Hamlet, Laertes, Claudius, and Gertrude – the ambassadors from England arrive and announce that "Rosencrantz and Guildenstern are dead." No one cares. A similar reaction might be produced among a group of physicists, or even among historians and philosophers of science, were someone to announce that "Lummer and Pringsheim are dead."

[21]Before the time of the cavity radiator, artificially blackened sheets of metal had been used but when heated these emitted radiation that was only very approximately black within a very narrow range of wavelengths.

[22]We will re-tell this chance encounter in greater detail in Chapter 7, but importantly for the moment what the work of Rubens and Kurlbaum showed was the major role the experiment had not only on Planck's work but on influencing a major new discovery in physics. It is in this wider context the work of Rubens and Kurlbaum is now seen as a classic of its type, emblematic of a new, closer interaction that was emerging between experimentalists and theoreticians by 1900.

[23]Die Plancksche Gleichung steht mit der Erfahrung in so guter Übereinstimmung, daß sie mindestens mit sehr großer Annäherung als der mathematische Ausdruck der Kirchhoffschen Funktion S_λ gelten kann. E. Pringsheim, 1904 'Über die Strahlungsgesetze,' *Archiv der Mathematik und Physik*, **7**(3), p. 253.

[24]For a number of years after Planck presented his form for the distribution law for a blackbody several other alternative forms continued to be proposed [131, 132, 176, 177, 178, 70, 117, 118, 104, 389, 663, 664, 530, 531, 532, 637, 152, 153]. The level of agreement between what each of these alternative forms predicted theoretically to what was observed experimentally, however, were unable to surpass those predictions made by Planck's law.

[25]The French mathematician and physicist Jean Baptiste Joseph Fourier (1768–1830) is considered the founder of dimensional analysis in the modern sense [475] as it was he who first applied its general methods to physical quantities [233]. Later, in 1871, James Clerk Maxwell introduced the symbolism still currently in use to denote the dimensions of a quantity such as M for the dimension of mass, L for the dimension of length, and T for the dimension of time [432]. But it is to Lord Rayleigh and his "method of dimensions" (or "method of similitude") the method finally found its greatest earlier supporter and advocate. Starting in 1871 Lord Rayleigh applied the methods of dimensional analysis in his study of the color of the sky [542] and would go on to make extensive use of its methods and extended its ideas considerably [543, 546, 551]. The first attempt at a general statement of what is now known as the Buckingham Π-theorem came in 1892 from the French mathematician A. Vaschy [655] though special cases of the theorem had been proved by J. Bertrand in 1878 who was also work working in France in connection to problems relating to heat conduction and electrodynamics [80]. Vaschy's work however, went unnoticed and it was later independently rediscovered by others. This can be seen most noticeably in the 1911 work of the two Russians, D. Riabouchinsky [557] and A. K. Federman [223], and in 1914 of

the American Edgar Buckingham [111] to whom the theorem now owes its name. Given that the work of Vaschy predates Buckingham's own work on his theorem by some twenty-two years, in the French literature it is often referred to as *le Théorème de Vaschy-Buckingham*. Further details on the historical development of dimensional analysis can be found in the work of Macagno [421], Görtler [265], and de A. Martins [174].

[26] Experimentally it is known the spectral distribution curve for a blackbody vanishes in the long ($\lambda \to \infty$) and short ($\lambda \to 0$) wavelength limits and so must have at least one maximum at some finite value for the wavelength.

[27] Day, A. L. and C. E. Van Orstrand, 1904 "The blackbody and the measurement of extreme temperatures," *The Astrophysical Journal*, **19**(1), p. 2.

Section II

Theoretical and numerical matters

2 Theoretical developments

In this chapter we consider a number of consequences stemming directly from Planck's law. Each is developed in some detail and many of the results given here will be needed for the numerical work that is to follow in the next chapter. A discussion on the important differences between two types of cases one can consider, be it the *radiometric* case or the *actinometric* case, is also given. It is shown how each case can be closely associated to the type of physical mechanism involved in the detection of thermal radiation and the advantages to be gained from doing so. Following an approach whereby Planck's law for thermal radiation is written in terms of a polylogarithm function; a special function that can be thought of as a generalisation of the natural logarithmic function; closed, analytic expressions in terms of this special function for the important class of functions known as the blackbody fractional functions of the first and second kinds will be found. But first we commence with a discussion on the nature of spectral representations and how they affect interpretations relating to Planck's law.

2.1 SPECTRAL REPRESENTATIONS

The spectral distribution of radiation emitted from a blackbody into a hemispherical region in space is given by Planck's law for thermal radiation. Typically this law is given in one of two common spectral representations — that of either wavelength or frequency. So far, whenever Planck's law has been referred to or considered it has been within the spectral context of a linear wavelength representation, though as we are about to see, many other alternative spectral representations are possible and depend on the spectral quantity chosen for the independent variable.

Depending on the spectral representation selected, the functional form for Planck's law will be different. Consequences stemming from these differences continue to not be fully appreciated by many and are responsible for no end of confusion. Misunderstandings in the interpretation of various features relating to different functional forms are often erroneously viewed as producing paradoxical results. It is not until one recognizes that Planck's law is an example of a distribution function which is differential in nature depending on the independent variable chosen does the confusion surrounding a host of seemingly paradoxical results begin to disappear. Appreciating the role that the choice in spectral representation plays is therefore essential in the correct interpretation of Planck's law.

The problem of how to represent spectra is a very old one. Dating back to the turn of the nineteenth century soon after the wave nature of light was first recognised, the work of Thomas Young (1773–1829), Joseph von

Fraunhofer (1787–1826), and others, quickly established wavelength as the spectral parameter of choice. The consequences stemming from this decision are still with us today, as for many, wavelength remains the spectral parameter of choice.

In 1847 the English-born American natural philosopher John William Draper (1811–1882) undertook a series of investigations into the radiation emitted from a platinum source as its temperature increased [190]. Draper pointed out the limitations in using a prism to analyze the radiation emitted from the surface of platinum compared to the use of a diffraction grating. While Draper clearly understood the advantages to be gained from the use of a diffraction grating over a prism he was forced to the later, as the intensity of radiation on passing through a diffraction grating was far too weak to be effectively measured. Compared to a prism, which causes the red end of the visible spectrum to crowd together while the blue end is unduly spread out, a grating arranges the colors of light side by side in order of their wavelength making it the obvious spectral variable of choice. And so spectra came to be universally presented in terms of wavelength, a practice that was to remain unchallenged for many years to come.

The slow mathematization of physics throughout the nineteenth century gradually saw the status of wavelength challenged. Arguments for spectral scales other than wavelength first began to appear in the later half of the nineteenth century. Commencing in 1871, at the *Forty-First Meeting of the British Association for the Advancement of Science* held in Edinburgh that year, the Anglo-Irish physicist George Johnstone Stoney (1826–1911) suggested it would be more advantageous if at least line spectra were presented on a scale equal to the reciprocal of wavelength. This he appropriately called wavenumbers; a term still with us today; and reasoned to do so was more convenient, at least from a theoretical point of view [636]. Some years later the German physicist Eugen von Lommel, in a paper on the theory of absorption and fluorescence, pointed out in a footnote that the spectrum applied along the abscissa in his present work was divided according to frequencies rather than the more familiar wavelength [394]. He reasoned

> This form of representation [frequency] is just as natural as by wavelengths, but has the advantage over it of being more closely suited to the prismatic spectrum.[1]

While Lommel was interested in advancing a theoretical understanding of absorption and fluorescence, it is a little unclear to us why he thought a scale based on frequency would be more suited to light measured using a prismatic spectrum. Despite this, he does appear to have been the first to advocate for the use of such a scale. The shift away from wavelength had began. It would be gradual at first, and would continue to rumble on for years to come, right down to the present day.

Commenting in the journal *Nature* several years later on the relative merits of the use of frequency (or wavenumber) for the representation of spectra compared to wavelength, Lord Rayleigh came down generally on the side of Stoney[2] but suggested a spectral representation intermediate between the two might be more appropriate [544]. In proposing the logarithm of the wavelength or of the frequency he was the first to do so and suggested that by employing such a scale one would obtain a mapping whereby every spectral octave, when drawn along the abscissa, would occupy the same amount of space and would perhaps give a fairer representation compared to either frequency or wavelength.

In the intervening years that followed after the initial suggestions of Stoney, Lommel, and Rayleigh, no conformity among practitioners regarding the most suitable choice in scale would be reached. Sixty years after Stoney's initial proposal William A. Shurcliff from the US lamented how general agreement between wavelength, frequency, or the logarithm of either of these quantities had not been achieved since the time of Rayleigh, he himself favoring a logarithmic frequency scale [599].

Nowadays many see frequency as the more fundamental quantity for light[3] as it remains unchanged on passing from one medium into another. Speaking in terms of frequency also has the added advantage, not realized until after Planck put forward his theory of the quanta, that the energy of a photon is directly proportional to its frequency. Despite these obvious advantages, the choice in scale made continues to be based largely on individual preference rather than on any considered theoretical basis and is how things remained until the 1960s.

The mid-1960s saw the first real concerted effort for change, calling for a shift away from wavelength to frequency. Led by a professor of biology at Harvard University in the US, George Wald was a fervent believer in the presentation of spectral data on a scale of frequency, or its equivalent such as wavenumber. Such was his conviction he met with the Board of Directors of the Optical Society of America in October 1964 and urged them to adopt the selection of frequency as official policy. He also insisted the editors of the society's two journals in print at the time, the *Journal of the Optical Society of America* and *Applied Optics*, only publish figures depicting spectral data which had been plotted using frequency along the abscissa. Not content with mere meetings, Wald openly followed his conviction through by writing a letter to the editor of the prestigious scientific journal *Science*. Subsequently published in December of the following year it forcefully expressed his many reasons for change to a frequency scale [661].

All this had its intended effect, creating a flurry of activity. First to respond was Stanley S. Ballard in his column "Optical Activities in the Universities," in the journal *Applied Optics* in February 1965, four months after Wald's initial meeting with the Optical Society of America's Board of Directors [62]. Ballard made a number of remarks on the various proposals put forward by

Wald and invited readers to send in their comments to the editor of the journal. Four months later in the June issue, the Editor of the journal John N. Howard, in his editorial wrote that while his office had not exactly been overwhelmed with responses to Ballard's request for comments, he none-the-less saw merit in Wald's suggestion and went on to reproduce Lord Rayleigh's short note from 1883 in its entirety [302]. January 1966 saw a number of letters to the editor appearing in *Science* responding mostly in the affirmative to Wald's suggestion [65]. A more extended response favouring the use of frequency appeared a few months later in a three page letter to the editor in *Applied Optics* by D. J. Baker and W. L. Brown [59]. Since this time the issue has occasionally been raised and discussed [390, 627] but to this day general agreement on the matter remains largely lacking.

2.2 TWO IMPORTANT SPECIAL FUNCTIONS

Before proceeding with a discussion of Planck's law and many of the consequences stemming from it, as the mathematical formulation to be developed depends largely on one being familiar with the polylogarithm and Lambert W functions, we begin by giving a brief propaedeutic to these two important special functions.

2.2.1 POLYLOGARITHMS

The polylogarithmic function, or polylogarithms for short, are one of the classical functions of mathematical physics. Unlike the usual approach presented for blackbody radiation, we intend to develop and follow a unique approach based on polylogarithms. As one will come to see, the polylogarithmic formulation is the most natural setting in which to consider the problem of blackbody radiation. Mathematically, it is as efficient as it is elegant. Due to the function's particular propensity to simplify integrals that otherwise are often thought to be intractable, to our mind sufficiently justifies the general polylogarithmic approach as a worthy successor to the existing mathematical formulation one may be currently familiar with for Planck's law.

Special cases of the polylogarithmic functions were first considered by the English mathematician John Landen (1719–1790) in 1760 [370] and eight years later by the great Swiss mathematician Leonhard Euler (1707–1783) [218]. They have been studied by some of the great mathematicians of the past including Abel, Lobachevsky, Kummer, and Ramanujan[4] to name but a few. Often buried deep in obscure texts on special functions, if they appeared at all, meant the polylogarithms were largely recondite and known only to the most dedicated of enthusiasts. Only in the last few decades have they began to rise in prominence, today making appearances across a broad range of disparate fields.[5]

The polylogarithm function is defined by the series [468]

$$\text{Li}_s(z) = \sum_{k=1}^{\infty} \frac{z^k}{k^s}, \quad |z| < 1, \tag{2.1}$$

and is absolutely convergent for all complex s and z inside the unit disc in the complex z-plane. For fixed order s the domain of the polylogarithms can be extended by analytic continuation using, for example, one of the function's known integral representations [157] to the whole complex z-plane with the exception being those points which lie on the branch cut along the real axis from one to infinite.

The function is appropriately named given its close connection with the natural logarithm. By restricting our attention to real arguments, when $s = 1$, from Eq. (2.1) it is apparent

$$\text{Li}_1(x) = \sum_{k=1}^{\infty} \frac{x^k}{k} = -\ln(1 - x), \tag{2.2}$$

and is just the Mercator series for the natural logarithm with x replaced by $-x$. In this way the polylogarithms can be thought of as one of the simplest generalisations of the natural logarithm. Second and third order polylogarithms are singled out for special attention since historically each was the object of extensive study in its own right [726]. They are known as the *dilogarithm* (of Euler) and the *trilogarithm* (of Landen) respectively.

From the series definition for the polylogarithmic function two important special values for the function follow. When $x = 0$, for all real orders s, one has $\text{Li}_s(0) = 0$. Also, when $s > 1$, at the point $x = 1$, since

$$\text{Li}_s(1) = \sum_{k=1}^{\infty} \frac{1}{k^s} = \zeta(s), \tag{2.3}$$

we immediately see the polylogarithm function for all orders greater than one reduces to the Riemann zeta function $\zeta(s)$ when its argument is equal to unity.

Derivatives for the polylogarithm function follow from Eq. (2.1). Here

$$\frac{d}{dx}\text{Li}_s(x) = \frac{\text{Li}_{s-1}(x)}{x}. \tag{2.4}$$

For example, setting $s = 1$ and using the known result for $\text{Li}_1(x)$ in terms of the natural logarithm as given by Eq. (2.2) gives $\text{Li}_0(x) = x/(1-x)$. Repeated application of Eq. (2.4) allows explicit expressions for the polylogarithms to all negative integer orders to be found. Eq. (2.4) implies that for all $s \in \mathbb{Z}^-$, $\text{Li}_s(x) \in \mathbb{Q}[x]$ is a rational function with a single singularity at $x = 1$ [706].

If n is a positive integer, a general expression for the polylogarithm of any negative integer order can be found. The result is[6]

$$\text{Li}_{-n}(x) = \frac{1}{(1-x)^{n+1}} \sum_{k=0}^{n} \left\langle {n \atop k} \right\rangle x^{n-k}, \quad n \geqslant 1. \tag{2.5}$$

Table 2.1

Euler's triangle showing the first few Eulerian numbers

n	$\left\langle {n \atop 0} \right\rangle$	$\left\langle {n \atop 1} \right\rangle$	$\left\langle {n \atop 2} \right\rangle$	$\left\langle {n \atop 3} \right\rangle$	$\left\langle {n \atop 4} \right\rangle$	$\left\langle {n \atop 5} \right\rangle$
0	1					
1	1	0				
2	1	1	0			
3	1	4	1	0		
4	1	11	11	1	0	
5	1	26	66	26	1	0

Here $\left\langle {n \atop k} \right\rangle$ are the Eulerian numbers [267].[7] The first few are shown in Table 2.1 with the resulting triangular array observed referred to as *Euler's triangle*.

From Eq. (2.5), in addition to the zeroth-order polylogarithm, the first few negative integer order polylogarithms are

$$\text{Li}_0(x) = \frac{x}{1-x}, \quad \text{Li}_{-1}(x) = \frac{x}{(1-x)^2}, \quad \text{Li}_{-2}(x) = \frac{x(1+x)}{(1-x)^3},$$
$$\text{Li}_{-3}(x) = \frac{x(1+4x+x^2)}{(1-x)^4}, \quad \text{Li}_{-4}(x) = \frac{x(1+11x+11x^2+x^3)}{(1-x)^5}, \dots \tag{2.6}$$

It is also apparent from Eq. (2.4) that the polylogarithms can be recursively defined as follows

$$\text{Li}_{s+1}(x) = \int_0^x \frac{\text{Li}_s(u)}{u} du. \tag{2.7}$$

Of particular interest to the present work will be polylogarithms with arguments of the form e^{-x}. For such arguments, derivatives and integrals for the polylogarithms reduce to simple ladder operations on the order index together with an associated change in sign. Here differentiating the polylogarithm with respect to x lowers its order by one

$$\frac{d}{dx}\text{Li}_s(e^{-x}) = -\text{Li}_{s-1}(e^{-x}), \tag{2.8}$$

while integrating the polylogarithm, to within an arbitrary constant, raises its order by one

$$\int \text{Li}_s(e^{-u})du = -\text{Li}_{s+1}(e^{-x}). \tag{2.9}$$

In view of what is to follow, evaluation of integrals of the form

$$\int u^k \text{Li}_0(e^{-u})du, \tag{2.10}$$

where k is a non-negative integer will often be needed. It can be evaluated using integration by parts k times together with the result for the integral of the polylogarithm function given by Eq. (2.9). When this is done, to within an arbitrary constant, the result is

$$\int u^k \mathrm{Li}_0(e^{-u}) du = -\sum_{n=0}^{k} u^{k-n} \mathrm{Li}_{n+1}(e^{-u}) \frac{\Gamma(k+1)}{\Gamma(k+1-n)}. \qquad (2.11)$$

Here $\Gamma(x)$ is the gamma function.[8]

Finally, a few important limits involving the polylogarithm function will be needed. For any real order s, two relatively simple ones are

$$\lim_{u \to \infty} \mathrm{Li}_s(e^{-u}) = \mathrm{Li}_s(0) = 0, \qquad (2.12)$$

and

$$\lim_{u \to 0} \mathrm{Li}_s(e^{-u}) = \mathrm{Li}_s(1) = \zeta(s). \qquad (2.13)$$

If k and n are integers such that $0 \leqslant n \leqslant k$, two other often needed limits which can readily be established are

$$\lim_{u \to \infty} u^{k-n} \mathrm{Li}_{n+1}(e^{-u}) = 0, \qquad (2.14)$$

and

$$\lim_{u \to 0} u^{k-n} \mathrm{Li}_{n+1}(e^{-u}) = \begin{cases} 0 & : 0 \leqslant n < k, \\ \zeta(n+1) & : n = k. \end{cases} \qquad (2.15)$$

2.2.2 THE LAMBERT W FUNCTION

The second of the special functions we intend to make use of is the so-called Lambert W function. As a recent arrival on the mathematical landscape, the Lambert W function has rapidly emerged as perhaps the most important special function to come along in the last fifty years or so. While not as old or familiar as the polylogarithms, as it is our intention to use the function in the solution to a particular transcendental equation, we introduce it here and consider some of its more important properties.

Denoted by $W(x)$, the Lambert W function is defined to be the inverse of the function $f(x) = xe^x$ satisfying

$$W(x) e^{W(x)} = x. \qquad (2.16)$$

Referred to as the *defining equation* for the Lambert W function, Eq. (2.16) is multivalued, and therefore has an infinite number of branches of which only two are real. By convention, the branch satisfying $W(x) \geqslant -1$ is taken to be the principal branch, denoted by $W_0(x)$, while that satisfying $W(x) < -1$ is known as the secondary real branch and is denoted by $W_{-1}(x)$.

For real arguments Eq. (2.16) can have either one unique positive real root $W_0(x)$ if $x \geq 0$ except for $W_0(0) = 0$; two negative real roots $W_0(x)$ and $W_{-1}(x)$ if $-1/e < x < 0$; one negative double real root $W_0(-1/e) = W_{-1}(-1/e) = -1$ if $x = -1/e$; and if $x < -1/e$ there are no real roots. The two real branches[9] share an order two branch point at $x = -1/e$. A plot of the Lambert W function showing its two real branches is shown in Fig. 2.1.

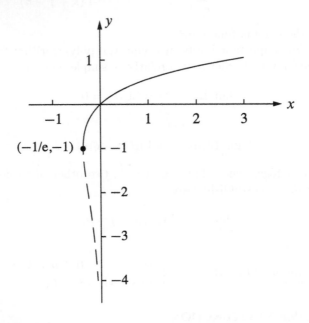

Figure 2.1   Plot of the Lambert W function for real arguments. The solid line shows the principal branch $W_0(x)$ while the dashed line shows the secondary real branch $W_{-1}(x)$. The branch point between the two real branches occurs at $x = -1/e$.

Expanding the Lambert W function about the origin, a Maclaurin series with rational coefficients can be found. Most simply, this is achieved using the Lagrange inversion theorem [49] which is a powerful way to revert series if the series for the inverse is known. Details of the procedure are given in Appendix A.1. The result is

$$W_0(z) = \sum_{n=1}^{\infty} \frac{(-n)^{n-1}}{n!} z^n, \quad |z| < \frac{1}{e}. \tag{2.17}$$

Some special values for the function are $W_0(0) = 0$, $W_0(-1/e) = W_{-1}(-1/e) = -1$, and $W_0(e) = 1$. Many other properties for the Lambert W function are known and all we can do here is direct the interested reader to the relevant literature [145, 284, 105, 468, 146].

The function is named in honor of the Swiss polymath Johann Heinrich Lambert (1728–1777) who in 1758 was the first to consider a problem requiring W(x) for its solution [368]. It was later picked up and studied by Euler [219]. The lending of attention by such a great name as Euler, however, did little to lift the function out of obscurity. Thereafter it languished just out of sight from mainstream mathematics for many years to come. With the passage of time the literature on the function became widely scattered and often obscure. Lacking a name, it was destine to be rediscovered by others [385, 386, 524, 526, 285, 527, 711, 179, 712, 242, 47, 193] and had to wait a further 230 years after its initial introduction before finally receiving a name [145]. Its immense popularity and use is a modern phenomenon driven largely by its early inclusion in computer algebra systems such as MAPLE [144] and MATHEMATICA. Having taken on the mantle of a bona fide function it now finds itself a respected place among the pantheon of special functions.[10]

2.3 TWO COMMON SPECTRAL SCALES USED TO REPRESENT BLACKBODY RADIATION

The very first spectral scale Planck himself chose to present his now famous law for thermal radiation, was a linear wavelength scale [511, 518]. That was in October of 1900. Two months later, he presented it using both linear wavelength and frequency scales [512, 519]. These two spectral scales came to dominate how radiation from a thermal source would be represented. As either one of these representations is typically the form in which one first encounters Planck's law, we consider each of these spectral scales in some detail while deferring discussion on other possible scales, how to convert between scales, and which, if any, of all the scales is the most appropriate to use, until later in the chapter.

The spectral radiate exitance in the linear wavelength representation $M_{e,\lambda}(\lambda, T)$ gives the amount of energy emitted by a body per unit time, per unit area, per unit wavelength interval, at a given wavelength λ. For a blackbody whose surface is at absolute temperature T radiating into a hemispherical envelope in space, it is given by Planck's well-known law for thermal radiation

$$M_{e,\lambda}^{b}(\lambda, T) = \frac{2\pi hc^2}{\lambda^5 \left[\exp\left(\dfrac{hc}{k_B \lambda T} \right) - 1 \right]}. \qquad (2.18)$$

For reasons which will become apparent later in the chapter, Planck's equation is an example of a distribution function and is the term we shall refer to it by. As is customary, the subscript λ for wavelength indicates a spectral (as opposed to a total) quantity being considered, while the superscript "b" is used to remind us that the body under consideration is that of a blackbody. The meaning of the suffix "e" will be given on page 59.

What is not so well known is that Planck's law can be rewritten in terms of a polylogarithm function. From the zeroth order polylogarithm, since

$$\text{Li}_0(e^{-x}) = \frac{e^{-x}}{1 - e^{-x}} = \frac{1}{e^x - 1}, \tag{2.19}$$

in terms of polylogarithms, Planck's law for thermal radiation in the linear wavelength representation can be rewritten as

$$M_{e,\lambda}^b(\lambda, T) = \frac{2\pi hc^2}{\lambda^5} \text{Li}_0 \left[\exp\left(-\frac{hc}{k_B \lambda T} \right) \right]. \tag{2.20}$$

While rarely seen in this form,[11] writing Planck's equation in terms of a polylogarithm function makes it more amenable to integration.

As is well known, Planck's distribution function within a linear wavelength representation contains a single peak in its spectrum. The wavelength at the peak can be found by taking the derivative of $M_{e,\lambda}^b(\lambda, T)$ with respect to the wavelength, setting the result equal to zero, and solving for the wavelength. The value for the wavelength at the peak we refer to as a *Wien peak*. Finding its value yields

$$\frac{d}{d\lambda}\left[M_{e,\lambda}^b(\lambda, T) \right] = 5\text{Li}_0(e^{-u}) - u\text{Li}_{-1}(e^{-u}) = \frac{5(e^u - 1) - ue^u}{(e^u - 1)} = 0. \tag{2.21}$$

Here $u = hc/(k_B \lambda T)$ is a dimensionless parameter corresponding to what is often referred to as the reduced wavelength while use of the explicit forms for the polylogarithms of orders zero and minus one listed in Eq. (2.6) has been made. Rearranging algebraically, Eq. (2.21) reduces to the following transcendental equation

$$(u - 5)e^{u-5} = -5e^{-5}. \tag{2.22}$$

The Wien peak is found on solving Eq. (2.22) for u and hence wavelength. Surprisingly, a number of closed-form expressions for the Wien peak can be found [604, 47, 652, 401, 628]. The simplest of these closed-form expressions we believe is the one written in terms of the Lambert W function. Recognising that Eq. (2.22) is exactly in the form for the defining equation of the Lambert W function, it can be readily solved in terms of this function. For the non-trivial solution[12] one has

$$u = 5 + W_0(-5e^{-5}). \tag{2.23}$$

After back substituting u for λ an expression for the Wien peak immediately follows. The result is

$$\lambda_{\text{max},\lambda} T = \frac{hc}{k_B[5 + W_0(-5e^{-5})]}. \tag{2.24}$$

The Lambert W function therefore provides an elegant approach to solving certain classes of transcendental equations involving exponentials in closed form.

The Wien peak occurs at the wavelength given by $\lambda_{\mathrm{max},\lambda}$. The appearance of a λ in the subscript is used to indicate the spectral scale, in this case a linear wavelength scale. It is done since when we come to consider other spectral scales, in order to avoid confusion, the use of an additional subscript should make it clear the spectral scale to which the result corresponds.

As already mentioned in Chapter 1, Eq. (2.24) is referred to as Wien's displacement law. The mathematical constant $5 + W_0(-5e^{-5})$ appearing in this law has a value equal to $4.965\,114\,231\ldots$[13] while the constant term appearing to the right of the equality in Eq. (2.24) is often referred to as Wien's displacement law constant. Denoted by the letter b its value is equal to

$$b = \frac{hc}{k_B[5 + W_0(-5e^{-5})]} = 2897.773\,\mu\mathrm{m}\cdot\mathrm{K}. \tag{2.25}$$

In Fig. 2.2 the spectral radiant exitance of a blackbody in the linear wavelength representation as a function of wavelength at six different temperatures between $500\,\mathrm{K}$ to $1000\,\mathrm{K}$ is given. In each of the four plots, slightly different combinations in the scales for the ordinate and the abscissa (either linear or log) have been used. The dashed curve appearing in each plot gives the locus of the Wien peak.

A feature of these plots is the very rapid increase observed in the value for the spectral radiant exitance at the Wien peak with temperature. It can be understood as follows. At the Wien peak, if the value for $\lambda_{\mathrm{max},\lambda}$ is substituted into Eq. (2.20) one finds

$$M_{e,\lambda}^{b}(\lambda_{\mathrm{max},\lambda}, T) = -\frac{2\pi k_B^5}{h^4 c^3}W_0(-5e^{-5})[5 + W_0(-5e^{-5})]^4 T^5 = aT^5. \tag{2.26}$$

Here a is a constant and has a value equal to

$$a = 1.286\,692 \times 10^{-5}\,\mathrm{W\cdot m^{-3}\cdot K^{-5}}. \tag{2.27}$$

At the Wien peak the spectral radiant exitance is therefore proportional to the fifth power of the absolute temperature. So a small increase in temperature leads to a very large increase in the value for the spectral radiant exitance. On normalizing the spectral radiant exitance relative to its peak value one finds

$$\frac{M_{e,\lambda}^{b}(\lambda, T)}{M_{e,\lambda}^{b}(\lambda_{\mathrm{max},\lambda}, T)} = -\left(\frac{hc}{k_B\lambda T}\right)^5 \frac{\mathrm{Li}_0\left[\exp\left(-\dfrac{hc}{k_B\lambda T}\right)\right]}{W_0(-5e^{-5})[5 + W_0(-5e^{-5})]^4} \tag{2.28}$$

A plot of the normalized spectral radiate exitance as given by Eq. (2.28) together with plots for several other normalized spectral curves involving other wavelength scales can be found later in Fig. 2.6.

For the second of the most common spectral scales in use, the spectral radiate exitance in the linear frequency representation, $M_{e,\nu}(\nu, T)$ gives the amount of energy emitted by a body per unit time per unit area per unit

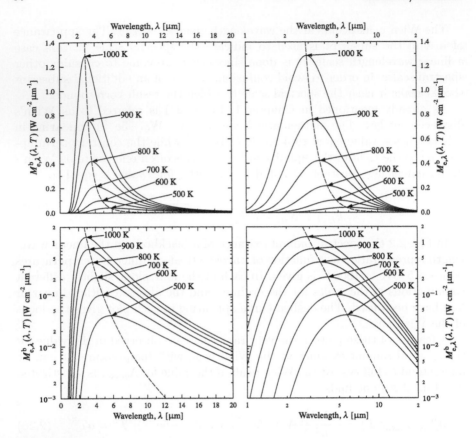

Figure 2.2 The spectral radiant exitance of a blackbody in the linear wavelength representation as a function of wavelength for six different temperatures. In each of the four plots, slightly different scales for the ordinate and abscissa have been used. Writing the abscissa first, followed by the ordinate, these are: TOP LEFT: linear–linear; TOP RIGHT: log–linear; BOTTOM LEFT: linear–log; BOTTOM RIGHT: log–log. The dashed curve shown in each figure gives the locus of the Wien peak.

frequency interval at a given frequency ν. For a blackbody at absolute temperature T radiating into a hemispherical envelope in space it is given by

$$M_{e,\nu}^{b}(\nu, T) = \frac{2\pi h \nu^3}{c^2 \left[\exp\left(\dfrac{h\nu}{k_B T} \right) - 1 \right]}. \tag{2.29}$$

Rewritten in terms of the zeroth order polylogarithm function it becomes

$$M_{e,\nu}^{b}(\nu, T) = \frac{2\pi h \nu^3}{c^2} \mathrm{Li}_0 \left[\exp\left(-\frac{h\nu}{k_B T} \right) \right]. \tag{2.30}$$

Within the linear frequency representation, as is well known, Planck's distribution function is again singly peaked. The peak value can be found in a similar fashion to how the peak was found for the linear wavelength case. Differentiating Eq. (2.30) with respect to frequency, setting the result equal to zero and solving, the following transcendental equation similar in form to Eq. (2.22) results

$$(u - 3)e^{u-3} = -3e^{-3}. \tag{2.31}$$

Here u is a dimensionless parameter corresponding to the reduced frequency given by $h\nu/(k_B T)$. On solving in terms of the Lambert W function, for the non-trivial solution, the Wien peak in the linear frequency representation is given by [693]

$$\frac{\nu_{\max,\nu}}{T} = \frac{k_B[3 + W_0(-3e^{-3})]}{h}. \tag{2.32}$$

Often Eq. (2.32) is referred to as Wien's displacement law for frequency. The mathematical constant $3 + W_0(-3e^{-3})$ appearing in the equation has a value equal to $2.821\,439\,372\ldots$. Rewritten in terms of a so-called Wien's displacement law constant for frequency b^*, one has

$$\frac{\nu_{\max,\nu}}{T} = b^*, \tag{2.33}$$

where

$$b^* = \frac{k_B[3 + W_0(-3e^{-3})]}{h} = 5.878\,924 \times 10^{10}\ \text{Hz·K}^{-1}. \tag{2.34}$$

To convert the value from frequency to wavelength, since $\lambda = c/\nu$ one has

$$\lambda_{\max,\nu} T = \frac{hc}{k_B[3 + W_0(-3e^{-3})]}. \tag{2.35}$$

At the Wien peak, the spectral radiant exitance for a blackbody in the linear frequency representation can once again be expressed in closed form. The result is

$$M_{e,\nu}^b(\nu_{\max,\nu}, T) = -\frac{2\pi k_B^3}{h^2 c} W_0(-3e^{-3})[3 + W_0(-3e^{-3})]^2 T^3 = a^* T^3, \tag{2.36}$$

where the constant a^* has a value equal to

$$a^* = 1.785\,762 \times 10^{-10}\ \text{W·m}^{-1}\text{·K}^{-3}. \tag{2.37}$$

Using a linear frequency scale, at its peak, the spectral radiant exitance for a blackbody does not increase as fast with temperature compared to the linear wavelength scale, increasing in proportion to only the cube of the temperature. Finally, normalizing relative to the peak value the spectral radiant exitance in the linear frequency representation can be written as

$$\frac{M_{e,\nu}^b(\nu, T)}{M_{e,\nu}^b(\nu_{\max,\nu}, T)} = -\left(\frac{h\nu}{k_B T}\right)^3 \frac{\text{Li}_0\left[\exp\left(-\frac{h\nu}{k_B T}\right)\right]}{W_0(-3e^{-3})[3 + W_0(-3e^{-3})]^2}. \tag{2.38}$$

From the Wien peak values within the linear wavelength and frequency representations it is interesting to note the product between the two does not give a value corresponding to the speed of light in a vacuum. Instead one has

$$\nu_{\text{max},\nu}\lambda_{\text{max},\lambda} = \frac{3 + \text{W}_0(-3\text{e}^{-3})}{5 + \text{W}_0(-5\text{e}^{-5})}c = 0.568\,257c \neq c. \tag{2.39}$$

Initially many are surprised by this seemingly contradictory result [60, 170]. There is however, nothing unusual about the result as each value for the Wien peak found came from considering a distribution function corresponding to different interval widths of either per unit frequency or per unit wavelength. The importance of differences in interval widths used to represent Planck's distribution function will be taken up and developed more completely in Section 2.5.

2.4 OTHER SPECTRAL SCALE REPRESENTATIONS

Considering that today no consensus has been reached on how spectra ought to be represented, it is hardly surprising to discover that when it comes to representing spectra arising from thermal radiation, a number of competing spectral representations are to be found. As already noted, the two most common are those given in terms of linear wavelength or frequency but they are by no means the only spectral representations possible. The logarithm of wavelength or frequency is often used and on occasion the independent variable in terms of either the wavelength or frequency squared may be selected or some other more exotic index involving either of these two quantities.

In this section we consider Planck's distribution function in the context of a more general spectral representation. We show how one can easily move between any two representations, and consider the source of confusion which arises when Planck's law is expressed using different spectral scales.

For a given selected spectral representation φ, if dM_φ is taken to represent the spectral exitance in the differential interval $d\varphi$ for a blackbody, one can write

$$dM_\varphi = M_{\varsigma,\varphi}^{\text{b}}(\varphi,T)d\varphi. \tag{2.40}$$

Here $M_{\varsigma,\varphi}^{\text{b}}(\varphi,T)$ is the well-known spectral exitance for a blackbody at absolute temperature T within the spectral representation denoted by φ while ς is a generalised suffix introduced to denote case type, be it radiometric or actinometric, to be discussed later (see Section 2.7). By writing Eq. (2.40) in this way it is clear the spectral exitance is a distribution function which is differential in form. The spectral exitance as given by Planck's radiation law therefore tells one how the power emitted by a blackbody from a unit area of its surface is distributed throughout the spectrum (on a unit spectral scale basis) and how it changes as a function of temperature.

The differential spectral exitance dM_φ in the differential interval $d\varphi$ has the property that when summed over the entire spectrum it gives the total exitance M_ς^b radiated by a blackbody at some fixed temperature T — a conserved quantity related to energy conservation. Energy conservation thus ensures that different spectral representations for dM_φ correspond to each other since the energy must be the same and is the means by which one can move between different spectral representations. If φ_1 and φ_2 are two different spectral representations such that φ_1 and φ_2 are functionally related to one another in some way, from energy conservation one must have

$$M_{\varsigma,\varphi_1}^b(\varphi_1, T)d\varphi_1 = M_{\varsigma,\varphi_2}^b(\varphi_2, T)d\varphi_2. \tag{2.41}$$

As an example, consider the case of the linear wavelength and frequency scales. For these two scales Eq. (2.41) becomes

$$M_{\varsigma,\lambda}^b(\lambda, T)d\lambda = M_{\varsigma,\nu}^b(\nu, T)d\nu. \tag{2.42}$$

In a vacuum the frequency of electromagnetic radiation is related to its wavelength by

$$\nu = \frac{c}{\lambda}, \tag{2.43}$$

where c is the speed of light in a vacuum and provides the necessary functional relation between the two scales. The corresponding interval widths between the two scales are therefore related by

$$d\nu = -\frac{c}{\lambda^2}d\lambda. \tag{2.44}$$

Simple substitution of Eq. (2.43) into the spectral exitance in moving between the two scales is therefore not enough. As one is dealing with a quantity which is differential in nature the prefactor, called the Jacobian of the transformation, seen in Eq. (2.44) also needs to be included when transforming between spectral representations.[14] Except for the negative sign which is an artifact related to the direction of integration used when finding the total exitance, one has

$$M_{\varsigma,\lambda}^b(\lambda, T)d\lambda = \frac{c}{\lambda^2}M_{\varsigma,\nu}^b(\nu, T)d\lambda. \tag{2.45}$$

Eq. (2.45) shows how to correctly move between the linear wavelength and frequency scales. Eight different spectral scales often used to represent blackbody radiation are shown in Table 2.2.

With many different spectral representations possible, it is natural to ask which scale ought to be selected, and is one particular scale over all others to be preferred? It is a very old question, and has often been asked, yet despite this it has no clear answer. Considering no general agreement regarding what quantity ought to be used when presenting spectra, this lack in definiteness has often led many to argue for the use of one scale over all others. Reasons given for doing so have tended to range from one of appropriateness or usefulness, down to one particular choice being somehow fundamentally superior to all others.

Table 2.2

Eight common spectral scales used to represent blackbody radiation.

φ	$M_{\varsigma,\varphi}^{b}(\varphi, T)d\varphi$	Spectral scale
ν^2	$2\nu M_{\varsigma,\nu^2}^{b}(\nu, T)d\nu$	frequency-squared
ν	$M_{\varsigma,\nu}^{b}(\nu, T)d\nu$	linear frequency
$\sqrt{\nu}$	$\frac{1}{2\sqrt{\nu}}M_{\varsigma,\sqrt{\nu}}^{b}(\nu, T)d\nu$	square root frequency
$\ln \nu$	$\frac{1}{\nu}M_{\varsigma,\ln \nu}^{b}(\nu, T)d\nu$	logarithmic frequency
$\ln \lambda$	$\frac{1}{\lambda}M_{\varsigma,\ln \lambda}^{b}(\lambda, T)d\lambda$	logarithmic wavelength
$\sqrt{\lambda}$	$\frac{1}{2\sqrt{\lambda}}M_{\varsigma,\sqrt{\lambda}}^{b}(\lambda, T)d\lambda$	square root wavelength
λ	$M_{\varsigma,\lambda}^{b}(\lambda, T)d\lambda$	linear wavelength
λ^2	$2\lambda M_{\varsigma,\lambda^2}^{b}(\lambda, T)d\lambda$	wavelength-squared

Historically, the distribution of spectra arising from thermal radiation has been presented as a function of wavelength. Clear early examples of this can be seen in the work of the Irish experimental physicist John Tyndall (1820–1893) from the late 1860s onwards [651] and particularly in the work performed in the mid-1880s by Captain William de Wiveleslie Abney (1843–1920) and Lieutenant-Colonel Edward Robert Festing (1839–1912), who were both Royal Engineers in the British army at the time [182]. The two gave a number of spectral distribution curves for various incandescent light sources at different temperatures as a function of wavelength.[15] Later, when in the late 1890s intense experimental work to establish the fundamental laws of thermal radiation had commenced, particularly from the group of contemporary German experimental physicists associated with the Physikalisch Technischen Reichsanstalt in Berlin, one finds the spectral curves they present are almost universally those as a function of wavelength.

As prevalent as the use of the wavelength scale was among experimentalists, not all people working on problems related to thermal radiation during this period used wavelength. One notable exception was the American experimental astrophysicist Samuel Langley. Having invented the bolometer; a device for measuring the incident power of radiation via the heating of a material with a temperature-dependent electrical resistance; around 1880 [371] allowed him for the first time to measure very accurately the distribution of radiant energy from a thermal source. Importantly, his bolometer could not only detect radiation in the visible portion of the electromagnetic spectrum down to 0.344 μm but also in the infrared out to a wavelength of 2.356 μm.

This he proceeded to do in 1884 [373]. In a paper of his from that year we find he is the first to give an account comparing a number of different spectral scales. After presenting his experimental data using a number of different spectral scales, Langley draws the conclusion "...nature has no law which *must* govern us in representing the distribution of energy, and all maps and charts of it are but conventions."[16]

Nature does indeed have no particular law governing the choice of spectral scale selected. As one is dealing with a distribution function, it is nothing more than a rule telling you how radiant energy is distributed throughout the spectrum as a function of the spectral variable chosen. While a freedom in the choice of scale resides with the practitioner, Langley writes

> While we remain at liberty, then, to represent the energy spectrum in terms of the wave frequency [wavenumber], or of the reciprocal of the square of the wave-length, or of any other function of it, and while we may often find occasion to use these scales for some special purposes, we are (and all the more especially that we habitually speak in terms of wavelength) led by considerations of a very practical kind to take as our normal or standard scale that of wave-length itself.[17]

Langley's selection of a linear wavelength scale was a pragmatic one. It conformed to what others at the time were familiar with and using, rather than from any deep sense of it being intrinsically superior or more fundamental than any other scale.

Within the linear wavelength spectral representation Langley gives precise details together with a charmingly drawn figure, reproduced in Fig. 2.3, illustrating graphically how the principle of transformation used to convert between the measured prismatic spectrum; a spectrum highly dependent on the particular prism used; to what he calls the "normal" spectrum, that is, the instrument-independent spectral curve. The measured prismatic spectrum, labelled as CD in the figure, is converted to a normal spectrum, labelled as AB, using a mapping that employs a known dispersion relation (refractive index as a function of wavelength and is the curve labelled EF in Fig. 2.3) for the prism used. Langley, however, went further. Not content with the use of wavelength only, in keeping with his conviction that in nature there was really no "standard," in addition to presenting a distribution curve in terms of wavelength one finds him presenting distribution curves as functions of three other spectral variables.

In the first, Langley chose the scale initially proposed by Stoney, that of wavenumber (Langley called this quantity wave-frequency). In the second the proposal of Lord Rayleigh's of a logarithmic wavelength scale is used, while in the third a far more interesting scale is chosen. Here he considers a scale adjusted so that the bounding spectral curve remains constant and parallel to the abscissa — a scale he thought had not previously been proposed before. Nor have we seen such a scale used since. We will refer to this unique

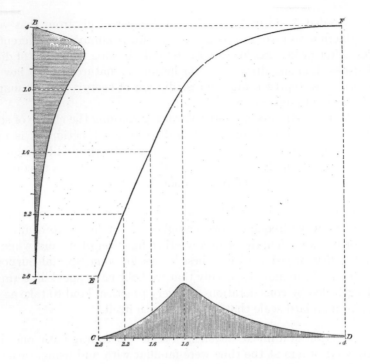

Figure 2.3 Depiction of the transformation from the measured prismatic spectrum to the instrument independent "normal" spectrum made by Langley in 1884. The measured prismatic spectrum is labelled CD while the normal spectrum obtained after the transformation is labelled AB. The mapping between the two spectra is achieved by using the known dispersion relation for the prism and is the curve labelled EF. *Source:* Langley, S. P., 1884 "Experimental determination of wavelengths in the visible prismatic spectrum," *American Journal of Science* [Third Series], **27**(159), p. 185.

representation as a "Langleysian scale." Each of the four spectra drawn by Langley is reproduced in Fig. 2.4 and they are labelled A through to D.

Langley did not suggest nor promote the use of one scale over any other and after the publication of his work almost all experimentalists working in the field of thermal radiation continued to present their spectral data as they had always done — using a linear wavelength scale. Change, however, was on the way. An early exception to the rule was Paschen, the German experimental physicist whom we met in the previous chapter. Starting in 1896 one finds him presenting spectral curves from blackbody sources in terms of a logarithmic wavelength scale [481]. Paschen was convinced to change from a linear over to a logarithmic wavelength scale by his colleague, the applied mathematician

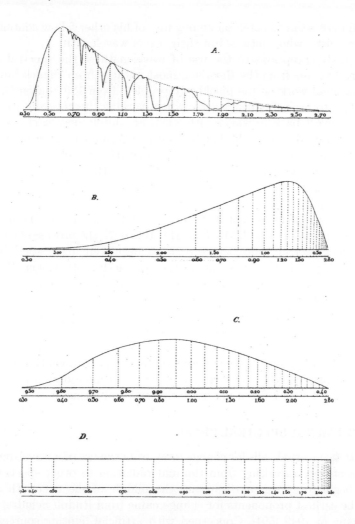

Figure 2.4 Four spectra drawn by Langley dating from the 1880s. Each depicts a different spectral scale representation. The first three are the now common linear wavelength (labelled *A*), linear frequency (labelled *B*), and the logarithm of the wavelength (labelled *C*), while the fourth used a uniquely Langleysian scale (labelled *D*). *Source:* Langley, S. P., 1884 "Experimental determination of wave-lengths in the visible prismatic spectrum," *American Journal of Science* [Third Series], **27**(159), p. 186.

and physicist Carl David Tolm e Runge (1856–1927) [482]. He wrote that doing so allowed him to more readily untangle the many complicated spectral curves he needed to analyses without which his task would have been made all the more difficult. Despite Paschen making heavy use of a logarithmic wavelength scale it did not seem to catch on among any of his other fellow contemporary experimentalists, who continued in their use of wavelength.

Other early exceptions to the use of wavelength as the spectral scale of choice was to come from the theorists, most notably from Planck himself. In his now seminal work on the blackbody problem from 1897 onwards, Planck was the first to use a frequency scale when working theoretically, only being forced to convert back to wavelength when absolutely necessary, as was the case when he needed a result that had been determined experimentally and presented within the context of a linear wavelength scale. In doing so, Planck appears to have been the first to convert between these two scales, doing so in May of 1899 in the fifth of his series of papers on irreversible radiation processes [507].

Judging from the treatment it received in early texts where the blackbody problem was considered [332, 516, 559, 517, 169, 667], the process of conversion between two spectral scales seems to have been well understood by those early practitioners in the early part of the twentieth century, something that appears to have been largely lost by many in the modern era. The loss is seen in the short reminders often given in the literature on how to correctly convert between scales [616, 322, 611, 612, 418]. Where it was once common practice to include this in almost all physics texts, today by contrast, beyond specialized texts dealing with radiometry, it is rarely seen. As we shall see, this modern day loss of a proper understanding of spectral width conversion is largely responsible for the constant source of confusion found in many people when Planck's law is written in alternative spectral forms.

2.5 EPHEMERAL SPECTRAL PEAKS

The earliest concerted calls for change away from the use of a wavelength scale to represent spectra arising from thermal radiation to other spectral scales did not begin to appear until around the middle of the twentieth century. One of the earliest proponents for change came from Rudolf Schulze working in Germany in 1949 [591]. Concerned with artificial lighting sources at the time, he suggested the linear wavelength scale was but a "special case" and commented in many instances it did not meet the requirements of practice. An extensive comparison between the linear wavelength, frequency, and the logarithmic frequency and wavelength scales in both energetic and photonic units followed. Of the three, he concluded a scale in terms of the logarithm was most appropriate.

Other suggestions for change were quick to follow [253, 88, 228, 567, 272, 585, 273]. The problem boiled down to one of selecting what many tried to argue was the "best" spectral scale. In many instances the selection was based on attributes associated with the spectral curve itself such as curve symmetry or peak location. Such criteria however, turn out to be completely arbitrary and tend to be based more on the individual user's personal predilections rather than having any claim to being intrinsically more fundamental.

Selection of a spectral scale based on peak location found in the spectral distribution curve is a case in point. The peak is often seen as having particular physical signifiance, but as we shall see, turns out to be completely ephemeral.

Following the development we presented for locating the spectral peaks in Section 2.3 when linear wavelength and frequency scales were used, if the idea is extended to any arbitrary spectral scale and the location of the resulting peak in the spectral curve determined, a transcendental equation similar to the transcendental equations found for the two linear cases results. For the general case we have

$$(u - s)e^{u-s} = -se^{-s}. \tag{2.46}$$

Here $u = hc/(k_B \lambda T)$ while s is a parameter related to the spectral scale selected. It will be seen that values greater than one for the parameter s are required if the spectral distribution function is to contain a peak. Comparisons with Eqs (2.22) and (2.31) show the linear frequency and wavelength representations are obtained by setting s equal to three and five respectively.

As we have seen, historically the peak in the spectrum occurring in the linear wavelength representation has been singled out as being particularly important with the product of the wavelength at the peak with temperature being referred to as Wien's displacement law. We will, however, show that it is no more important nor fundamental than any other spectral scale one may care to consider.

Returning to Eq. (2.46), as it is exactly in the form for the defining equation of the Lambert W function it can once again be solved in terms of this function. For the non-trivial solution one has

$$u = s + W_0(-se^{-s}), \tag{2.47}$$

or in terms of the Wien peak

$$\lambda_{\text{max},\varphi} T = \frac{hc}{k_B[s + W_0(-se^{-s})]}. \tag{2.48}$$

When a solution to the transcendental equation given by Eq. (2.46) exists, Planck's distribution function will be singly peaked. A non-trivial solution to the equation exists provided the argument for the Lambert W function term appearing in Eq. (2.47) is greater than the value of the argument of the Lambert W function at its branch point, namely

$$-se^{-s} > -\frac{1}{e}. \tag{2.49}$$

As the left-hand side of the above inequality is exactly in the form for the defining equation for the Lambert W function, the inequality can be solved for s. Noting the principal branch of the Lambert W function increases monotonically with increasing argument, on solving for s in terms of W one can write

$$-s < W_0(-1/e) = -1 \quad \text{or} \quad s > 1. \tag{2.50}$$

A peak in Planck's distribution function is therefore no longer found for spectral representations which turn out to be equal to frequency-cubed or beyond. The correspondence between spectral scale parameter s to the type of spectral representation will be taken up in Section 2.12. For the moment we remain content by simply stating that $s \leqslant 1$ corresponds to spectral scales equal to frequency-cubed or greater.

In Fig. 2.5 the product of the wavelength at the Wien peak with temperature as a function of the spectral scale parameter is plotted while Table 2.3 gives a list of several Wien peaks with their associated spectral scale parameters for a number of common spectral representations.

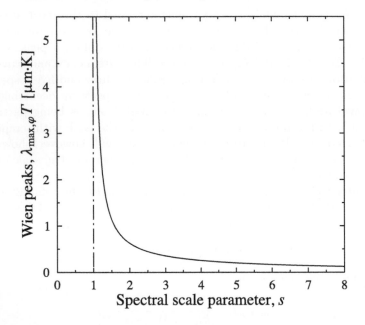

Figure 2.5 Plot of the wavelength–temperature product at the Wien peak as a function of the spectral scale parameter s. The dot-dashed line represents a vertical asymptote at $s = 1$.

Inspecting Table 2.3, at first glance the case of $s = 4$ in energetic units (the radiometric case) seems interesting. Here the Wien peaks for two different spectral scales coincide. It represents the crossover point in the spectral representation from one of frequency $(1 < s \leqslant 4)$ to one of wavelength $(s \geqslant 4)$ and on this basis alone has often been singled out as the preferred scale of choice.[18] As we will show, while it can be useful, it is no more significant nor privileged than any other spectral scale one may care to consider.

All too often, the appearance of a peak in a spectral distribution curve is seen as having particular importance. While it is true a peak in a curve provides one with a convenient point of reference, it is nothing more than this, a

Table 2.3
Wien peaks for several common spectral representations.

Spectral scale	φ	s	$\lambda_{\mathrm{max},\varphi}T\ [\mu\mathrm{m}\cdot\mathrm{K}]^{a}$
		Energetic units	
frequency-cubed	ν^3	1	—
frequency-squared	ν^2	2	9028.335
linear frequency	ν	3	5099.445
square root frequency	$\sqrt{\nu}$	3 1/2	4255.546
logarithmic frequency	$\ln\nu$	4	3669.704
logarithmic wavelength	$\ln\lambda$	4	3669.704
square root wavelength	$\sqrt{\lambda}$	4 1/2	3235.167
linear wavelength	λ	5	2897.773
wavelength-squared	λ^2	6	2404.012
		Photonic units^{b}	
frequency-squared	ν^2	1	—
linear frequency	ν	2	9028.335
square root frequency	$\sqrt{\nu}$	2 1/2	6447.256
logarithmic frequency	$\ln\nu$	3	5099.445
logarithmic wavelength	$\ln\lambda$	3	5099.445
square root wavelength	$\sqrt{\lambda}$	3 1/2	4255.546
linear wavelength	λ	4	3669.704
wavelength-squared	λ^2	5	2897.773

^{a}The values for the wavelength–temperature products given here are correct to seven significant figures. These products can be expressed in exact form in terms of the Lambert W function by substituting the appropriate value for the spectral scale parameter s into Eq. (2.48), but for brevity we have instead chosen to give each value in decimal form.
^{b}See Section 2.7 starting on page 58.

convenient point to which no physical significance should be attached. While Fig. 2.5 shows the expected shift in peak wavelength in Planck's distribution function at a fixed temperature from shorter wavelengths in any of the wavelength representations ($s \geqslant 4$) to longer wavelengths for representations in frequency ($1 < s \leqslant 4$), any suitable choice in spectral scale can result in a maximum being found in any part of the electromagnetic spectrum ranging from radio waves all the way through to x-rays and beyond. And not only is it possible to arbitrarily shift the spectral peak by a change in scale, but for frequency representations equal to frequency-cubed or beyond, to entirely suppress it all together. The appearance of a vertical asymptote at the cut-off point where the spectral parameter is equal to unity simply tells one that in such spectral representation Planck's distribution function no longer contains a peak.

Attempting to attach physical significance to the peak found in the spectral curve, as Langley observed more than a century ago, is therefore meaningless. In fact, the arbitrariness in peak location lead the American William D. Ross in 1954 to suggest the importance generally assigned to the peak in Wien's displacement law, or for that matter any of these so-called "spectral displacement laws," should be de-emphasized [567].

Not that this has stopped others from trying. Since the early 1950s a spectral representation in terms of either a logarithmic frequency or wavelength scale has held a particular allure to many. In the linear wavelength and frequency scales the non-linear relation between wavelength and frequency via the wave equation $\lambda = c\nu^{-1}$ has a profound effect on the shape and peak location of each spectral curve. As we have already seen, the interval widths between the two scale are related by $d\nu = -c\lambda^{-2}d\lambda$, while energy conservation requires the area under each spectral curve per unit interval width to be equal regardless of the spectral scale used. For increasing wavelength, intervals of constant $d\lambda$ in the linear wavelength representation correspond to increasingly narrower (that is, decreasing, and accounts for the negative sign) intervals of $d\nu$ in the linear frequency representation. The non-constancy in interval widths between the two scales accounts for the differences observed in spectral curve shapes and corresponding peak locations between the two representations. As Bernard H. Soffer and David K. Lynch adroitly pointed out only recently [611], the shift in the Wien peak between the linear wavelength and frequency scales is due not just to the change in variable of $\nu = c/\lambda$ made on moving between the two scales but also to a Jacobian weighting factor of c/λ^2 resulting from the differential nature of the Planck distribution function.

Despite this, differences in peak location found between the linear wavelength and frequency representations continue to be a source of unending confusion and an apparent paradox to many [653, 135, 136, 393, 100, 611, 612, 322, 418, 470, 291, 419, 81]. The ephemeral nature of the Wien peak stems from having to consider a spectral distribution function, a function that is differential in character. Since radiation from a spectrally continuous source such as a blackbody is distributed over the entire spectrum, the area under the spectral curve gives the total radiant exitance radiated by the blackbody. As any interval width within the spectrum goes to zero the spectral radiant exitance at that point must also go to zero. Rather than the distribution function itself, it is the differential form which has physical meaning. Only after this is recognized does the seemingly paradoxical behavior of shifting spectral peaks on moving between different spectral scales disappear.

On account of its differential behavior, the spectral radiant exitance only makes sense when it is specified within a spectral band of finite width. When viewed in this way, the spectral radiant exitance acts as nothing more than an auxiliary quantity enabling one to find the power radiated per unit area into a selected spectral band. The spectral radiant exitance is fundamentally an area preserving quantity and has the property that the ordinates vary with the type of spectral scale chosen for the abscissa. As first noted by Lord Rayleigh [544], and later by others [135], different spectral representations are simply different ways of partitioning and viewing the same total amount of power radiated per unit area. Mathematically each spectral representation contains the same amount of information and transformationally are equivalent even though spectrally their shapes and peak locations can appear to be quite different. The source of the ambiguity stems from trying to attach physical meaning and importance to a distribution function which is nothing more than a convenient mathematical tool used to describe the spectral distribution of radiation from a thermal source.

2.6 LOGARITHMIC SPECTRAL SCALES

In use, the problem with a linear wavelength or frequency scale is seen in the asymmetric form the spectral distribution takes on in either representation about its peak. Here the spectral curve is skewed towards either longer wavelengths (shorter frequencies) in the linear frequency representation or towards higher frequencies (shorter wavelengths) in the linear wavelength representation.

Skewness in the shape of the distribution can be reduced by a more appropriate choice in spectral scale. Spectral representations which are interval width preserving on moving between wavelength and frequency scales have the effect of changing the shape of the spectral curve so that it is more symmetric about its peak. Such scales are logarithmic in form. Within a logarithmic representation, as

$$d(\ln \lambda) = \frac{d\lambda}{\lambda} = -\frac{d\nu}{\nu} = -d(\ln \nu), \qquad (2.51)$$

except for the change in sign, it can be seen as spectral intervals when expressed in terms of the logarithm of the wavelength or frequency are equivalent. Proportionality between the two scales means the areas under each spectral curve per unit interval width are the same. For either spectral curve the shape and the corresponding peak locations will also be the same. As the location of the Wien peak in either of the two logarithmic spectral representations coincide it is often referred to as the *wavelength–frequency neutral peak* and represents the crossover point from one of frequency to wavelength.

The unexpected equality in peak location found when using a logarithmic scale for either frequency or wavelength has often lead many to single it out as the spectral scale of choice [88, 96, 272, 273, 596, 681, 450, 734, 428]. And while a logarithmic scale leads to spectral curves which are clearly more symmetric about their coincident peaks compared to any of the other more commonly used spectral scales, such a choice in scale is completely arbitrary. As has been widely recognised [253, 567, 585, 611, 287, 135, 630], the logarithmic scale is no more significant nor privileged that any other scale one may care to choose. To reiterate the observation made by Langley well over a century ago, there simply is no standard spectral representation in nature, as nature has no law governing such a choice.

To find the spectral radiant exitance in terms of a logarithmic wavelength scale, from Eq. (2.40), setting $\varphi = f(\lambda) = \ln \lambda$ gives

$$dM_{\ln \lambda} = M^{b}_{\varsigma,\ln \lambda}(\lambda, T) d(\ln \lambda) = \frac{1}{\lambda} M^{b}_{\varsigma,\ln \lambda}(\lambda, T) d\lambda, \qquad (2.52)$$

while in a linear wavelength representation

$$dM_{\lambda} = M^{b}_{\varsigma,\lambda}(\lambda, T) d\lambda. \qquad (2.53)$$

As energy within each differential interval is conserved one has

$$dM_{\lambda} = dM_{\ln \lambda} \qquad (2.54)$$

giving

$$M^{b}_{\varsigma,\lambda}(\lambda, T) d\lambda = \frac{1}{\lambda} M^{b}_{\varsigma,\ln \lambda}(\lambda, T) d\lambda \qquad (2.55)$$

In energetic units, as the form for the spectral radiant exitance within the wavelength representation is known (see Eq. (2.20)), in the logarithmic wavelength scale, from Eq. (2.55) one finds

$$M^{b}_{e,\ln \lambda}(\lambda, T) = \frac{2\pi h c^2}{\lambda^4} \mathrm{Li}_0 \left[\exp \left(-\frac{hc}{k_B \lambda T} \right) \right]. \qquad (2.56)$$

We return to the discussion of logarithmic spectral scales after the importance of two types of cases known as the radiometric and the actinometric cases have been introduced.

2.7 THE RADIOMETRIC AND ACTINOMETRIC CASES

Before proceeding further, an important distinction regarding how thermal radiation emitted from the surface of an object is measured needs to be made. While we have already hinted at this in various places, we now wish to make the distinction definite. The difference is important because what one measures determines the type of quantity under consideration. Radiation emitted from a body can be characterized in one of two different ways. In the first instance, radiation is an electromagnetic wave of wavelength λ. For the vast majority of objects at commonly encountered temperatures (say between 200 to 10 000 K) most of the thermal radiation emitted falls within the ultraviolet, visible, and infrared portions of the electromagnetic spectrum.[19] In the second instance, radiation exhibits particle-like properties due to its quantum nature. The particles of light are referred to as photons. Using a particle description the radiation emitted from a body due to its temperature can be thought of as a "leakage" of photons from the surface of a body into its surroundings.

When it comes to differentiating between the two different ways one may use to characterize thermal radiation, separation into two distinct cases of either radiometric or actinometric turns out to be very useful. The characterization of thermal radiation by type depends largely on how the radiation is detected. Broadly speaking, detectors can be divided into one of two common types depending on the mechanism used in the detection process. The first are the so-called *thermal* detectors [705]. Here a heating effect caused by the incident radiation results in a variation in some physical parameter (usually electrical) of the detector with temperature. The second are the so-called *quantum* detectors [361]. Here there is a direct interaction between the incident photons and the electrons of the detector. In the former case the detector response is proportional to the energy absorbed while in the later case it is proportional to the number of photons absorbed. As a way of connecting particular units in a natural way to the type of detector used it turns out to be useful when discussing various aspects associated with thermal radiation.

Radiometry is measurement associated with the energy content or power of electromagnetic radiation. It is based on units of joules or watts. It gives rise to the so-called "energetic" system of units and is the origin of the suffix "e" that has been used so far. The radiometric case corresponds to those quantities used to describe radiant energy and is the natural unit of choice for thermal detectors. *Actinometry*, on the other hand, is measurement associated with photon counting. It corresponds to those quantities related to the measure of photons such as photon flux (photons per second) or number of photons (dose). It leads to the so-called "photonic" system of units and is the more appropriate choice for quantum detectors [727]. For this case the suffix "q," for quantum, will be used.

So far, whenever Planck's law has been considered it has been done using the radiometric case. It serves as the form for his law met in the vast majority of cases. A form for Planck's law for the actinometric case can also be defined. While an obvious corollary to Planck's radiation law, it is somewhat surprising to see it was not formally defined for almost four decades after Planck first introduced his law in 1900. Published as a short one-page note in the May issue of the *Journal of the Optical Society of America* of 1938, the Austrian-born physicist Arthur Erich Haas (1884–1941) who emigrated to the US in 1935 appears to have been the first to give the form for the spectral distribution of photons emitted from a blackbody [274]. The work was based on a short presentation he and his colleague Eugene Guth delivered before the *Thirty-Ninth Annual Meeting of the American Physical Society* which was held in Indianapolis, Indiana, in December of the previous year [275]. In a succession of papers shortly after its publication, many of the consequences stemming from the idea of a spectral distribution of photons from a blackbody were quickly worked out by Haas' compatriot, the US physicist Archie Garfield Worthing (1881–1949) [708, 709, 710].

At around the same time as Haas was considering and developing his ideas on this matter, Zen'emon Miduno, who was working at Kyusyu Imperial University in Hukuoka, Japan, arrived independently at the same idea. In a paper marked as read before a meeting of the *Physico-Mathematical Society of Japan* held on October 15, 1938, and subsequently received for publication in the society's *Proceedings* two weeks later on October 31, 1938, Miduno presented tabulations and a nomogram[20] for the purposes of calculating values for the spectral photon exitance [446]. At the time of publication the journal was not widely known outside of Japan. As a result Miduno's work remained largely unknown to those working in the West and it would be some years before the wider international community would finally learn of his work.

In obtaining the spectral distribution for photons emitted from a blackbody, as the energy carried by a single photon is given by hc/λ, on dividing Eq. (2.20) by this amount, Planck's law for the spectral photon exitance $M_{q,\lambda}^{b}$ for a blackbody immediately follows. The result is

$$M_{q,\lambda}^{b}(\lambda, T) = \frac{2\pi c}{\lambda^4}\mathrm{Li}_0\left[\exp\left(-\frac{hc}{k_B \lambda T}\right)\right].\qquad(2.57)$$

The peak in the curve for the spectral photon exitance is found in a similar manner to how peaks in the curves for the spectral radiant exitance were found. Setting the derivative of Eq. (2.57) with respect to wavelength equal to zero and solving yields

$$(u-4)e^{u-4} = -4e^{-4},\qquad(2.58)$$

a transcendental equation similar in form to Eq. (2.22). Here $u = hc/(k_B\lambda T)$. On moving from a linear wavelength representation to some more general spectral representation, an equation similar in form to Eq. (2.58) is found. The result is

$$(u-s)e^{u-s} = -se^{-s}.\qquad(2.59)$$

Once more s is a parameter related to the spectral scale chosen but as it applies to the actinometric case, this time the crossover point from one of frequency to wavelength occurs at $s = 3$ rather than at $s = 4$ as previously found for the radiometric case.

Given Eq. (2.59) is once more exactly in the form for the defining equation for the Lambert W function it can be solved in terms of this function. The result is

$$\lambda_{\mathrm{max,q}}T = \frac{hc}{k_B[s + \mathrm{W}_0(-se^{-s})]}.\qquad(2.60)$$

Table 2.3 lists a number of Wien peaks in photonic units for several of the more common spectral representations. Note in photonic units the wavelength–frequency neutral peak that gives the division between the frequency and wavelength scales occurs when $s = 3$. A peak in the spectral distribution function still only exists when $s > 1$. So for the actinometric case, Planck's distribution function in photonic units no longer contains a peak for spectral representations equal to frequency-squared or beyond.

2.8   NORMALIZED SPECTRAL EXITANCE

Since the wavelength at the Wien peak can be found in closed form in terms of the Lambert W function it allows the spectral radiant and photon exitances normalized relative to their corresponding peak value to be expressed in a compact, analytic form. We already considered this briefly in Section 2.3 where we considered spectral scales corresponding to linear wavelength and frequency in energetic units. Now we move our attention to the more general case for any spectral scale in either energetic or photonic units.

The spectral radiant exitance normalized relative to its peak value in the logarithmic wavelength scale will be

$$\frac{M^{b}_{e,\ln\lambda}(\lambda, T)}{M^{b}_{e,\ln\lambda}(\lambda_{\max,\ln\lambda}, T)} = -\left(\frac{hc}{k_B\lambda T}\right)^4 \frac{\mathrm{Li}_0\left[\exp\left(-\dfrac{hc}{k_B\lambda T}\right)\right]}{W_0(-4\mathrm{e}^{-4})[4 + W_0(-4\mathrm{e}^{-4})]^3}, \quad (2.61)$$

while for a general wavelength representation of the form $\varphi = f(\lambda) = \lambda^\gamma$ where $\gamma > 0$ one has

$$\frac{M^{b}_{e,\lambda^\gamma}(\lambda, T)}{M^{b}_{e,\lambda^\gamma}(\lambda_{\max,\lambda^\gamma}, T)} = -\left(\frac{hc}{k_B\lambda T}\right)^s \frac{\mathrm{Li}_0\left[\exp\left(-\dfrac{hc}{k_B\lambda T}\right)\right]}{W_0(-s\mathrm{e}^{-s})[s + W_0(-s\mathrm{e}^{-s})]^{s-1}}. \quad (2.62)$$

Here $s = 4 + \gamma$. For frequency scales, for a logarithmic frequency scale one has

$$\frac{M^{b}_{e,\ln\nu}(\nu, T)}{M^{b}_{e,\ln\nu}(\nu_{\max,\ln\nu}, T)} = -\left(\frac{h\nu}{k_B T}\right)^4 \frac{\mathrm{Li}_0\left[\exp\left(-\dfrac{h\nu}{k_B T}\right)\right]}{W_0(-4\mathrm{e}^{-4})[4 + W_0(-4\mathrm{e}^{-4})]^3}, \quad (2.63)$$

while for a general frequency representation of the form $\varphi = f(\nu) = \nu^\beta$ where $0 < \beta < 3$ one has

$$\frac{M^{b}_{e,\nu^\beta}(\nu, T)}{M^{b}_{e,\nu^\beta}(\nu_{\max}, T)} = -\left(\frac{h\nu}{k_B T}\right)^s \frac{\mathrm{Li}_0\left[\exp\left(-\dfrac{h\nu}{k_B T}\right)\right]}{W_0(-s\mathrm{e}^{-s})[s + W_0(-s\mathrm{e}^{-s})]^{s-1}}. \quad (2.64)$$

Here $s = 4 - \beta$. Note the use of s in both the wavelength and frequency representations corresponds to the familiar spectral scale parameter used previously when discussing Wien peaks for various spectral scales. We see the spectral scale parameter s relative to a general spectral representation can be written as

$$s = \begin{cases} 4 - \beta, & \text{for} \quad 0 < \beta < 3, \\ 4 + \gamma, & \text{for} \quad \gamma > 0. \end{cases} \qquad (2.65)$$

In a similar manner, normalized expressions for the spectral photon exitance at its peak can be found but are not given here. Plots for the normalized spectral radiate exitance for a number of different wavelength scales as a function of the wavelength–temperature product are given in Fig. 2.6. The skewness in the shape of each spectral curve is gradually reduced as the spectral scale chosen approaches a logarithmic one. A similar plot for the normalized spectral radiate exitance for a number of different frequency scales as a function of the frequency–temperature quotient is given in Fig. 2.7.

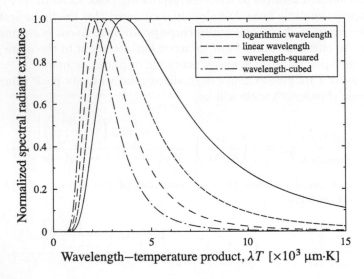

Figure 2.6 The normalized spectral radiant exitance for the following four spectral wavelength representations of: logarithmic wavelength (solid line); linear wavelength (dashed line); wavelength-squared (chained line); and wavelength-cubed (dot-dashed line) as a function of the wavelength–temperature product.

2.9 THE STEFAN–BOLTZMANN LAW

Even though the Stefan–Boltzmann law antedates Planck's law it is a direct consequence of the later's law. In this section we will derive the law for both the radiometric and actinometric cases directly from Planck's spectral distribution function, using what we term the traditional approach and using a more recent approach based on polylogarithms.

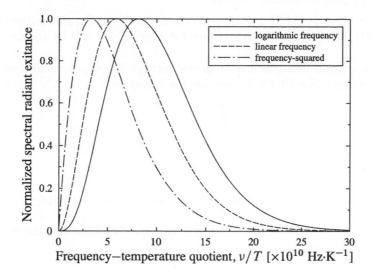

Figure 2.7 The normalized spectral radiant exitance for three spectral frequency representations of: logarithmic frequency (solid line); linear frequency (dashed line); and frequency-squared (dot-dashed line) as a function of the frequency–temperature quotient.

2.9.1 THE TRADITIONAL APPROACH

The total radiant exitance is found from the spectral radiant exitance by integrating over all wavelengths

$$M_e^b(T) = \int_0^\infty M_{e,\lambda}^b(\lambda, T)d\lambda = 2\pi h c^2 \int_0^\infty \frac{d\lambda}{\lambda^5 \left[\exp\left(\frac{hc}{k_B \lambda T} \right) - 1 \right]}. \quad (2.66)$$

In the first of these approaches, which we refer to as the tradition approach, the form for Planck's law is left in a form as it was first given by Planck himself; that with the exponential term appearing in the denominator of his law. After the change of variable $x = hc/(k_B \lambda T)$ is made, Eq. (2.66) can be rewritten in terms of the following dimensionless integral

$$M_e^b(T) = \frac{2\pi k_B^4}{h^3 c^2} T^4 \int_0^\infty \frac{x^3}{e^x - 1} dx. \quad (2.67)$$

The value for the integral appearing in Eq. (2.67) can be determined in a number of ways [646, 110, 262]. Due to a need when we come to the actinometric case, its evaluation is best considered within the context of the following a slightly more general case of

$$\int_0^\infty \frac{x^{s-1}}{e^x - 1} dx, \quad s > 1, \quad (2.68)$$

an integral first found by the German mathematician Bernhard Riemann in 1859 [561]. Besides the $s = 4$ case needed here, $s = 3$ will be required later on.

As the integral we are interested in evaluating is improper, we first establish its convergence. For all $s > 1$, consider the integrals

$$\int_0^1 \frac{x^{s-1}}{e^x - 1} dx, \tag{2.69}$$

and

$$\int_1^\infty \frac{x^{s-1}}{e^x - 1} dx, \tag{2.70}$$

both of which are improper and sum to give the integral appearing in Eq. (2.68).

The convergence of the first integral can be established in the following manner. Observing

$$e^x = 1 + x + \frac{x^2}{2!} + \frac{x^3}{3!} + \cdots > 1 + x, \tag{2.71}$$

for all $x \in (0, 1]$, $e^x - 1 > x$. For $\varepsilon > 0$ sufficiently small one has

$$\int_0^1 \frac{x^{s-1}}{e^x - 1} dx < \int_\varepsilon^1 \frac{x^{s-1}}{x} dx = \int_\varepsilon^1 x^{s-2} dx = \frac{x^{s-1}}{s-1}\Big|_\varepsilon^1 = \frac{1}{s-1} - \frac{\varepsilon^{s-1}}{s-1}. \tag{2.72}$$

Since $\varepsilon^{s-1}/(s-1) \to 0$ as $\varepsilon \to 0^+$ this shows the first of the integrals converges.

For the second improper integral, observing that

$$\frac{x^{s-1}}{e^x - 1} = \frac{x^{s-1}}{\sum_{k=0}^\infty \frac{x^k}{k!} - 1} = \frac{x^{s-1}}{\sum_{k=1}^\infty \frac{x^k}{k!}} \leqslant \frac{x^{s-1}}{\frac{x^n}{n!}}, \tag{2.73}$$

for all $n \in \mathbb{N}$ (the set of natural numbers) and $x \geqslant 1$. Here the Maclaurin series expansion for the exponential function has been used. Accordingly, one can see

$$\int_1^\Lambda \frac{x^{s-1}}{e^x - 1} dx \leqslant n! \int_1^\Lambda \frac{1}{x^{n-s+1}} dx = n! \frac{x^{s-n}}{s-n}\Big|_1^\Lambda = \frac{n!}{s-n}\left(\frac{1}{\Lambda^{n-s}} - 1\right), \tag{2.74}$$

provided Λ is sufficiently large and n is chosen so that $n \geqslant s + 1$. Thus $1/\Lambda^{n-s} \to 0$ as $\Lambda \to \infty$ and shows the second of the integrals also converges. We therefore conclude the improper integral we started with, namely Eq. (2.68), converges.

As the integral given by Eq. (2.68) converges it has a finite value. Moreover, in this particular case it is possible to evaluate the integral in closed form in terms of known functions of mathematical physics. After rearranging the integrand appearing in the integral given by Eq. (2.68) it can be rewritten as

$$\int_0^\infty \frac{e^{-x} x^{s-1}}{1 - e^{-x}} dx. \tag{2.75}$$

Recognising the term $1/(1-e^{-x})$ as the sum of the convergent geometric series $\sum_{n=0}^{\infty} e^{-nx}$ for all $x > 0$, on shifting the summation index and interchanging the order of the integration and summation,[21] the integral becomes

$$\int_0^{\infty} \frac{x^{s-1}}{e^x - 1} dx = \sum_{n=1}^{\infty} \int_0^{\infty} x^{s-1} e^{-nx} dx. \tag{2.76}$$

Applying the change of variable $u = nx$, the integral appearing in Eq. (2.76) can be written as

$$\int_0^{\infty} \frac{x^{s-1}}{e^x - 1} dx = \sum_{n=1}^{\infty} \frac{1}{n^s} \int_0^{\infty} e^{-u} u^{s-1} du. \tag{2.77}$$

Recognising

$$\Gamma(s) = \int_0^{\infty} e^{-u} u^{s-1} du, \quad s > 0, \tag{2.78}$$

as the Gamma function and

$$\zeta(s) = \sum_{n=1}^{\infty} \frac{1}{n^s}, \quad s > 1, \tag{2.79}$$

as the Riemann zeta function, in terms of these two functions one has

$$\int_0^{\infty} \frac{x^{s-1}}{e^x - 1} dx = \Gamma(s)\zeta(s), \quad s > 1. \tag{2.80}$$

Setting $s = 4$ in Eq. (2.80), the value for the integral appearing in Eq. (2.67) will be equal to $\Gamma(4)\zeta(4)$. It is, however, possible to express this number in terms of more familiar mathematical constants. For positive integers n, it is well know the Gamma function can be expressed in terms of the factorial, namely

$$\Gamma(n+1) = n!. \tag{2.81}$$

So $\Gamma(4) = 3! = 6$. On the other hand for positive even integers, in Appendix A.2 we show the Riemann zeta function is equal to the remarkable formula

$$\zeta(2n) = \frac{(-1)^{n+1} B_{2n} (2\pi)^{2n}}{2(2n)!}, \quad n \geqslant 1. \tag{2.82}$$

Here B_{2n} are the Bernoulli numbers. We will meet these numbers again in the chapters to follow. The first few are:

$$B_0 = 1, \ B_1 = -\frac{1}{2}, \ B_2 = \frac{1}{6}, \ B_3 = 0, \ B_4 = -\frac{1}{30}, \ B_5 = 0, \ B_6 = \frac{1}{42}, \ \cdots \tag{2.83}$$

So we find $\zeta(4) = -B_4\pi^4/3 = \pi^4/90$. The value for the integral appearing in Eq. (2.67) is $\pi^4/15$ and the total radiant exitance becomes

$$M_e^b(T) = \frac{2\pi^5 k_B^4}{15 h^3 c^2}T^4 = \sigma T^4, \tag{2.84}$$

which is the forementioned Stefan–Boltzmann law.[22] The constant σ appearing in the Stefan–Boltzmann law is know appropriately enough as the Stefan–Boltzmann constant.

The Stefan–Boltzmann law appearing in Eq. (2.84) is the form of the law most commonly encountered. It is given in terms of energetic units and corresponds to the radiometric case. For the actinometric case, a law similar in form to the Stefan–Boltzmann law except in photonic units can be found. It gives the total photon exitance, and to differentiate it from the law corresponding to the more familiar case, is referred to as the Stefan–Boltzmann law for photon exitance.

From Planck's law, for the spectral photon exitance $M_{q,\lambda}^b$ for a blackbody the total photon exitance is found from the spectral photon exitance by integrating over all wavelengths. When this is done one finds

$$M_q^b(T) = \int_0^\infty M_{q,\lambda}^b(\lambda, T)d\lambda = 2\pi c \int_0^\infty \frac{d\lambda}{\lambda^4 \left[\exp\left(\dfrac{hc}{k_B \lambda T}\right) - 1\right]}. \tag{2.85}$$

On making the change of variable $x = hc/(k_B \lambda T)$ the integral appearing in Eq. (2.85) can be rewritten as

$$M_q^b(T) = \frac{2\pi k_B^3}{h^3 c^2}T^3 \int_0^\infty \frac{x^2}{e^x - 1}dx. \tag{2.86}$$

Comparing with Eq. (2.80), the dimensionless integral appearing in the total photon exitance is found by setting $s = 3$. We have already shown improper integrals of this form converge for all $s > 1$ while its value will be equal to $\Gamma(3)\zeta(3) = 2\zeta(3)$. So for the total photon exitance one has

$$M_q^b(T) = \frac{4\pi k_B^3 \zeta(3)}{c^2 h^3}T^3 = \sigma_q T^3. \tag{2.87}$$

The units for $M_q^b(T)$ are photons per second per square meter [photon·s^{-1}·m^{-2}] while σ_q is the Stefan–Boltzmann constant in photonic units whose currently accepted value is $1.520\,459 \times 10^{15}\,\text{s}^{-1}\cdot\text{m}^{-2}\cdot\text{K}^{-3}$. The law in this form was first given by both Hass [274] and Miduno [446] in 1938.

2.9.2   A POLYLOGARITHMIC APPROACH

In this section we reconsider the deviation of the Stefan–Boltzmann law from Planck's law, this time however using the polylogarithmic approach developed in Section 2.2.1.

By integrating the spectral radiate exitance over all wavelengths, as before, the total radiant exitance can be found

$$M_e^b(T) = \int_0^\infty M_{e,\lambda}^b(\lambda, T)\, d\lambda. \tag{2.88}$$

Substituting Planck's spectral distribution function in the linear wavelength representation written in terms of polylogarithms (see Eq. (2.20)) into Eq. (2.88), after the change of variable $u = hc/(k_B\lambda T)$ is made, one has

$$M_e^b(T) = \frac{2\pi k_B^4 T^4}{h^3 c^2} \int_0^\infty u^3 \mathrm{Li}_0(e^{-u})\, du. \tag{2.89}$$

The dimensionless improper integral appearing in Eq. (2.89) is just that given by Eq. (2.68) written in an alternative mathematical form. It has already been shown to converge. To find its value we write the improper integral as

$$M_e^b(T) = \frac{2\pi k_B^4 T^4}{h^3 c^2} \left[\lim_{\alpha \to \infty} \int_a^\alpha u^3 \mathrm{Li}_0(e^{-u})\, du + \lim_{\alpha \to 0^+} \int_\alpha^a u^3 \mathrm{Li}_0(e^{-u})\, du \right],$$
$$\tag{2.90}$$

for some finite, positive number a. The integrals are evaluated using Eq. (2.11) by setting $k = 3$. The result is

$$M_e^b(T) = \frac{2\pi k_B^4 T^4}{h^3 c^2} \left(\mathfrak{J}_1 - \mathfrak{J}_2 \right), \tag{2.91}$$

where

$$\mathfrak{J}_1 = \lim_{\alpha \to 0^+} \left[\alpha^3 \mathrm{Li}_1(e^{-\alpha}) + 3\alpha^2 \mathrm{Li}_2(e^{-\alpha}) + 6\alpha \mathrm{Li}_3(e^{-\alpha}) + 6\mathrm{Li}_4(e^{-\alpha}) \right], \tag{2.92}$$

and

$$\mathfrak{J}_2 = \lim_{\alpha \to \infty} \left[\alpha^3 \mathrm{Li}_1(e^{-\alpha}) + 3\alpha^2 \mathrm{Li}_2(e^{-\alpha}) + 6\alpha \mathrm{Li}_3(e^{-\alpha}) + 6\mathrm{Li}_4(e^{-\alpha}) \right]. \tag{2.93}$$

For the first term $\mathfrak{J}_1$, as $\lim_{\alpha \to 0^+} \mathrm{Li}_s(e^{-\alpha}) = 1$ for all orders s, the only non-zero term comes from the fourth term appearing in Eq. (2.92). Thus

$$\mathfrak{J}_1 = 6 \lim_{\alpha \to 0^+} \mathrm{Li}_4(e^{-\alpha}) = 6\mathrm{Li}_4(1). \tag{2.94}$$

For the second term $\mathfrak{J}_2$, from Eq. (2.14) we see each term appearing in the limit is equal to zero. Thus $\mathfrak{J}_2 = 0$. So the total radiant exitance becomes

$$M_e^b(T) = \frac{2\pi k_B^4 T^4}{h^3 c^2} 6\mathrm{Li}_4(1). \tag{2.95}$$

But as $\mathrm{Li}_4(1) = \zeta(4) = \pi^4/90$, one finally has

$$M_e^b(T) = \frac{2\pi^5 k_B^4}{15 h^3 c^2} T^4 = \sigma T^4, \tag{2.96}$$

as expected.

A similar calculation for the total photon exitance using the polylogarithmic form for the spectral photon exitance as given by Eq. (2.57) leads to

$$M_q^b(T) = \frac{2\pi k_B^3 T^3}{h^3 c^2} 2\text{Li}_3(1) = \frac{4\pi k_B^3 \zeta(3)}{h^3 c^2} T^3 = \sigma_q T^3. \qquad (2.97)$$

The resulting improper integral in terms of polylogarithms which appears in the calculation is taken care of by setting $k = 2$ in Eq. (2.11) while the two resulting limits involving a sum of polylogarithmic functions are dealt with in a manner similar to the treatment done for the radiometric case. Lastly, use of the result $\text{Li}_3(1) = \zeta(3)$ has been made.

2.10   FRACTIONAL FUNCTIONS OF THE FIRST KIND

In applications the amount of energy radiated by a blackbody into a given spectral band is often required. Between wavelengths λ_1 and λ_2 ($\lambda_2 > \lambda_1$) this in-band exitance is found from

$$M_{\varsigma,\lambda_1 \to \lambda_2}^b = \int_{\lambda_1}^{\lambda_2} M_{\varsigma,\lambda}^b(\lambda, T) d\lambda. \qquad (2.98)$$

Here ς is a generalised suffix introduced to denote either the radiometric ($\varsigma = \text{e}$) or the actinometric ($\varsigma = \text{q}$) case.

A more convenient form, particularly when used as an aid to calculation, is the fractional amount normalized relative to the total exitance $M_\varsigma^b(T)$. Given by the ratio

$$\mathfrak{F}_{\varsigma,\lambda_1 \to \lambda_2} = \frac{\displaystyle\int_{\lambda_1}^{\lambda_2} M_{\varsigma,\lambda}^b(\lambda, T) d\lambda}{\displaystyle\int_0^\infty M_{\varsigma,\lambda}^b(\lambda, T) d\lambda}, \qquad (2.99)$$

its value depends on the type of case considered, be it radiometric or actinometric. It is a dimensionless quantity between zero and one and gives a measure of the amount of radiation radiated by a blackbody into a finite wavelength interval compared to the total amount radiated over all wavelengths.

Restricting for the moment our attention to the radiometric case, from Eq. (2.99) one has

$$\mathfrak{F}_{\text{e},\lambda_1 \to \lambda_2} = \frac{\displaystyle\int_{\lambda_1}^{\lambda_2} M_{\text{e},\lambda}^b(\lambda, T) d\lambda}{\displaystyle\int_0^\infty M_{\text{e},\lambda}^b(\lambda, T) d\lambda}. \qquad (2.100)$$

The integral appearing in the denominator for the fractional function, being nothing more than the Stefan–Boltzmann law, has already been found. The result of its evaluation is σT^4 where σ corresponds to the Stefan–Boltzmann constant.

From properties of the definite integral over a contiguous interval the integral appearing in the numerator of Eq. (2.100) can be written as

$$\mathfrak{F}_{e,\lambda_1 \to \lambda_2} = \frac{1}{\sigma T^4} \left[\int_0^{\lambda_2} M_{e,\lambda}^b(\lambda, T) d\lambda - \int_0^{\lambda_1} M_{e,\lambda}^b(\lambda, T) d\lambda \right] = \mathfrak{F}_{e,0 \to \lambda_2} - \mathfrak{F}_{e,0 \to \lambda_1}.$$

(2.101)

Note the advantage gained by setting the lower limit of integration equal to zero in the fractional function is largely one of convenience.

Historically, as it was thought to be not possible to evaluate the integral appearing in the numerator of the fractional function in closed form, for the purposes of tabulation, rather than having to find $\mathfrak{F}_{e,0 \to \lambda}$ at each temperature T, since the radiant spectral exitance can be written as a function of the single variable λT as follows

$$\frac{M_{e,\lambda}^b(\lambda T)}{T^5} = \frac{2\pi hc^2}{(\lambda T)^5} \mathrm{Li}_0 \left[\exp \left(-\frac{hc}{k_B \lambda T} \right) \right],$$

(2.102)

this added complexity in the evaluation of the fractional function can be removed by writing it in terms of a single variable. When this is done one has

$$\mathfrak{F}_{e,0 \to \lambda_*} = \frac{1}{\sigma T^4} \int_0^{\lambda_*} M_{e,\lambda}^b(\lambda, T) d\lambda = \frac{1}{\sigma} \int_0^{\lambda_*} \frac{M_{e,\lambda}^b(\lambda T)}{T^5} d(\lambda T) = \mathfrak{F}_{e,0 \to \lambda_*}.$$

(2.103)

Here λ_* is an arbitrary finite wavelength. To avoid further use of λ_*, which can be rather cumbersome, for an arbitrary wavelength we will simply use λ on the understanding that it should not be confused with the use of λ for the dummy variable of integration appearing in Eq. (2.103).

The function $\mathfrak{F}_{e,0 \to \lambda T}$ is referred to as the *blackbody fractional function of the first kind* corresponding to the radiometric case. The term was first introduced by Marinus Czerny in 1954 [160] and subsequently adopted and used by others [164, 203, 471, 632], though the origin of the function can be traced back to the late nineteenth century.[23] In its evaluation a polylogarithmic approach will be taken. Here the integral will be evaluated in closed form in terms of polylogarithm functions.

Substituting the polylogarithmic form for the radiant spectral exitance into Eq. (2.103) gives

$$\mathfrak{F}_{e,0 \to \lambda T} = \frac{15}{\pi^4} \left(\frac{hc}{k_B T} \right)^4 \int_0^{\lambda} \frac{1}{\lambda^5} \mathrm{Li}_0 \left[\exp \left(-\frac{hc}{k_B \lambda T} \right) \right] d\lambda.$$

(2.104)

After the change of variable $u = hc/(k_B \lambda T)$ is made, Eq. (2.104) can be rewritten more compactly in terms of the following dimensionless integral

$$\mathfrak{F}_{e,0 \to \lambda T} = \frac{15}{\pi^4} \int_z^\infty u^3 \mathrm{Li}_0(e^{-u}) du,$$

(2.105)

where $z = hc/(k_B \lambda T)$.

Setting $k = 3$ in Eq. (2.11), after taking care of the limits of integration, a closed-form expression for the blackbody fractional function of the first kind corresponding to the radiometric case in terms of polylogarithms follows. The result is

$$\mathfrak{F}_{e,0 \to \lambda T} = \frac{15}{\pi^4} \left[z^3 \mathrm{Li}_1(\mathrm{e}^{-z}) + 3z^2 \mathrm{Li}_2(\mathrm{e}^{-z}) + 6z \mathrm{Li}_3(\mathrm{e}^{-z}) + 6\mathrm{Li}_4(\mathrm{e}^{-z}) \right].$$

(2.106)

The formulation followed here was recently developed by one of the authors [629], though the expression for the blackbody fractional function of the first kind corresponding to the radiometric case in terms of polylogarithms dates back to earlier work first performed in the late 1980s. This work however, does not seem to be widely known. Working with MACSYMA, one of the first early computer algebra systems,[24] Bradley A. Clark, who in 1986 was working at the Los Alamos National Laboratory in New Mexico in the US, appears to have been the first to have made the computer-aided discovery for the closed-form expression for the fractional function of the first kind for the radiometric case in terms of polylogarithms [138, 139]. Nor would he be the last to make such a serendipitous discovery using a computer algebra system.[25] Fifteen years later, V. Lampret, J. Peternelj and A. Krainer, who were all working at the University of Ljubljana in Slovenia, when faced with what they thought to be an intractable integral in their work on the luminous flux and luminous efficacy of blackbody radiation, after dropping it into MATHEMATICA also managed to independently stumble upon the polylogarithmic form for the blackbody fractional function of the first kind [369]. Either chance discovery however, belies a simplicity the polylogarithmic formulation brings to the problem of blackbody radiation.

The cognate to the fractional function of the first kind corresponding to the radiometric case is that for the actinometric case. Known appropriately enough as the fractional function of the first kind corresponding to the actinometric case, while not as common as the former, its origins can none-the-less be traced back to the early 1930s and the Dutch astronomer Herman Zanstra and his work on planetary nebulae [730]. Given by

$$\mathfrak{F}_{q,0 \to \lambda T} = \frac{\displaystyle\int_0^\lambda M_{q,\lambda}^{\mathrm{b}}(\lambda, T)d\lambda}{\displaystyle\int_0^\infty M_{q,\lambda}^{\mathrm{b}}(\lambda, T)d\lambda},$$

(2.107)

it can be evaluated in closed form in terms of polylogarithms in a manner similar to what was done for the radiometric case.

Starting with Eq. (2.57), Planck's law for the spectral photon exitance in the linear wavelength representation in terms of polylogarithms, substituting into Eq. (2.107), after recognising the denominator is just the Stefan–

Boltzmann law for photon exitance, one has

$$\mathfrak{F}_{q,0\to\lambda T} = \frac{1}{2\zeta(3)}\left(\frac{hc}{k_B T}\right)^3 \int_0^\lambda \frac{1}{\lambda^4}\mathrm{Li}_0\left[\exp\left(-\frac{hc}{k_B\lambda T}\right)\right]d\lambda. \qquad (2.108)$$

Introduction of the change of variable $u = hc/(k_B\lambda T)$ will once more reduce the integral appearing in Eq. (2.108) to dimensionless form. The result is

$$\mathfrak{F}_{q,0\to\lambda T} = \frac{1}{2\zeta(3)}\int_z^\infty u^2\mathrm{Li}_0(e^{-u})du, \qquad (2.109)$$

where $z = hc/(k_B\lambda T)$.

Setting $k = 2$ in Eq. (2.11) the integral appearing in Eq. (2.109) can be evaluated in terms of polylogarithms. After taking care of the limits of integration we find

$$\mathfrak{F}_{q,0\to\lambda T} = \frac{1}{2\zeta(3)}\left[z^2\mathrm{Li}_1(e^{-z}) + 2z\mathrm{Li}_2(e^{-z}) + 2\mathrm{Li}_3(e^{-z})\right]. \qquad (2.110)$$

The above closed-form result first appeared in the literature only very recently and was given by one of the authors [632, 634], though an appearance under a slightly different guise in the work of Heetae Kim, Seong Chu Lim, and Young Hee Lee on the size effect of two-dimensional thermal radiation a few years earlier can be found [351].

Some simple properties for either of the fractional functions of the first kind immediately follow from properties for definite integrals. These are

$$\mathfrak{F}_{\varsigma,\lambda T\to 0} = -\mathfrak{F}_{\varsigma,0\to\lambda T}, \qquad (2.111)$$

and

$$\mathfrak{F}_{\varsigma,\lambda T\to\infty} = 1 - \mathfrak{F}_{\varsigma,0\to\lambda T}. \qquad (2.112)$$

In Fig. 2.8 three fractional functions of the first kind are shown as a function of the wavelength–temperature product using a semi-log plot for the abscissa. The third fractional function for the so-called eta case is to be introduced in Section 2.11. The points on the fractional curves where its concavity changes sign correspond to a point of inflexion. From a necessary condition for a point of inflexion, namely that its second derivative with respect to wavelength be equal to zero, one has

$$\frac{\partial^2}{\partial\lambda^2}\left[\mathfrak{F}_{\varsigma,0\to\lambda T}\right] = \frac{1}{\sigma T^4}\frac{\partial}{\partial\lambda}\left[M_{\varsigma,\lambda}^{\mathrm{b}}(\lambda, T)\right] = 0, \qquad (2.113)$$

and is precisely the condition used to find the peak in the corresponding spectral distribution curve. Each inflexion point in the fractional function therefore occurs at its Wien peak.

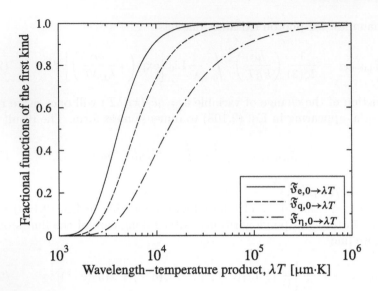

Figure 2.8 Fractional functions of the first kind for the radiometric (solid line), actinometric (dashed line), and eta cases (dot-dashed line) as a function of the wavelength–temperature product. A semi-log plot for the abscissa has been used.

Fractional functions of the first kind in terms of frequency can also be defined. For the radiometric case the fractional function of the first kind is defined by

$$\mathfrak{F}_{e,0\to\nu} = \frac{\displaystyle\int_0^\nu M_{e,\nu}^b(\nu,T)d\nu}{\displaystyle\int_0^\infty M_{e,\nu}^b(\nu,T)d\nu} = \frac{1}{\sigma T^4}\int_0^\nu M_{e,\nu}^b(\nu,T)d\nu. \qquad (2.114)$$

Since the spectral radiant exitance is able to be written in terms of a single variable ν/T as follows

$$\frac{M_{e,\nu}^b(\nu/T)}{T^3} = \frac{2\pi h}{c^2}\left(\frac{\nu}{T}\right)^3 \mathrm{Li}_0\left[\exp\left(-\frac{h}{k_B}\frac{\nu}{T}\right)\right], \qquad (2.115)$$

in terms of a single variable, Eq. (2.114) becomes

$$\mathfrak{F}_{e,0\to\nu} = \frac{1}{\sigma T^4}\int_0^\nu M_{e,\nu}^b(\nu,T)d\nu = \frac{1}{\sigma}\int_0^{\nu/T}\frac{M_{e,\nu}^b(\nu/T)}{T^3}d\left(\frac{\nu}{T}\right) = \mathfrak{F}_{e,0\to\nu/T}. \qquad (2.116)$$

Substituting Eq. (2.115) into Eq. (2.116), after the change of variable $u = h\nu/(k_B T)$ is made, the following dimensionless integral results

$$\mathfrak{F}_{e,0\to\nu/T} = \frac{15}{\pi^4}\int_0^z u^3\mathrm{Li}_0(e^{-u})\,du. \qquad (2.117)$$

Here $z = h\nu/(k_B T)$. From properties of the definite integral over a contiguous interval, the fractional function of the first kind for frequency can be related to its corresponding function for wavelength by means of

$$\mathfrak{F}_{e,0\to\nu/T} = \frac{15}{\pi^4} \int_0^z u^3 \mathrm{Li}_0(e^{-u})\, du = 1 - \frac{15}{\pi^4} \int_z^\infty u^3 \mathrm{Li}_0(e^{-u})\, du = 1 - \mathfrak{F}_{e,0\to\lambda T}.$$
(2.118)

The evaluation for the fractional function of the first kind in terms of frequency immediately follows. Using the previously found result for $\mathfrak{F}_{e,0\to\lambda T}$ one has

$$\mathfrak{F}_{e,0\to\nu/T} = 1 - \frac{15}{\pi^4} \left[z^3 \mathrm{Li}_1(e^{-z}) + 3z^2 \mathrm{Li}_2(e^{-z}) + 6z \mathrm{Li}_3(e^{-z}) + 6\mathrm{Li}_4(e^{-z}) \right].$$
(2.119)

In a similar manner, for the actinometric case, the fractional function of the first kind in terms of frequency can be readily shown to be given by

$$\mathfrak{F}_{q,0\to\nu/T} = 1 - \frac{1}{2\zeta(3)} \left[z^2 \mathrm{Li}_1(e^{-z}) + 2z \mathrm{Li}_2(e^{-z}) + z \mathrm{Li}_3(e^{-z}) \right].$$
(2.120)

2.11 FRACTIONAL FUNCTIONS OF OTHER KINDS

Fractional functions of other kinds do exist and it is these we plan to discuss in this section. While fractional functions of the first kind may be the most commonly encountered, occasionally one has a need for other closely related fractional functions.

Fractional functions of the second kind are defined in a similar way to the two fractional functions of the first kind already met. First introduced by Czerny in 1954 [160], it is once more a ratio between two integrals. The spectral exitance term which appears in the integrals found in the fractional function for the first kind, on replacing it with the spectral exitance partially differentiated with respect to temperature, leads to the fractional functions of the second kind, viz.,

$$\mathfrak{F}^*_{\varsigma,0\to\lambda T} = \frac{\displaystyle\int_0^\lambda \frac{\partial M^b_{\varsigma,\lambda}}{\partial T}\, d\lambda}{\displaystyle\int_0^\infty \frac{\partial M^b_{\varsigma,\lambda}}{\partial T}\, d\lambda}.$$
(2.121)

In its evaluation, for the fractional function of the second kind for the radiometric case one has

$$\mathfrak{F}^*_{e,0\to\lambda T} = \frac{\displaystyle\int_0^\lambda \frac{\partial M^b_{e,\lambda}}{\partial T}\, d\lambda}{\displaystyle\int_0^\infty \frac{\partial M^b_{e,\lambda}}{\partial T}\, d\lambda}.$$
(2.122)

The integral appearing in the denominator of Eq. (2.122) can be readily found. From Leibniz's rule for differentiating under the integral sign one can write

$$\int_0^\infty \frac{\partial M_{e,\lambda}^b}{\partial T} d\lambda = \frac{d}{dT} \int_0^\infty M_{e,\lambda}^b(\lambda, T) d\lambda. \tag{2.123}$$

The integral on the right side of Eq. (2.123) corresponds to the Stefan–Boltzmann law and yields

$$\int_0^\infty \frac{\partial M_{e,\lambda}^b}{\partial T} d\lambda = \frac{d}{dT}(\sigma T^4) = 4\sigma T^3. \tag{2.124}$$

As for the integral appearing in the numerator of Eq. (2.122), its evaluation is a little more involved but can also be found in closed form in terms of polylogarithm functions. Starting with the partial derivative for the spectral radiant exitance with respect to temperature, namely

$$\frac{\partial M_{e,\lambda}^b}{\partial T} = \frac{2\pi h^2 c^3}{k_B T^2 \lambda^6} \text{Li}_{-1}\left[\exp\left(-\frac{hc}{k_B \lambda T}\right)\right], \tag{2.125}$$

the integral appearing in the numerator of Eq. (2.122) can be written as

$$\int_0^\lambda \frac{\partial M_{e,\lambda}^b}{\partial T} d\lambda = \frac{2\pi k_B^4 T^3}{h^3 c^2} \int_z^\infty u^4 \text{Li}_{-1}(e^{-u}) du, \tag{2.126}$$

where the change of variable $u = hc/(k_B \lambda T)$ has been employed so as to make the integral dimensionless. Again $z = hc/(k_B \lambda T)$.

The fractional function of the second kind for the radiometric case can now be written as

$$\mathfrak{F}_{e,0\to\lambda T}^* = \frac{15}{4\pi^4} \int_z^\infty u^4 \text{Li}_{-1}(e^{-u}) du. \tag{2.127}$$

Note the expression for the Stefan–Boltzmann constant in terms of its more fundamental constants has been used (see page 66).

The integral appearing in Eq. (2.127) can be reduced to a form given by Eq. (2.11) using integrating by parts. Doing so, after the upper limit of integration has been evaluated by repeated application of de l'Hôpital's rule, yields

$$\mathfrak{F}_{e,0\to\lambda T}^* = \frac{15}{\pi^4}\left[\frac{z^4}{4}\text{Li}_0(e^{-z}) + \int_z^\infty u^3 \text{Li}_0(e^{-u}) du\right]. \tag{2.128}$$

Recognizing the integral in Eq. (2.128) as that which arose in the evaluation of the fractional function of the first kind, in closed form in terms of polylogarithms one has

$$\mathfrak{F}_{e,0\to\lambda T}^* = \frac{15}{\pi^4}\left[\frac{z^4}{4}\text{Li}_0(e^{-z}) + z^3\text{Li}_1(e^{-z}) + 3z^2\text{Li}_2(e^{-z}) + 6z\text{Li}_3(e^{-z}) + 6\text{Li}_4(e^{-z})\right]. \tag{2.129}$$

For reference, places where the function has made an appearance in the literature can be found [164, 203, 471, 441, 133].

From Eq. (2.128) an interesting connection between the fractional fractions of the first and second kinds can be made. As

$$\frac{d}{dz}\left[\mathfrak{F}_{e,0\to\lambda T}\right] = \mathfrak{F}'_{e,0\to\lambda T} = -\frac{15}{\pi^4}z^3\text{Li}_0(e^{-z}),\tag{2.130}$$

which results from the direct application of the first Fundamental Theorem of Calculus [255] to Eq. (2.105), Eq. (2.128) becomes

$$\mathfrak{F}^*_{e,0\to\lambda T} = \mathfrak{F}_{e,0\to\lambda T} - \frac{z}{4}\mathfrak{F}'_{e,0\to\lambda T},\tag{2.131}$$

a result first given by Czerny [161].

In an analogous manner to the radiometric case, a fractional function of the second kind for the actinometric case can also be defined. Here one has

$$\mathfrak{F}^*_{q,0\to\lambda T} = \frac{\displaystyle\int_0^\lambda \frac{\partial M^{\text{b}}_{q,\lambda}}{\partial T}d\lambda}{\displaystyle\int_0^\infty \frac{\partial M^{\text{b}}_{q,\lambda}}{\partial T}d\lambda}.\tag{2.132}$$

In a similar way as was done for the radiometric case, one can once again show it is related to the corresponding fractional function of the first kind. The result is

$$\mathfrak{F}^*_{q,0\to\lambda T} = \mathfrak{F}_{q,0\to\lambda T} - \frac{z}{3}\mathfrak{F}'_{q,0\to\lambda T},\tag{2.133}$$

where the prime once again denotes the derivative of the fractional function of the first kind with respect to z. In terms of polylogarithms, the result is

$$\mathfrak{F}^*_{q,0\to\lambda T} = \frac{1}{2\zeta(3)}\left[\frac{z^3}{3}\text{Li}_0(e^{-z}) + z^2\text{Li}_1(e^{-z}) + 2z\text{Li}_2(e^{-z}) + 2\text{Li}_3(e^{-z})\right].\tag{2.134}$$

Figure 2.9 shows the two fractional functions of the second kind for the radiometric and actinometric cases as a function of the wavelength–temperature product using a semi-log plot for the abscissa.

Interestingly, alternative definitions leading to the same fractional function for the actinometric case can be given. For example, the fractional fraction of the first kind for the actinometric case can be defined in terms of the spectral radiant exitance as follows

$$\mathfrak{F}_{q,0\to\lambda T} = \frac{\displaystyle\int_0^\lambda \lambda M^{\text{b}}_{e,\lambda}d\lambda}{\displaystyle\int_0^\infty \lambda M^{\text{b}}_{e,\lambda}d\lambda},\tag{2.135}$$

and has often caught out unwary authors. From the consideration of a problem where the spectral emissivity of a body was modeled using a wavelength dependent term which is linear, Wei Jing and Tiegang Fang from North

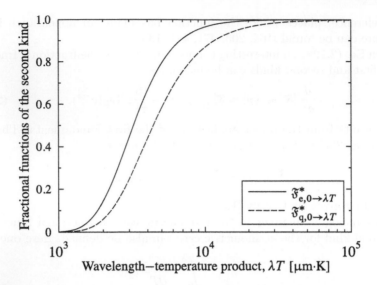

Figure 2.9 Fractional functions of the second kind for the radiometric (solid line) and actinometric cases (dashed line) as a function of the wavelength–temperature product.

Carolina State University in the US were recently lead to defining Eq. (2.135) as a "new" fractional function [333]. What was not recognised at the time by either author was their newly defined fractional function was nothing more than the already defined, though less commonly used, fractional function of the first kind for the actinometric case, a fact duly pointed out recently by one of us [632].

Just as it is possible to write the fractional function of the first kind for the actinometric case in terms of integrals involving both the spectral photon exitance (see Eq. (2.107)) and the spectral radiant exitance (see Eq. (2.135)), the same is true for the fractional function of the second kind for the actinometric case. In terms of integrals involving the spectral radiant exitance one has

$$\mathfrak{F}^*_{q,0\to\lambda T} = \frac{\displaystyle\int_0^\lambda \lambda \frac{\partial M^b_{e\lambda}}{\partial T} d\lambda}{\displaystyle\int_0^\infty \lambda \frac{\partial M^b_{e\lambda}}{\partial T} d\lambda}. \tag{2.136}$$

Other fractional functions may also be defined. For example, from an integrand which consists of the product between the spectral radiant exitance and the square of the wavelength, a new fractional function results and can once more be evaluated in closed form in terms of polylogarithms. The result is

$$\mathfrak{F}_{\eta,0\to\lambda T} = \frac{\displaystyle\int_0^\lambda \lambda^2 M^b_{e,\lambda} d\lambda}{\displaystyle\int_0^\infty \lambda^2 M^b_{e,\lambda} d\lambda} = \frac{6}{\pi^2}\left[z\text{Li}_1(e^{-z}) + \text{Li}_2(e^{-z}) \right]. \tag{2.137}$$

Here the eta suffix has been introduced so as not to confuse it with either fractional functions of the first kind corresponding to the radiometric or actinometric cases. To date, however, the need for such a formulation is quite rare [633]. A plot of the fractional function of the first kind for the eta case can be found in Fig. 2.8.

2.12 CENTROID AND MEDIAN WAVELENGTHS

From time to time, authors have defined other points of reference in the distribution curve which give a measure of central tendency. Statistically, a measure of central tendency is the central or typical value for a probability distribution function. For blackbody radiation the probability distribution function is of course Planck's distribution function. Here we intend to discuss two such measures.

The first is what we will refer to as a *centroid* wavelength. Introduced by the Dutch physicist Hendrik Antoon Lorentz (1853–1928) in a paper he presented before the *Koninklijke Akademie van Wetenschappen te Amsterdam* (Royal Academic of Sciences in Amsterdam) on December 29, 1900 [395, 396] he never proceeded with its formal evaluation. That was left to the American physicist Paul D. Foote (1888–1971), who unaware of Lorentz's work, introduced it in 1916 as a new relation that could be derived from Planck's equation [229, 230].

The centroid is a geometric concept. It gives the geometric center of a region in the plane. It is not a peak as such but a measure of central tendency in the distribution, but for convenience we will continue with the practice of referring to it as a "peak." From Planck's distribution function, the centroid wavelength in the linear wavelength representation is defined as

$$\lambda_{c,\lambda} = \frac{\displaystyle\int_0^\infty \lambda M_{e,\lambda}^b(\lambda, T)\, d\lambda}{\displaystyle\int_0^\infty M_{e,\lambda}^b(\lambda, T)\, d\lambda}. \tag{2.138}$$

Substituting for the spectral radiant exitance and applying the change of variable $u = hc/(k_B \lambda T)$, Eq. (2.138) becomes

$$\lambda_{c,\lambda} = \frac{hc}{k_B T} \frac{\displaystyle\int_0^\infty u^2 \mathrm{Li}_0(e^{-u})\, du}{\displaystyle\int_0^\infty u^3 \mathrm{Li}_0(e^{-u})\, du}. \tag{2.139}$$

Each integral appearing in the numerator and denominator of Eq. (2.139) can be evaluated in closed form by setting $k = 2$ and 3 in Eq. (2.11) respectively. The result is the displacement-like law of

$$\lambda_{c,\lambda} T = \frac{30\zeta(3)}{\pi^4} \frac{hc}{k_B} = 5326.481 \ \mu\mathrm{m\cdot K}. \tag{2.140}$$

This little known result for the centroid peak in the linear wavelength representation was first given by Foote in 1916 [229, 230] and later considered by Frank Benford [76, 77, 78]. The last reference to the wavelength–temperature product at the center of gravity, namely at the centroid peak, we have managed to find appears in Figure 4 on page 60 of the third edition of William H. McAdams' text *Heat Transmission* from 1954 [299], though no explanation of it either in the main body of the text or in the corresponding figure caption is provided. Until the work of one of the present authors [629], references to the centroid peak have been relatively few and far between in the literature [521], suggesting the concept has not proven particularly useful.

In a similar manner, the centroid peak in the linear frequency representation can be shown to be

$$\lambda_{c,\nu} = \frac{\pi^4}{360\zeta(5)} \frac{hc}{k_B} = 3754.413 \, \mu\text{m·K}. \tag{2.141}$$

Table 2.4 contains a summary of Wien and centroid peaks for six common spectral representations in energetic units. Since the spectral distribution curves in both the wavelength and frequency representations rise rapidly to their peak value before gradually falling away asymptotically to zero, the centroid peak, being the geometric center, will be found to lie to the right of the Wien peak in all cases. In the wavelength representation, for fixed temperature, this means the centroid peak falls to the longer wavelength side of its corresponding Wien peak. In the frequency representation however, at fixed temperatures, since the centroid frequency peak falls to the higher frequency side of the corresponding Wien frequency peak, on converting to wavelength the higher centroid frequency peak must fall to shorter wavelengths. So at fixed temperatures the wavelength of the centroid peak in the spectral frequency representation occurs at shorter wavelengths compared to its corresponding Wien peak.

A fractional centroid peak, denoted by $\lambda_{c,\lambda_1 T \to \lambda_2 T}^{\{s\}}$, can also be defined. For a portion of the blackbody radiation spectrum between the wavelength limits λ_1 and λ_2 ($\lambda_2 > \lambda_1$) the geometric centre for such a region corresponds to the fractional centroid peak. If the spectral band is extended to the entire spectrum the fractional centroid peak becomes the centroid peak.

From the definition for the centroid, the fractional centroid peak for any spectral representation can be found. It is given by

$$\lambda_{c,\lambda_1 T \to \lambda_2 T}^{\{s\}} = \frac{\displaystyle\int_{\lambda_1}^{\lambda_2} \lambda M_{\varsigma,\varphi}^{\text{b}}(\lambda, T) \, d\lambda}{\displaystyle\int_{\lambda_1}^{\lambda_2} M_{\varsigma,\varphi}^{\text{b}}(\lambda, T) \, d\lambda}. \tag{2.142}$$

Substituting for Planck's distribution function in the linear wavelength representation and applying the change of variable $u = hc/(k_B \lambda T)$, one

Table 2.4
Comparison between the Wien and centroid peaks found for six common spectral representations.

Spectral scale	Wien peak [μm·K]		Centroid peak [μm·K]	
ν^2	$\dfrac{hc}{k_B[2 + W_0(-2e^{-2})]}$	9028.335	$\dfrac{hc}{k_B}\dfrac{30\zeta(3)}{\pi^4}$	5326.481
ν	$\dfrac{hc}{k_B[3 + W_0(-3e^{-3})]}$	5099.445	$\dfrac{hc}{k_B}\dfrac{\pi^4}{360\zeta(5)}$	3754.413
$\ln \nu$	$\dfrac{hc}{k_B[4 + W_0(-4e^{-4})]}$	3669.704	$\dfrac{hc}{k_B}\dfrac{189\zeta(5)}{\pi^6}$	2932.950
$\ln \lambda$	$\dfrac{hc}{k_B[4 + W_0(-4e^{-4})]}$	3669.704	$\dfrac{hc}{k_B}\dfrac{\pi^2}{12\zeta(3)}$	9844.351
λ	$\dfrac{hc}{k_B[5 + W_0(-5e^{-5})]}$	2897.773	$\dfrac{hc}{k_B}\dfrac{30\zeta(3)}{\pi^4}$	5326.481
λ^2	$\dfrac{hc}{k_B[6 + W_0(-6e^{-6})]}$	2404.012	$\dfrac{hc}{k_B}\dfrac{\pi^4}{360\zeta(5)}$	3754.413

finds

$$\lambda^{\{\lambda\}}_{c,\lambda_1 T \to \lambda_2 T} = \frac{hc}{k_B T} \frac{\int_{z_2}^{z_1} u^2 \text{Li}_0(e^{-u})\, du}{\int_{z_2}^{z_1} u^3 \text{Li}_0(e^{-u})\, du}. \tag{2.143}$$

Here $z_1 = hc/(k_B \lambda_i T)$ for $i = 1, 2$. Employing Eq. (2.11) each integral appearing in Eq. (2.143) can be evaluated in terms of polylogarithms. The result is [629]

$$\lambda^{\{\lambda\}}_{c,\lambda_1 T \to \lambda_2 T} = \frac{hc}{k_B} \frac{\displaystyle\sum_{n=0}^{2} \left[z_2^{2-n} \text{Li}_{n+1}(e^{-z_2}) - z_1^{2-n} \text{Li}_{n+1}(e^{-z_1}) \right] \frac{\Gamma(3)}{\Gamma(3-n)}}{\displaystyle\sum_{n=0}^{3} \left[z_2^{3-n} \text{Li}_{n+1}(e^{-z_2}) - z_1^{3-n} \text{Li}_{n+1}(e^{-z_1}) \right] \frac{\Gamma(4)}{\Gamma(4-n)}}. \tag{2.144}$$

Over the entire spectral range, $\lambda_1 \to 0$ and $\lambda_2 \to \infty$. In this limit $z_1 \to \infty$ while $z_2 \to 0$ and the expression given for the fractional centroid peak in Eq. (2.144) reduces to the centroid peak value as given by Eq. (2.140), as expected. Similarly, within a linear frequency representation the fractional centroid peak will be

$$\lambda_{c,\lambda_1 T \to \lambda_2 T}^{\{\nu\}} = \frac{hc}{k_B} \frac{\displaystyle\sum_{n=0}^{3} \left[z_2^{3-n} \mathrm{Li}_{n+1}(e^{-z_2}) - z_1^{3-n} \mathrm{Li}_{n+1}(e^{-z_1}) \right] \frac{\Gamma(4)}{\Gamma(4-n)}}{\displaystyle\sum_{n=0}^{4} \left[z_2^{4-n} \mathrm{Li}_{n+1}(e^{-z_2}) - z_1^{4-n} \mathrm{Li}_{n+1}(e^{-z_1}) \right] \frac{\Gamma(5)}{\Gamma(5-n)}}.$$

(2.145)

Extending the limit over the entire spectral range the value for the fractional centroid peak once more reduces to the centroid peak value found for the linear frequency representation.

In Fig. 2.10 the ratio of the fractional centroid peak to the centroid peak in the linear wavelength and frequency spectral representations as a function of the wavelength–temperature product is shown. We see that as the spectral band with zero at one of its end-points increases, each fractional centroid peak increases and rapidly approaches the limiting value given by the centroid peak.

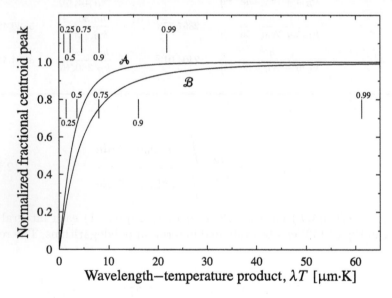

Figure 2.10 Ratio of the fractional centroid peak to the centroid peak in the linear frequency (curve **A**) and wavelength (curve **B**) representations as a function of the wavelength–temperature product. The two sets of five short vertical lines running across the top of the plot indicate, from left to right, the centroid peak fractional amounts $0.25, 0.5, 0.75, 0.9$ and 0.99 relative to its corresponding centroid peak value.

A second slightly more common measure for central tendency is the so-called *median* case. It corresponds to the wavelength $\tilde{\lambda}$ which divides the total radiate exitance into two equal halves. First suggested by Gershun [253], and later by others [299, 292, 300, 311, 287], as a more appropriate "peak" compared to any of the Wien peaks, in energetic units it is found on setting the fractional function of the first kind corresponding to the radiometric case equal to one-half and solving for λ. When this is done, on solving numerically one finds

$$\widetilde{\lambda}_{\bar{p}}T = 4107.250\,\mu\text{m·K}. \tag{2.146}$$

and can be viewed as a displacement-like law for the median wavelength. If one must speak of a "peak" in Planck's distribution function, it is often preferred by some authors as its spectral scale independence means it is the most physically meaningful of all the displacement-like laws for blackbody radiation.

The type of spectral scale one ought to consider if one wishes the Wien peak to coincide with the median peak can be found. In energetic units Wien peaks are the solutions to the transcendental equation given by Eq. (2.46). If we rewrite this later equation as

$$(s - x_\varphi)\exp(x_\varphi) - s = 0, \tag{2.147}$$

where s is the spectral parameter used previously while x_φ corresponds to the non-trivial solution previously found and takes the form

$$x_\varphi = \frac{hc}{k_B \lambda_{\max,\varphi} T}. \tag{2.148}$$

For the median case, as we can rewrite Eq. (2.146) as

$$\lambda_{\bar{p}}T = 4107.250\,\mu\text{m·K} = \frac{hc}{k_B(3.503\,019)}, \tag{2.149}$$

for equality in wavelength between Eqs (2.148) and (2.149), allows one to identify $x_\varphi = 3.503\,019$. Substituting the value for x_φ into Eq. (2.147) and solving for the spectral scale parameter yields $s = 3.611\,756$. Since $s < 4$ the median case falls to the frequency representation side of the wavelength–frequency neutral peak divide and can be represented by

$$\varphi = f(\nu) = \nu^{0.388\,244} \approx \nu^{0.4}. \tag{2.150}$$

Therefore an approximately fifth root squared spectral frequency scale gives a Wien peak coincident with the median peak.

In writing Eq. (2.149) one may have noticed a subscript $\bar{p}$ was used. The reason is that the concept of a median case can be readily generalised as division of the total radiant exitance into an amount other than two equal halves. The fraction by which a wavelength divides the total radiant exitance into a certain percentile amount $(0 \leqslant \bar{p} \leqslant 1)$ can be found from the fractional function of the first kind. The median peak is therefore just a special case of what we will call a *percentile* "peak" $\lambda_{\bar{p}}$. It should be remembered that these percentile peaks are not real peaks in the sense they correspond to some maximum point on a curve, but instead are just a convenience label to use when referring to the resulting wavelengths found.

From the expression for the fractional function of the first kind, setting $\mathfrak{F}_{e,0\to\lambda T} = \bar{p}$ and solving for wavelength leads to

$$\alpha^3 \text{Li}_1(e^{-\alpha}) + 3\alpha^2 \text{Li}_2(e^{-\alpha}) + 6\alpha \text{Li}_3(e^{-\alpha}) + 6\text{Li}_4(e^{-\alpha}) = \frac{\pi^4}{15}\bar{p}, \quad (2.151)$$

an equation which must be solved numerically to find α. Once α is known, the percentile peak wavelength follows and is given by

$$\lambda_{\bar{p}} T = \frac{hc}{k_B \alpha}. \quad (2.152)$$

The percentile peak required to coincide with any given Wien peak in Planck's distribution function can also be found. Matching the percentile peak wavelength to an equivalent Wien peak wavelength allows such a determination to be made. From Eq. (2.46), as $u = \alpha$ is the solution we seek, solving for the spectral parameter s yields $s = \alpha e^{\alpha}/(e^{\alpha} - 1)$. The equivalent spectral scale can now be found as follows. In the frequency representation we set $\varphi = f(\nu) = \nu^{\beta}$ so that

$$dM^b_{\varsigma,\nu^{\beta}} = M^b_{\varsigma,\nu^{\beta}}(T)\, d(\nu^{\beta}) = \beta\nu^{\beta-1} M^b_{\varsigma,\nu^{\beta}}(T)\, d\nu. \quad (2.153)$$

Similarly, in the wavelength representation we set $\varphi = f(\lambda) = \lambda^{\gamma}$ so that

$$dM^b_{\varsigma,\lambda^{\gamma}} = M^b_{\varsigma,\lambda^{\gamma}}(T)\, d(\lambda^{\gamma}) = \gamma\lambda^{\gamma-1} M^b_{\varsigma,\lambda^{\gamma}}(T)\, d\lambda. \quad (2.154)$$

In terms of the spectral scale parameter s, its relation to the indices β and γ has been given in Eq. (2.65).

As an example, for the quarter percentile case ($\bar{p} = 0.25$), on solving Eq. (2.151) numerically one finds $\alpha = 4.965\,526$ which gives $s = 5.000\,400$. Since $s > 4$, the quarter percentile case falls on the wavelength representation side of the wavelength–frequency neutral peak divide and thus can be represented by a wavelength scale $\varphi = \lambda^{\gamma}$ where $\gamma = 1.000\,400$, an approximately linear wavelength scale. This confirms the widely known general rule of thumb that for the linear wavelength scale the region between zero and its peak wavelength one-quarter of the total radiant exitance is to be found [75, 77, 78, 358]. The case for other representative percentiles, and their associated spectral scales are summarised in Table 2.5.

Finally, from Eq. (2.151) it is also possible to find the fraction of the total radiant exitance, namely the percentile $\bar{p}$, for wavelength intervals ranging from zero up to the the Wien peak for any of the various spectral representations one may care to consider. Results for eight different spectral scales are summarised in Table 2.6. Note that while it is possible to express each percentile amount in exact form in terms of polylogarithms and the Lambert W function, for brevity we give each value for $\bar{p}$ as a decimal correct to six decimal places.

Table 2.5

Equivalent spectral scales associated with eleven different percentile amounts as found on matching the percentile peak wavelength to its corresponding spectral scale dependent Wien peak.

$\bar{p}$	α	s	φ
0.01	9.937 050	9.937 530	$\lambda^{5.937\,530}$
0.1	6.554 228	6.563 575	$\lambda^{2.563\,575}$
0.2	5.376 478	5.401 455	$\lambda^{1.401\,455}$
0.3	4.613 189	4.659 411	$\lambda^{0.659\,411}$
0.4	4.016 206	4.089 911	$\lambda^{0.089\,911}$
0.5	3.503 018	3.611 755	$\nu^{0.388\,244}$
0.6	3.032 090	3.185 687	$\nu^{0.814\,312}$
0.7	2.573 955	2.786 369	$\nu^{1.213\,630}$
0.8	2.096 264	2.390 034	$\nu^{1.609\,965}$
0.9	1.534 548	1.956 216	$\nu^{2.043\,783}$
0.99	0.628 717	1.347 084	$\nu^{2.652\,915}$

Table 2.6

Equivalent percentile amounts for wavelength intervals ranging from zero up to the the Wien peak for eight different spectral scales.

φ	s	$\bar{p}$
ν^2	2	0.890 787
ν	3	0.646 006
$\sqrt{\nu}$	7/2	0.525 338
$\ln \nu$	4	0.417 710
$\ln \lambda$	4	0.417 710
$\sqrt{\lambda}$	9/2	0.325 876
λ	5	0.250 054
λ^2	6	0.141 088

2.13 THE STANDARD PROBABILITY DISTRIBUTION AND CUMULATIVE PROBABILITY DISTRIBUTION FUNCTIONS FOR BLACKBODY RADIATION

It is instructive to interpret the spectral radiant exitance for a blackbody and its associated fractional function of the first kind using the language of probability theory. For a spectral representation in terms of a dimensionless parameter x equal to $hc/(k_B \lambda T)$, within a differential interval dM_φ, as energy is conserved one has

$$M_{e,\lambda}^{b}(\lambda, T)d\lambda = M_e(x)dx. \tag{2.155}$$

When correctly normalized, the Planck distribution function $M_e(x)$ in standard form can be written as

$$M_e^*(x) = \begin{cases} \dfrac{15}{\pi^4}x^3\,\mathrm{Li}_0(e^{-x}) & \text{for } x \geqslant 0, \\ 0 & \text{for } x < 0. \end{cases} \tag{2.156}$$

Recognising $M_e^*(x)$ as a *probability distribution function*, this accounts for our use of the term "distribution function" when referring to Planck's law. The factor of $15/\pi^4$ arises from correctly normalizing the distribution since in order for it to be a valid probability distribution function one requires

$$\int_{-\infty}^{\infty} M_e^*(x)dx = 1. \tag{2.157}$$

For the fractional fraction of the first kind, writing Eq. (2.105) in terms of the probability distribution function $M_e^*(x)$, one has

$$\mathfrak{F}_{e,0\to\lambda T} = \int_z^{\infty} M_e^*(x)\,dx = 1 - \int_{-\infty}^{z} M_e^*(x)\,dx. \tag{2.158}$$

Since the second of the integral terms appearing in Eq. (2.158) corresponds to the *cumulative probability distribution function* for $M_e^*(x)$, one can interpret the fractional function of the first kind corresponding to the radiometric case as a *complementary* cumulative probability distribution function.

From the standard probability distribution function for blackbody radiation, other statistical measures for the distribution can be readily found. In terms of measures of central tendency, if x is the value assigned to the continuous random variable X, the mean μ is found from

$$\mu = \int_{-\infty}^{\infty} xM_e^*(x)\,dx. \tag{2.159}$$

The mean or expectation value gives the weighted average value of a random variable. In the language of Section 2.12 it corresponds to the centroid and has a value of

$$\mu = \frac{360\zeta(5)}{\pi^4} = 3.832\,229. \tag{2.160}$$

The mode of a continuous probability distribution function is the value $\hat{x}$ at which its probability distribution function has its maximum value. It corresponds to the peak value in the distribution curve and is found by solving

$$\frac{d}{dx}\left[M_e^*(x)\right] = 0. \tag{2.161}$$

The result is

$$\widehat{x} = 3 + W_0(-3e^{-3}) = 2.821\,439. \tag{2.162}$$

The mode is of course nothing more than the location of the Wien peak for the standard probability distribution function for blackbody radiation.

The median of a continuous probability distribution function is the value of x which divides the area under the probability distribution curve into two equal halves. It is found by the number $\widetilde{x}$ which satisfies the equation

$$\int_{-\infty}^{\widetilde{x}} M_{\mathrm{e}}^*(x)\,dx = \frac{1}{2}. \tag{2.163}$$

On solving

$$\frac{15}{\pi^4} \int_0^{\widetilde{x}} x^3 \mathrm{Li}_0(e^{-x})\,dx = \frac{1}{2}, \tag{2.164}$$

yields

$$\widetilde{x}^3 \mathrm{Li}_1(e^{-\widetilde{x}}) + 3\,\widetilde{x}^2 \mathrm{Li}_2(e^{-\widetilde{x}}) + 6\,\widetilde{x}\,\mathrm{Li}_3(e^{-\widetilde{x}}) + 6\mathrm{Li}_4(e^{-\widetilde{x}}) = \frac{\pi^4}{30}. \tag{2.165}$$

As no closed-form solution for Eq. (2.165) is currently known to exist, it must be solved for numerically. The result is

$$\widetilde{x} = 3.503\,019. \tag{2.166}$$

The median for the standard probability distribution function for blackbody radiation is the same as the median wavelength found for Planck's distribution function in Section 2.12 (see Eq. (2.149)).

Figure 2.11 shows a plot of the standard probability distribution function for blackbody radiation as a function of x for the value assigned to the continuous random variable X. The three measures of central tendency of mode, median, and mean are indicated in the figure as $\widehat{x}$, $\widetilde{x}$, and μ respectively.

In terms of the various measures of spread, the variance is the most important. Defined by

$$\sigma^2 = \int_{-\infty}^{\infty} (x - \mu)^2 M_{\mathrm{e}}^*(x)\,dx, \tag{2.167}$$

for the standard probability distribution function for blackbody radiation, it can be shown to be equal to

$$\sigma^2 = \frac{40\pi^2}{21} - \left(\frac{360\zeta(5)}{\pi^4}\right)^2 = 4.113\,264. \tag{2.168}$$

One measure of the shape of the distribution function is found in the skewness [187]. Skewness gives a measure of the asymmetry in the curve of the probability distribution function. It is defined by

$$\gamma_1 = \frac{\nu_3}{\sigma^3} = \frac{1}{(\sigma^2)^{3/2}} \int_{-\infty}^{\infty} (x - \mu)^3 M_{\mathrm{e}}^*(x)\,dx. \tag{2.169}$$

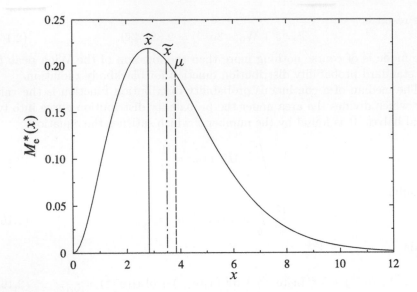

Figure 2.11 Plot showing the standard probability distribution function for black-body radiation as a function of x for the value assigned to the continuous random variable X. The three measures of central tendency indicated in the figure as $\widehat{x}$, $\widetilde{x}$, and μ correspond to the mode, median, and mean, respectively.

Here ν_3 is known as the third central moment [562]. Evaluating, we find

$$\gamma_1 = \frac{\frac{15}{\pi^4}\left[720\zeta(7) - \frac{960}{7}\pi^2\zeta(5) + \frac{6\,220\,800}{\pi^8}\zeta(5)^3\right]}{\left[\frac{40}{21}\pi^2 - \left(\frac{360}{\pi^4}\zeta(5)\right)^2\right]^{3/2}} = 0.986\,474. \tag{2.170}$$

The small positive value in the skewness accounts for the tail to the right observed in the curve for the distribution function. A perfectly symmetric distribution on the other hand, such as the normal (Gaussian) distribution, will have a skewness of zero.

2.14   INFRARED, VISIBLE, AND ULTRAVIOLET COMPONENTS IN THE SPECTRAL DISTRIBUTION OF BLACKBODY RADIATION

A question often asked is what fraction of the total radiation emitted by a blackbody at a given temperature is radiated into different portions of the electromagnetic spectrum. At low temperatures one expects most of the emitted radiation to fall in the infrared portion of the spectrum and would account for why objects at room temperatures do not appear self-luminous. With increasing temperatures the portion radiated into the visible and the ultraviolet regions gradually increases until at very high temperatures one expects the majority of the radiation to fall within the ultraviolet part of the spectrum and beyond.

Using the blackbody fractional function of the first kind corresponding to the radiometric case, the amount of flux radiated into a given spectral interval can be readily calculated. Given by $\mathfrak{F}_{\lambda_1 T \to \lambda_2 T}$, in terms of polylogarithms, the result is

$$\mathfrak{F}_{\lambda_1 T \to \lambda_2 T} = \frac{15}{\pi^4} \sum_{n=0}^{3} \left[z_2^{3-n} \mathrm{Li}_{n+1}(e^{z_2}) - z_1^{3-n} \mathrm{Li}_{n+1}(e^{z_1}) \right] \frac{\Gamma(4)}{\Gamma(4-n)}, \quad (2.171)$$

where $z_i = hc/(k_B \lambda_i T)$ for $i = 1, 2$.

Radiation which causes a response in the human eye is customarily defined as the visible portion of the electromagnetic spectrum. Depending on the individual and the type of light used, the human eye can respond to light in the region of the spectrum ranging in wavelength from about 380 to 760 nm. If the infrared is considered to consist of all those wavelengths beyond the red end while the ultraviolet is taken to consist of all wavelengths less than those for violet, it allows the electromagnetic spectrum to be divided into three major spectral components — infrared, visible, and ultraviolet. Taking $\lambda_1 = 380$ nm and $\lambda_2 = 760$ nm Table 2.7 shows the portion of radiation radiated by a blackbody into the three regions of the spectrum for temperatures ranging from 500 to 15 000 K while Fig. 2.12 plots these fractional amounts for each of the three different spectral regions as a function of temperature.

As expected, for temperatures below 500 K we see almost all radiation emitted from a blackbody falls into the infrared region of the spectrum while little is found in either the visible or ultraviolet portions of the spectrum. As temperature increases, the amount radiated into the infrared gradually decreases while the amount radiated into the ultraviolet gradually increases. Sitting between the infrared and the ultraviolet bands, the amount of radiation radiated as visible light increases rapidly at first with increasing temperature, reaches a maximum, before gradually decreasing with increasing temperature as more of the emitted radiation moves into the ultraviolet portion of the spectrum.

The temperature which maximises the amount of radiation radiated into a given waveband with finite wavelength limits, as for example in the visible portion of the spectrum, can be found and depends on the lower and upper wavelength limits selected for the waveband. Starting with

$$\mathfrak{F}_{\lambda_1 T \to \lambda_2 T} = \frac{15}{\pi^4} \int_{z_2}^{z_1} u^3 \mathrm{Li}_0(e^{-u}) du, \quad (2.172)$$

where $z_i = hc/(k_B \lambda_i T)$, for $i = 1, 2$ and is a temperature dependent parameter. Differentiating with respect to the temperature, application of Leibniz's rule yields

$$\frac{d}{dT} \left[\mathfrak{F}_{\lambda_1 T \to \lambda_2 T} \right] = \frac{15}{\pi^4 T} \left[z_2^4 \mathrm{Li}_0(e^{-z_2}) - z_1^4 \mathrm{Li}_0(e^{-z_1}) \right]. \quad (2.173)$$

Table 2.7
Portion of radiation radiated by a blackbody into the infrared, visible, and ultraviolet portions of the spectrum for temperatures ranging from 500 K to 15 000 K.

Temperature [K]	Fraction infrared	Fraction visible	Fraction ultraviolet
500	1.000	2.515×10^{-14}	3.528×10^{-30}
1000	1.000	7.374×10^{-6}	2.515×10^{-14}
1500	0.999	1.307×10^{-3}	3.048×10^{-8}
2000	0.986	0.014	7.374×10^{-6}
2500	0.948	0.052	1.735×10^{-4}
3000	0.884	0.115	1.307×10^{-3}
3500	0.804	0.191	5.209×10^{-3}
4000	0.718	0.268	0.014
4500	0.634	0.336	0.030
5000	0.557	0.391	0.052
6000	0.427	0.457	0.116
7000	0.328	0.475	0.196
8000	0.256	0.462	0.282
9000	0.201	0.433	0.366
10 000	0.161	0.396	0.443
11 000	0.130	0.357	0.513
12 000	0.107	0.320	0.573
13 000	0.088	0.286	0.626
14 000	0.074	0.254	0.672
15 000	0.062	0.227	0.711

Stationary points occur when the derivative in Eq. (2.173) is set equal to zero. Doing so one has

$$z_1^4 \mathrm{Li}_0(e^{z_1}) = z_2^4 \mathrm{Li}_0(e^{z_2}), \qquad (2.174)$$

or equivalently, after both zeroth-order polylogarithms are written in terms of their explicit rational form in terms of the exponential function as

$$\left(\frac{\lambda_1}{\lambda_2}\right)^4 = \frac{\exp\left(\dfrac{c_2}{\lambda_2 T}\right) - 1}{\exp\left(\dfrac{c_2}{\lambda_1 T}\right) - 1}. \qquad (2.175)$$

For a given wavelength interval Eq. (2.175) can be solved for the temperature T. Unfortunately, as Eq. (2.175) is a transcendental equation in terms of T, it cannot in general be solved in closed form for T but instead must be found numerically. Denoting the temperature which solves Eq. (2.175) as T_{opt}, it represents the optimal temperature for maximum efficiency in the production of radiation from a blackbody into a given spectral band.[26]

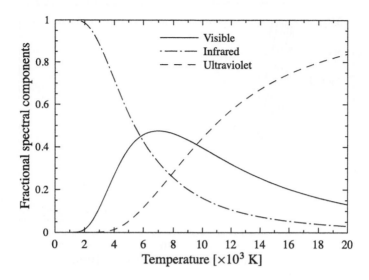

Figure 2.12 Fractional amount of radiation radiated by a blackbody into the visible (solid line), infrared (broken chain line), and ultraviolet (dashed line) regions of the spectrum as a function of temperature. The upper and lower bounds used for the visible region were $\lambda_1 = 380\,\text{nm}$ and $\lambda_2 = 760\,\text{nm}$ respectively.

Despite the root to Eq. (2.175) being dependent on a numerical evaluation for its solution, it is still possible to obtain considerable information about the behavior of the optimal temperature for arbitrary sized spectral bands. Setting $\lambda_2 = \alpha\lambda_1$, the parameter α represents the ratio of the upper wavelength bound λ_2 relative to the lower bound λ_1. As $\lambda_2 > \lambda_1$, one has $\alpha > 1$. Substituting λ_2/α for λ_1 into Eq. (2.175), after rearranging algebraically, the following equation results

$$x^\alpha - \alpha^4 x + \alpha^4 - 1 = 0, \qquad (2.176)$$

where $x = \exp\left(hc/(k_B\lambda_2 T)\right)$. Physically, as the argument of the exponential term must be positive, $x \geqslant 1$.

Clearly one real root to Eq. (2.176) is $x_1 = 1$ for all $\alpha > 1$. In Appendix A.3 we show that there exists only one other real root on the interval $x > 1$ for all $\alpha > 1$. Calling this root ξ^*, in terms of the upper wavelength bound λ_2 the optimal temperature within any given spectral band with finite wavelength limits can be expressed as

$$T_{\text{opt}} = \frac{hc}{k_B\lambda_2 \ln(\xi^*)}. \qquad (2.177)$$

Depending on the value selected for α, cases exist where it is possible to explicitly find the root ξ^*. For example, for the selected wavelength bounds used in Fig. 2.12, as the upper bound is twice the lower bound, it is possible

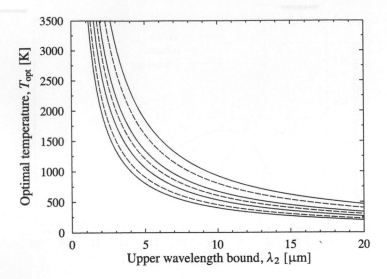

Figure 2.13 Optimal temperature which maximises the amount of radiation eradi-
ated by a blackbody into a given waveband as a function of wavelength for the upper
waveband for a number of spectral waveband sizes. Seven spectral waveband sizes
are shown. Starting from the bottom curve and moving through to the top curve
these are: $\alpha = 1.2, 1.5, 2.0, 2.5, 3.0, 4.0$, and 5.0.

to find a closed-form expression for $T_{\rm opt}$ [381]. In this case Eq. (2.176) reduces
to the quadratic equation

$$x^2 - 16x + 15 = (x - 1)(x - 15) = 0. \tag{2.178}$$

On solving, we see $\xi^* = 15$ is the only root to the equation on the interval
$x > 1.$[27] In this particular case an exact, analytic expression for the optimal
temperature will be

$$T_{\rm opt} = \frac{hc}{k_B \lambda_2 \ln(15)}. \tag{2.179}$$

If we set an upper bound of $\lambda_2 = 760\,{\rm nm}$ corresponding to deep red at the
visible end of the spectrum the optimal temperature will be

$$T_{\rm opt} = 6990.742\,{\rm K}. \tag{2.180}$$

This temperature is far too high for any material object and will only apply
to Sun-like objects. In the infrared however, far cooler optimal temperatures
can be achieved for wavebands of technological importance. For example, the
two most important atmospheric windows in the infrared occur at 3–5 μm and
8–12 μm and are known as the midwave band and longwave band respectively.
For the midwave band, $\alpha = 5/3$ giving $\xi^* = 20.053\,264$ corresponding to an

optimal temperature of 959.699 K. For the longwave band, $\alpha = 3/2$ giving $\xi^* = \frac{65}{512}(97 + \sqrt{8385}) = 23.939\,506$ and corresponds to the far cooler optimal temperature of 377.569 K.

In Fig. 2.13 the optimal temperature as a function of the upper wavelength bound is plotted for a number of different values for the parameter α. The optimal temperature is seen to decrease as the value of the upper wavelength band increases, which we should expect since a greater portion of radiation from a blackbody will be emitted into the long wavelength region of the spectrum as the temperature decreases. On the other hand, the optimal temperature increases as the size of the spectral bandwidth increases. A larger value in the parameter α means a larger spectral bandwidth size. As the bandwidth increases in size, the amount of radiation emitted into the bandwidth will obviously be greater. A greater amount of the emitted radiation appearing within a given bandwidth means far higher temperatures are needed before the fractional amount is maximised after which it gradually falls away.

Notes

[1]Lommel, E., 1878 "Theorie der Absorption und Fluorescenz," *Annalen der Physik und Chemie*, **3**(2), p. 259. In the original German it reads: Diese Darstellung ist ebenso naturgemäss wie die nach Wellenlängen, hat aber gegenüber dieser den Vortheil, dass sie sich dem prismatischen Spectrum enger anschliesst.

[2]Rayleigh does not appear to have been aware of Lommel's work that seems to have been the first to suggest the use of frequency. Stoney suggested the use of wavenumber which, while being equal to the reciprocal of wavelength, is proportional to frequency but not identically equal to it.

[3]We use the term *light* to include electromagnetic radiation ranging from the near ultraviolet, visible, and into the near infrared portions of the electromagnetic spectrum.

[4]In the order of their appearance, we have the Norwegian mathematician Niels Henrik Abel (1802–1829), the Russian mathematician Nikolai Ivanovich Lobachevsky (1792–1858), the German mathematician Ernst Eduard Kummer (1810–1893), and the Indian mathematician Srinivasa Ramanujan (1887–1920).

[5]The lifting of the polylogarithms out of obscurity in modern times is due largely to the work of the English electrical engineer Leonard Lewin (1919–2007). In a 1986 review to his book *Polylogarithms and associated functions*, published in 1981 as a follow up to his earlier *Dilogarithms and associated functions* from 1958, Gian-Carlo Rota, as Editor of the journal *Advances in Mathematics* wittily wrote [569]

> Dilogarithms would not have been given a chance thirty years ago, not even by someone who contributed heavily to the Salvation Army. Along came I. M. Gel'fand and connected them to Grassmannians, so people started to take notice. Now they are about to break into High Society, and the bets are on that they will be seen at the next debutantes' ball.

Here the I. M. Gel'fand Rota refers to was the Soviet mathematician Israïl Moyseyovich Gel'fand who made major contributions to many branches of mathematics while a Grassmannian is a rather abstract concept in mathematics connected with a space which parametrizes all linear subspaces of a vector space of some given dimension, an idea that turns out to be useful in numerous fields of mathematics and theoretical physics.

On the remarkableness of the polylogarithmic functions Don Zagier, then of the *Max-Planck-Institut für Mathematik* in Bonn, Germany, in a paper based on a lecture he gave on the occasion of F. Hirzebruch's 60th birthday wrote

...the dilogarithm is one of the simplest non-elementary functions one can imagine. It is also one of the strangest. It occurs not quite often enough, and in not quite an important enough way, to be included in the Valhalla of the great transcendental functions – the gamma function, Bessel and Legendre functions, hypergeometric series, or Riemann's zeta function. And yet it occurs too often, and in far too varied contexts, to be dismissed as a mere curiosity.

And if this were not enough, to highlight their remarkableness even further, Zagier continued by writing

Almost all of its [the polylogarithm function] appearances in mathematics, and almost all the formulas relating to it, have something of the fantastical in them, as if this function alone among all others possessed a sense of humor.

Zagier, D. (2007) "The dilogarithm function," in *Frontiers in number theory, physics, and geometry II – On conformal field theories, discrete groups and renormalization*, P. Cartier, P. Moussa, B. Julia and P. Vanhove (Editors), Berlin: Springer-Verlag. p. 6.

[6]Alternatively, in terms of an operator of the form $z\,d/dz$, the polylogarithms for any negative integer order can be found using

$$\mathrm{Li}_{-n}(z) = \left(z\frac{d}{dz}\right)^n \left[\frac{z}{1-z}\right], \quad n \geqslant 0. \tag{2.181}$$

Here n is a positive integer.

[7]The Eulerian numbers are not to be confused with the slightly more common special numbers known as *Euler numbers* and denoted by E_n. While the two sets of special numbers may have similar sounding names, they are not the same.

[8]Recall the gamma function is defined by

$$\Gamma(x) = \int_0^\infty \mathrm{e}^{-t} t^{x-1} dt.$$

[9]The canonical form for indicating branches of the Lambert W function with complex argument z is to write these as $\mathrm{W}_k(z)$. Here k is an integer with $k = -1, 0$ being the only two branches when $\mathrm{W}_k(z)$ becomes real. Only the principal branch is analytic at zero while all other branches have a branch point at the origin.

[10]In a 1993 article which first introduced the function as the "Lambert W function," Robert M. Corless, Gaston H. Gonnet, D. E. G. Hare and David J. Jeffrey remarked that "For a function, getting your own name is rather like Pinocchio getting to be a real boy." Corless, R. M., Gonnet, G. H., Hare, D. E. G., and Jeffrey, D. J., 1993 "Lambert's W function in Maple," *The Maple Technical Newsletter* **9**, p. 12.

An indication of its acceptance as one of the special functions of mathematical physics one need look no further than the *NIST handbook of mathematical functions*. Published in 2010, an entry for the Lambert W function can be found. This text is an updated version of the 1964 publication *Handbook of mathematical functions with formulas, graphs, and mathematical tables* by Milton Abramowitz and Irene A. Stegun. At the time of its publication this text was considered by many to be the definitive work of reference on special functions. Such has been its influence, if a function did not appear in Abramowitz and Stegun it was not worth knowing about. Inclusion of the Lambert W function in the updated version of Abramowitz and Stegun is enough to convince many the function is now considered worth knowing.

[11]In fact, as far as we are aware, the only time the polylogarithmic form for Planck's law has ever been seen is in the work of the present authors [629, 632, 633].

[12]The trivial solution corresponding to the second real solution to Eq. (2.22) is often not given by many authors [607]. The solution $u = 0$ corresponds to a wavelength that tends to infinity and therefore is not physically meaningful. In the formulation presented, if the secondary real branch for the Lambert W function, $\mathrm{W}_{-1}(x)$, were to be chosen the trivial solution follows and is apparent on recognising the simplification $\mathrm{W}_{-1}(-se^{-s}) = -s$ for $s > 1$.

[13]The value for the mathematical constant $\xi^* = 5 + W_0(-5e^{-5})$ can be evaluated in a variety of ways. For example, as the argument appearing in the Lambert W function is less than $1/e$ in absolute value, the power series expansion for the function about the origin can be used (see Appendix A.1). Doing so yields

$$\xi^* = 5 + W_0(-5e^{-5}) = 5 - \sum_{n=1}^{\infty} \frac{(5n)^n}{n!n} e^{-5n}. \qquad (2.182)$$

If the infinite sum is truncated after the first two terms have been summed one has [47, 656]

$$\xi^* \simeq 5 - 5e^{-5} - 25e^{-10} = 4.965\,175\,266\,760\ldots \qquad (2.183)$$

This result is able to recover the first four decimal places of the true value. For a second approach, writing the Lambert W function as

$$W_0(x) = \ln(1+x)M(x), \qquad (2.184)$$

where $M(x)$ is an auxiliary function, when a $(2,2)$-Padé approximant is found for $M(x)$ one finds [417]

$$\xi^* \simeq 5 + \ln(1 - 5e^{-5}) \frac{1 - \frac{123}{8}e^{-5} + \frac{105}{2}e^{-10}}{1 - \frac{143}{8}e^{-5} + \frac{3565}{48}e^{-10}} = 4.965\,114\,231\,797\ldots \qquad (2.185)$$

a result which is able to recover the first ten decimal places of the true value. If precision to an arbitrary number of decimal places is required, by far the best approach to use is a numerical root finding technique based on Halley's method [145].

[14]The interpretation of Planck's distribution function has a long and venerable reputation for leading many an author astray. In the past many have made the egregious mistake of referring to Planck's equation as a "function of either wavelength or frequency," often belied by its true differential nature. Instead, its correct differential interpretation as a "function of wavelength or frequency *per unit wavelength or frequency*," is pertinent in understanding why it is not possible to simply substitute $\nu = c/\lambda$ into Planck's distribution function when moving between linear wavelength and frequency representations. We will go to great pains to make this point in the discussion which is to follow in the text.

[15]What Abney and Festing actually present in their paper of 1884 is a plot of what they called "total radiation" on the ordinate versus number of screw turns on the abscissa. Here their so-called total radiation was a measure for the intensity of the radiation detected and was related to the size of a deflection as measured by a galvanometer. The number of screw turns on the other hand was calibrated to wavelength. Here radiation from an incandescence light source passed through a prism made from either glass or rock salt and was dispersed before passing through a very narrow slit. By turning the screw, the slit stepped through different parts of the spectrum with the position of the slit within the spectrum after each screw turn corresponding to a known wavelength. Different temperatures for the incandescent light source were achieved by passing currents of various sizes through the circuit connected to the light source.

[16]Langley, S. P., 1884 "Experimental determination of wave-lengths in the invisible prismatic spectrum," *American Journal of Science* [Third Series] **27**(159), p. 170.

[17]Langley, S. P., 1884 'Experimental determination of wave-lengths in the invisible prismatic spectrum,' *American Journal of Science* [Third Series] **27**(159), p. 170.

[18]If a reciprocal scale in either wavelength or frequency were to be chosen, as wavelength and frequency are reciprocally proportional to one another such a choice passes one from a wavelength scale over to a frequency scale, or vice verse, with an associated change in sign occurring.

[19]See Section 2.14 where the claim that most of the thermal radiation emitted by the vast majority of objects at commonly encountered temperatures falls within the ultraviolet, visible, and the infrared portions of the electromagnetic spectrum is addressed.

[20]A nomogram is also known as a nomograph or alignment chart.

[21]A criterion which tells when one is allowed to interchange the order of the integration and summation is given by Tonelli's theorem [648]. Named after the Italian mathematician

Leonida Tonelli (1885–1946), under very general conditions satisfied by the function f states that if $\{f_n(x)\}_n$ is a positive $(f_n(x) \geqslant 0)$ sequence of integrable functions for all n, x then

$$\int \sum_n f_n(x)dx = \sum_n \int f_n(x)dx.$$

For the case given in the text, as $f_n(x) = x^{s-1}e^{-nx} > 0$ for all $x > 0, n \geqslant 1$ Tonelli's theorem applies with the interchange of the integration with the summation being a permissible operation.

[22]This result should be compared to the result obtained using dimensional analysis on pages 19–20.

[23]Around the turn of the twentieth century it was common practice amongst illumination engineers to speak of the efficiency of a light source as the amount of radiation radiated into the visible portion of the spectrum divided by the total power radiated over all wavelengths [52, 622, 623, 467, 442, 319, 66, 89]. This notion of efficiency was first introduced by Langley in 1883 [372, 374] and was later termed "radiant efficiency" by the US physicist Edward Leamington Nichols in 1894 [464, 463]. An experimental determination for the radiant efficiency for a visible source was made by measuring the area under the spectral curve within the visible portion of the spectrum using a mechanical measuring device known as a planimeter. On dividing this quantity by the total amount of radiation radiated over all wavelengths the value for the radiant efficiency could be found for any thermal light source. For light sources that are well approximated as grey-bodies the radiant efficiency corresponds to what we now call the fractional function of the first kind for the radiometric case. In 1907 C. V. Drysdale calculated the percentage of radiation radiated by a blackbody into the visible part of the spectrum using Wien's equation as an approximation to Planck's law [195]. It was not until the late 1920s that L. L. Holladay performed the first full calculation for the radiant efficiency of a blackbody without the use of any such approximation [294].

[24]Project MACs SYmbolic MAnipulator, or MACSYMA for short, is a computer algebra system initially developed in the late 1960s at the Massachusetts Institute of Technology in Cambridge, Massachusetts, as part of the university's Project on Mathematics and Computation (Project MAC). It is considered to be the world's first comprehensive symbolic mathematics system with many of its ideas later adopted by more recent computer algebra system arrivals such as MATHEMATICA and MAPLE.

[25]In his initial announcement to finding a closed form solution to the fractional function of the first kind for the radiometric case Clark writes [138]

[T]he fact that the Planck integral problem persisted for so long, and was solved so soon after MACSYMA was applied to it, is an indication that symbolic algebra packages are just beginning to make an impact in computational physics.

And indeed they were.

[26]The idea of an optimal temperature for maximum spectral efficiency was first proposed by the Danish astronomer Ejnar Hertzsprung (1873–1967) in 1906 [289] and independently some time later by the German physicist J. Salpeter [582] and the American physicist Frank Benford [75, 77, 78] of "Benford's law" fame [74]. Here a measure of the spectral efficiency η of production of radiation from a blackbody source at a selected wavelength was characterised by

$$\eta_{e,\lambda} = \frac{M_{e,\lambda}^{b}(\lambda, T)}{M_e^b(T)} = \frac{15c_2^4}{\pi^4 T^4 \lambda^5} \mathrm{Li_0}\left[\exp\left(-\frac{hc}{k_B \lambda T}\right)\right]. \tag{2.186}$$

The optimal temperature T_{opt} for maximum spectral efficiency is then found on differentiating the above equation with respect to the temperature and solving the resulting equation $\partial \eta_{e,\lambda}/\partial T = 0$ for T. Doing so yields the following displacement-like law of

$$\lambda T_{\mathrm{opt}} = \frac{hc}{k_B[4 + W_0(-4e^{-4})]}. \tag{2.187}$$

However, as is the case with any differential quantity, the spectral efficiency suffers from a clear, unambiguous physical interpretation being attached to it. Instead, the temperature that maximises the efficiency of production within a given spectral band has real physical meaning and is the optimal temperature considered in the text.

[27] For an upper wavelength bound which is finite, the second solution to Eq. (2.178) is rejected on physical grounds as it corresponds to the case of infinite temperature.

3 Computational and numerical developments

The complicated mathematical form with which Planck's law presents itself makes working with it or other closely related functions derived from it far from easy. Without the modern day benefit of a calculator or computer the task of finding values for the spectral exitance is a lengthy and tedious affair. Several well-known approximations, valid within various limits, allow for considerable simplification in Planck's law. Perhaps the two best known of these are the laws of Rayleigh–Jeans and Wien, though other approximate forms exist. We commence this chapter by considering these two well-known approximations together with a number of other lesser known approximations that have been developed to approximate Planck's law.

The remainder of the chapter is devoted to the problem of the accurate numerical evaluation of the blackbody fractional function of the first kind.[1] Until quite recently the definite integral appearing in the expression for the fractional function was not thought to be expressible in closed form. As an often required quantity, its importance meant historically a number of approximations were developed to deal specifically with this integral. In order to handle the apparent intractability, various quadrature methods were used, but as we shall see are by no means the best methods available. Computationally, rapidly converging infinite series expansions are far superior and offer the most efficient means available for the evaluation of this integral. These and other methods are considered in some detail in this chapter.

3.1 APPROXIMATIONS TO THE SPECTRAL EXITANCE

We commence by considering a number of approximations to the spectral radiant exitance.[2] While computationally no longer as important as they once were, at least two of these approximations are very well known. Within the limit where each applies they allow for considerable simplification, accounting for their continued use and popularity.

3.1.1 THE LAWS OF WIEN AND RAYLEIGH–JEANS

Two very well-known approximations to Planck's law for the spectral radiant exitance are the laws of Rayleigh–Jeans and Wien. Mathematically each of the these laws is far simpler in form compared to Planck's law, and in the limit where each applies, allow the spectral exitance for a blackbody source to be estimated with far less computational effort.

Historically, the first of the laws to be given was that of Wien's. Briefly touched upon in Chapter 1, in June of 1896, using thermodynamic arguments Lord Rayleigh would later describe as little more than conjecture [548], Wien derived the law that now bears his name [687, 688]. Despite the questionable theoretical underpinnings leading to the law it quickly gained both experimental and theoretical support. Wien's new law for thermal radiation seemed to fit the existing experimental data better than any law previously proposed while theoretically it gained further support from an unlikely quarter. Shortly after Planck decided to turn his full attention to the problem of blackbody radiation and its relation to the second law of thermodynamics in the late 1890s, using arguments quite different and more plausible than those used by Wien, Planck's first major theoretical advance came from his confirmation of Wien's law in 1899 [507]. That year however, saw significant progress made by the experimentalists. For the first time they were able to extend their measurements deep into the infrared. Discrepancies between what Wien's law predicated compared to what they measured had far reaching consequences for all of physics, leading Planck to having his now famous "quantum" change of heart.

While both the laws of Rayleigh–Jeans and Wien precede those of Planck, we will derive these two approximate laws from the latter's law.[3] In the low temperature, short wavelength (high frequency) limit, the exponential term appearing in the denominator of Planck's law dominates. In this case the negative one term can be ignored and the well-known Wien form for the spectral radiant exitance follows

$$M_{e,\lambda}^{b}(\lambda, T) \simeq \frac{2\pi h c^2}{\lambda^5} \exp\left(-\frac{hc}{k_B \lambda T}\right), \qquad \frac{hc}{k_B \lambda T} \gg 1. \qquad (3.1)$$

The Wien approximation is accurate to within 1% when $\lambda T \lesssim 3100 \ \mu m \cdot K$.

At the other limit, for high temperatures and long wavelengths (low frequencies) one arrives at the now equally well-known Rayleigh–Jeans law. If the exponential term appearing in the denominator of Planck's equation is expanded as a Maclaurin series, namely

$$\exp\left(\frac{hc}{k_B \lambda T}\right) = 1 + \frac{1}{1!}\left(\frac{hc}{k_B \lambda T}\right) + \frac{1}{2!}\left(\frac{hc}{k_B \lambda T}\right)^2 + \cdots \qquad (3.2)$$

in the limit $hc/(k_B \lambda T) \ll 1$ all terms above the first degree will be small so can be ignored. Doing so yields

$$M_{e,\lambda}^{b}(\lambda, T) \simeq \frac{2\pi h c^2}{\lambda^5}\left[\left(1 + \frac{hc}{k_B \lambda T}\right) - 1\right]^{-1} = \frac{2\pi c k_B T}{\lambda^4}, \qquad (3.3)$$

the well-known Rayleigh–Jeans approximation for the spectral radiate exitance in the high temperature, long wavelength limit. The Rayleigh–Jeans approximation is accurate to within 1% when $\lambda T \gtrsim 7.2 \times 10^5 \ \mu m \cdot K$.

The story of how the Rayleigh–Jeans law came about is an interesting one and is worth re-telling. First put forward by Lord Rayleigh in 1900 in a paper entitled "Remarks upon the law of complete radiation" [548] as an attempt to reconcile blackbody radiation with the classical law of the equipartition of energy,[4] the distribution law in this limit was initially given as being proportional to $T\lambda^{-4}$. Five years later, still troubled by the inability of the equipartition theorem to correctly describe blackbody radiation, one again finds Lord Rayleigh returning to the question. In a letter to *Nature* dated 16 May 1905 concerning "The dynamical theory of gases and radiation," [549] Rayleigh re-derives his law once more except this time he includes an explicit value for the constant of proportionality. At the time Rayleigh's letter appeared, James Jeans, who was then a Lecturer of Applied Mathematics at Cambridge University, was also concerned with problems related to the equipartition theorem. In a paper entitled "On the partition of energy between matter æther" he was preparing for submission to the *Philosophical Magazine* he read Rayleigh's letter where he raised a number of questions centering on the failure of the law of the equipartition of energy when applied to blackbody radiation. Responding to Rayleigh's questions, in a postscript dated 7 June 1905 to the paper Jeans had by now already prepared, he pointed out the numerical factor Rayleigh had found was in error by a factor of eight [328],[5] a mistake Rayleigh readily acknowledged in a short letter published in the July 13 1905 edition of *Nature* [550]. Thereafter both the names of Rayleigh and Jeans would come to be associated with this approximate law even though the form of the law given by Rayleigh five years prior to the work of Jeans was essentially correct.

By 1900 it was known that the spectral distribution of a blackbody must satisfy four limiting conditions. These were:

(i)   At a fixed temperature, the spectral radiant exitance tends to zero as the wavelength tends to zero.

(ii) At a fixed temperature, the spectral radiant exitance tends to zero as the wavelength tends to infinity.

(iii) For fixed wavelength, the spectral radiant exitance tends to zero as the temperature tends to zero.

(iv) For fixed wavelength, the spectral radiant exitance tends to infinity as the temperature tends to infinity.

Rayleigh recognised that Wien's approximation satisfied the first three of the above four limiting conditions [548]. For the fourth, as the temperature is raised, for a given wavelength, the spectral radiant exitance approaches a non-zero limiting value equal to $2\pi hc^2/\lambda^5$ and is the principal reason why Lord Rayleigh by 1900 could no longer accept Wien's law as the final law for thermal radiation. On the other hand, the Rayleigh–Jeans approximation satisfies three of the above four limiting conditions, these being the second, third, and fourth. For the first, for fixed temperature as one moves towards the short wavelength region of the spectrum the spectral radiant exitance increases without bound, a result in direct discord with experiment. Eleven years later the Austrian-born Dutch physicist Paul Ehrenfest coined the term *ultraviolet catastrophe*[6] to describe the failure of the Rayleigh–Jeans law at short wavelengths [207]. As dramatic sounding as the term may be, one should recognise Lord Rayleigh and Jeans both understood their law was only valid in the limit of large wavelength–temperature products.

In addition to these two well-known approximations, several other approximations to Planck's law, valid in various limits, have been proposed in the literature. We now consider a number of these in turn.

3.1.2 EXTENDED WIEN AND RAYLEIGH–JEANS APPROXIMATIONS

The simplest type of approximation beyond the Rayleigh–Jeans and Wien approximations are the so-called "extended" forms. First introduced formally by D. E. Erminy in 1967 [215] the idea behind them is an extension of the Rayleigh–Jeans and Wien approximations into the intermediate wavelength–temperature product regions where the former simpler approximations are no longer valid.

For the extended Rayleigh–Jeans approximation, starting from Planck's law, when the argument $hc/(k_B \lambda T)$ is small. On expanding the exponential term one has

$$M_{e,\lambda}^{b}(\lambda, T) = \frac{2\pi hc^2}{\lambda^5} \left[\left(1 + \frac{1}{1!} \left(\frac{hc}{k_B \lambda T} \right) + \frac{1}{2!} \left(\frac{hc}{k_B \lambda T} \right)^2 + \cdots \right) - 1 \right]^{-1}.$$

$$(3.4)$$

Retaining up to second order terms gives

$$M_{e,\lambda}^{b}(\lambda, T) \simeq \frac{2\pi ck_B T}{\lambda^4} \left[1 + \frac{1}{2} \frac{hc}{k_B \lambda T} \right]^{-1}.$$

$$(3.5)$$

This is known as the extended Rayleigh–Jeans approximation. It is accurate to within 1% for $\lambda T \gtrsim 5.7 \times 10^4 \mu\text{m·K}$.

For $hc/(k_B \lambda T) \ll 1$, the extended form can be further approximated using the binomial expansion. Since

$$(1 + x)^{-1} = 1 - x + x^2 - x^3 + \cdots, \quad |x| < 1,$$

$$(3.6)$$

Eq. (3.5) becomes

$$M_{e,\lambda}^{b}(\lambda, T) \simeq \frac{2\pi ck_B T}{\lambda^4} \left[1 - \frac{1}{2} \frac{hc}{k_B \lambda T} \right].$$

$$(3.7)$$

In this form we will refer to it as the extended binomial Rayleigh–Jeans approximation. It is accurate to within 1% for $\lambda T \gtrsim 4.5 \times 10^4 \mu\text{m·K}$.

For the extended form of Wien's approximation, starting from Planck's law and rearranging, one can write

$$M^b_{e,\lambda}(\lambda, T) = \frac{2\pi hc^2}{\lambda^5} \frac{1}{\exp\left(\dfrac{hc}{k_B \lambda T}\right) - 1} = \frac{2\pi hc^2}{\lambda^5} \frac{\exp\left(-\dfrac{hc}{k_B \lambda T}\right)}{1 - \exp\left(-\dfrac{hc}{k_B \lambda T}\right)}, \quad (3.8)$$

or

$$M^b_{e,\lambda}(\lambda, T) = \frac{2\pi hc^2}{\lambda^5} \exp\left(-\frac{hc}{k_B \lambda T}\right) \left[1 - \exp\left(-\frac{hc}{k_B \lambda T}\right)\right]^{-1}. \quad (3.9)$$

In the limit where Wien's approximation applies, $hc/(k_B \lambda T) \gg 1$ meaning that the argument appearing in each exponential term will be small. The binomial series expansion can be applied to the term appearing in the denominator of Eq. (3.9). Doing so yields

$$M^b_{e,\lambda}(\lambda, T) \simeq \frac{2\pi hc^2}{\lambda^5} \exp\left(-\frac{hc}{k_B \lambda T}\right) \left[1 + \exp\left(-\frac{hc}{k_B \lambda T}\right)\right], \quad (3.10)$$

and is the so-called extended Wien approximation. It is accurate to within 1% for $\lambda T \lesssim 6200 \mu\text{m·K}$.

In Fig. 3.1 plots showing the relative error between the Rayleigh–Jeans and Wien approximations and the extended Rayleigh–Jeans and Wien approximations compared to Planck's equation are given. We see the extended forms do indeed extend the range of validity of each approximation but notice for wavelength–temperature products around $1 \times 10^4 \mu\text{m·K}$ neither approximation nor their extended forms give a particularly accurate approximation to Planck's equation.

3.1.3 POLYNOMIAL INTERPOLATION AND LOGARITHMIC CORRECTION FACTORS

In 1946 Parry H. Moon and Domina E. Spencer [461] suggested that while it seemed doubtful that a simple approximation to Planck's equation over the entire spectral range was possible, a reasonably accurate approximation valid at least within the visible region ($\lambda = 0.40$ to 0.70 μm) at any temperature was possible using a three term polynomial interpolation of the form

$$M^b_{e,\lambda}(\lambda, T) = K_0(T) + K_1(T)\lambda + K_2(T)\lambda^2. \quad (3.11)$$

Here $K_0(T)$, $K_1(T)$, and $K_2(T)$ are temperature dependent constants to be found. At a given temperature these three constants were determined by making Eq. (3.11) coincide with Planck's equation at the three wavelengths 0.45, 0.55, and 0.65 μm. Agreement between the two was good except at the ends

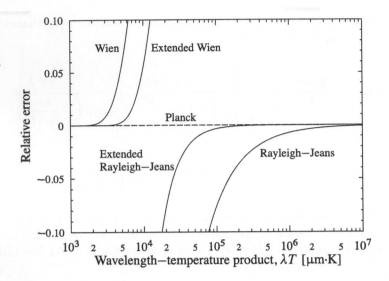

Figure 3.1 Relative error between the Rayleigh–Jeans and Wien approximations and the extended Rayleigh–Jeans and Wien's approximations compared to Planck's equation.

of the visible region. To help overcome such discrepancies, if a more accurate representation was needed, a seven term polynomial in place of the three was suggested. The values for the extended polynomial were interpolated with Planck's equation at wavelengths equal to 0.40, 0.45, 0.50, 0.55, 0.60, 0.65, and 0.70 μm. When this was done, for temperatures between 2000 and 20 000 K, the error between the two within the visible region was found to be no larger than 0.02%. The increase in computational effort required to gain greater accuracy however, was significant and meant such an approximation was not suitable for most applications. A further problem was that if many different temperatures needed to be considered it meant the computational effort required to find each of the temperature dependent constants would be great.

A year later, in the course of preparing tables for Planck's function made some years earlier [41], Elliot Q. Adams, then from the Lamp Development Laboratory at General Electric in the US, found he could use Wien's approximation with the addition of a single logarithmic correction term in place of Planck's equation [40]. These he tabulated and allowed Wien's approximation to be readily corrected. The approximation only applies where Wien's approximation is valid, it giving a slightly improved correction to the result found using Wien's law alone.

Writing Planck's equation for the spectral radiant exitance as

$$M_{e,\lambda}^{b}(\lambda, T) = \frac{2\pi ch^2}{\lambda^5} \frac{\exp\left(-\dfrac{hc}{k_B \lambda T}\right)}{1 - \exp\left(-\dfrac{hc}{k_B \lambda T}\right)}, \tag{3.12}$$

if the logarithm of both sides is taken, one can write

$$\ln M_{e,\lambda}^{b}(\lambda, T) = \ln c_1 - 5\ln\lambda - \frac{c_2}{\lambda T} - \ln\left[1 - \exp\left(-\frac{c_2}{\lambda T}\right)\right]. \tag{3.13}$$

For brevity we have introduced the constants c_1 and c_2 for the first and second radiation constants respectively.[7] The first three terms on the right-hand side of Eq. (3.13) gives the Wien approximation. The fourth and final term needs to be approximated in some manner and leads to what will be referred to as a logarithmic correction.

In the limit where Wien's law holds, namely $c_2/(\lambda T) \gg 1$, the exponential term appearing in the final log term to the right of Eq. (3.13) will be small. The log term can therefore be approximated using its Maclaurin series expansion. Noting that

$$\ln(1 - x) = -x - \frac{x^2}{2} - \frac{x^3}{3} - \cdots, \quad |x| < 1, \tag{3.14}$$

if only the first term in the series expansion is retained, Eq. (3.13) can be approximated as

$$\ln\left[M_{e,\lambda}^{b}(\lambda, T)\right] \simeq \ln c_1 - 5\ln\lambda - \frac{c_2}{\lambda T} + \exp\left(-\frac{c_2}{\lambda T}\right), \tag{3.15}$$

giving

$$M_{e,\lambda}^{b}(\lambda, T) \simeq \frac{2\pi hc^2}{\lambda^5} \exp\left(-\frac{hc}{k_B \lambda T}\right) \exp\left[\exp\left(-\frac{hc}{k_B \lambda T}\right)\right]. \tag{3.16}$$

This so-called logarithmic corrected form for Wien's law gives values for the spectral radiant exitance which turn out to be more accurate than those given by either Wien's law or its extended form. The logarithmic corrected form of Wien's law maintains accuracies to within 1% for $\lambda T \lesssim 7100\,\mu\text{m}\cdot\text{K}$.

While the logarithmic corrected form for Wien's law may be more accurate than either the Wien or the extended Wien approximations, as it involves function composition between exponentials, computationally it cannot be said to be very efficient for the gain in accuracy made. Being more difficult to use compared to the exact form of Planck's law meant it was rarely considered and seems to have only ever been used by Adams himself.

3.1.4 LAURENT POLYNOMIALS AND NON-RATIONAL APPROXIMATIONS OF ERMINY

An extensive treatment of the problem of approximating Planck's equation was given by D. E. Erminy in 1967 [215]. Erminy was concerned with finding suitable approximations to Planck's equation in the intermediate region where the Wien and the Rayleigh–Jeans approximations are no longer valid. As we saw earlier, at first he suggested the approximations of Wien and Rayleigh–Jeans could simply be extended by retaining the first two terms in the power series expansions leading to each approximation. However, when this is done the mathematical form for each extended approximation is hardly any simpler than Planck's equation itself. Erminy next observed that approximating Planck's equation by a polynomial over some intermediate region was also quite unsuitable as in order to achieve a satisfactory fit, as Moon and Spencer had already found some years earlier [461], a polynomial of high degree was required and again hardly constituted a simple approximation.

Faced with these challenges, Erminy posed an approximation to the Planck function based on a Laurent polynomial; a "polynomial" with non-positive integer exponents in the variable which Erminy mistakenly termed a "multinomial." Following Erminy's procedure, he begun by writing Planck's equation as

$$M_{e,\lambda}^{b}(\lambda, T) = \frac{2\pi hc^2}{\lambda^5}\phi_P(v), \tag{3.17}$$

where v equals the dimensionless parameter $hc/(k_B\lambda T)$ while $\phi_P(v) = (e^v - 1)^{-1}$. The task now is to approximate the function $\phi_P(v)$ with a Laurent polynomial function $\phi(v)$. Writing

$$\phi(v) = \sum_{k=0}^{n} \frac{a_k}{v^k}, \tag{3.18}$$

where n is a small positive integer (usually either two or three), setting $\phi(v)$ equal to $\phi_P(v)$ the coefficients a_k are found by demanding that at a given point the two phi functions and their first (for two terms) and second (for three terms) derivatives are equal. That is

$$\phi(v) = \phi_P(v), \quad \phi'(v) = \phi_P'(v), \quad \phi''(v) = \phi_P''(v). \tag{3.19}$$

For small v (the Rayleigh–Jeans limit) a two term Laurent polynomial of the form $\phi(v) = b_0 + b_1/v$ gives an adequate approximation to the Planck function $\phi_P(v)$. The two coefficients are found by matching the value of the approximate function $\phi(v)$ and its derivative to the Planck function and its derivative at a suitably chosen point. When this is done, one finds

$$b_0(v) = \phi_P(v) + v\phi_P'(v) \quad \text{and} \quad b_1(v) = -v^2\phi_P'(v). \tag{3.20}$$

For larger v (the Wien limit) a three term Laurent polynomial of the form $\phi(v) = h_0 + h_1/v + h_2/v^2$ gives a more suitable approximation to $\phi_P(v)$. Matching the values for the Laurent polynomial and its first and second derivatives at a suitably chosen point with the corresponding Planck function and its first and second derivatives one finds

$$h_0(v) = \phi_P(v) + 2v\phi'_P(v) + \frac{v^2}{2}\phi''_P(v),$$
$$h_1(v) = -v^2[3\phi'_P(v) + v\phi''_P(v)],$$
$$h_2(v) = v^3\left(\phi'_P(v) + \frac{v}{2}\phi''_P(v)\right). \tag{3.21}$$

Using either two or three term Laurent polynomials, their fit to Planck's function over the intermediate region is found to be reasonable for the computational effort expended in employing the approximation. As an example of the improvement gained on the Rayleigh–Jeans approximation, for the two term Laurent polynomial selecting the point $v = 0$ one finds for the coefficients $b_0 = -1/2$ and $b_1 = 1$ respectively. A Laurent polynomial approximation for the spectral radiant exitance will therefore be

$$M_{e,\lambda}^b(\lambda, T) \simeq \frac{2\pi c k_B T}{\lambda^4} - \frac{\pi h c^2}{\lambda^5}, \tag{3.22}$$

an approximation more accurate than either the Rayleigh–Jeans or extended Rayleigh–Jeans approximations.

As a final approach to finding a suitable approximation in the intermediate region Erminy found a non-rational approximation to Planck's equation by approximating its derivative using a simple polynomial and then integrating the resulting differential equation. From

$$\frac{d\phi_P(v)}{dv} = -e^v\phi_P^2(v), \tag{3.23}$$

a satisfactory polynomial fit to the derivative requires a degree of higher order leading to an approximation which can hardly be said to be simple. Instead, rewriting the derivative as

$$\frac{d\phi_P(v)}{dv} = -\psi_P(v)\frac{\phi_P(v)}{v}, \tag{3.24}$$

where $\psi_P(v) = v(1 - e^{-v})^{-1}$, as $\psi_P(v)$ is a smooth, slowly varying function of v it can be more readily and accurately approximated using a polynomial of low degree of the form

$$\psi_P(v) \simeq \psi(v) = \sum_{k=0}^{n} \alpha_k v^k. \tag{3.25}$$

Applying the polynomial approximation to Eq. (3.24) one has

$$\frac{d\phi_P(v)}{dv} \simeq -\psi(v)\frac{\phi_P(v)}{v}, \tag{3.26}$$

a separable differential equation which can be readily solved. Its solution is

$$\phi_P(v) \simeq \exp\left(-\int \frac{\psi(v)}{v} dv\right). \tag{3.27}$$

Note the constant of integration has been set equal to zero.

For small v (the Rayleigh-Jeans limit) a function of the form $\psi(v) = \beta_0 + \beta_1 v$ gives an adequate approximation to $\psi_P(v)$. Performing the integration the result is

$$\phi_P(v) \simeq v^{-\beta_0} e^{-\beta_1 v}. \tag{3.28}$$

For $v > 1$ (the Wien limit) a function of the form $\phi(v) = \gamma_1 v$ is found to be adequate. After performing the necessary integration one has

$$\phi_P(v) \simeq e^{-\gamma_1 v}. \tag{3.29}$$

The two beta coefficients in Eq. (3.28) are found by fitting Planck's equation and its first derivative at a point with each of the approximating functions found. When this is done one finds

$$\beta_0(v) = \frac{\psi_P(v) + \ln \phi_P(v)}{1 - \ln v} \quad \text{and} \quad \beta_1(v) = -\frac{\psi_P(v) \ln v + \ln \phi_P(v)}{v(1 - \ln v)}. \tag{3.30}$$

For the gamma coefficient, fitting Planck's equation at a point to the approximating function will suffice. Doing so yields

$$\gamma_1(v) = -\frac{\ln \phi_P(v)}{v}. \tag{3.31}$$

The accuracy of each fit is reasonable given the computational expense involved in finding a non-rational approximation.

As an example of the improvement gain over the Rayleigh–Jeans approximation, if the point $v = 0$ is selected one finds for the two beta coefficients $\beta_0 = 1$ and $\beta_1 = 1/2$ respectively. In this particular case a non-rational approximation to the spectral radiant exitance will be

$$M_{e,\lambda}^b(\lambda, T) \simeq \frac{2\pi c k_B T}{\lambda^4} \exp\left(-\frac{hc}{2k_B \lambda T}\right), \tag{3.32}$$

an approximation more accurate than the previous two-term Laurent polynomial approximation found using the same point for v.

In Fig. 3.2 plots showing the relative error between all the various approximations to Planck's equation we have discussed are given. It will be noticed for approximations beyond either the Rayleigh–Jeans and Wien approximations and their extended forms, while an improvement in accuracy is gained over these more common approximations, none of the approximations discussed are accurate over all wavelength–temperature products of interest.

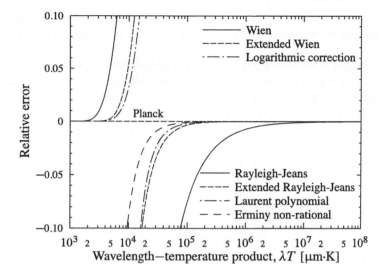

Figure 3.2 The relative error for various approximations developed for Planck's equation as a function of the wavelength–temperature product.

3.2 COMPUTATION OF THE FRACTIONAL FUNCTION OF THE FIRST KIND

For a long time the question of the integrability of the fractional function of the first kind in closed form in terms of any of the known functions of mathematical physics remained open. When faced with an integral involving the Planck distribution function over finite limits authors simply remarked it was not possible to express such an integral in closed form. Despite the prolonged failure to carry out the integration analytically, mathematically there is nothing formally precluding a closed form expression in terms of any of the special functions of mathematics being found.

In the past when attempting to evaluate the fractional function of the first kind for the radiometric and actinometric cases, integrals of the type

$$\mathfrak{F}_{e,0\to\lambda T} = \frac{15}{\pi^4} \int_z^\infty \frac{u^3}{e^u - 1} du, \tag{3.33}$$

and

$$\mathfrak{F}_{q,0\to\lambda T} = \frac{1}{2\zeta(3)} \int_z^\infty \frac{u^2}{e^u - 1} du, \tag{3.34}$$

were encountered. Outwardly these two forms may seem to differ from those given in Chapter 2 where their integrand were expressed in terms of a zeroth-order polylogarithm (see Eqs (2.105) and (2.109) for a comparison). The two alternative forms are however, equivalent if one recalls the zeroth-order poly-logarithm can be expressed as a rational function of its argument (see page 38).

Surprisingly, it was not until 1980 that Richard Pavelle, then at the Lincoln Laboratory at the Massachusetts Institute of Technology in the US, formally showed integrals of the fractional type could not be expressed in closed form in terms of any of the elementary functions of mathematics.[8] He was able to do this by exploiting a very powerful theoretical tool now known as the Risch algorithm [492, 493]. The Risch algorithm, named after the American mathematician Robert Henry Risch, who developed it in 1969 [563], is a systematic method for determining if a function has an elementary function as an indefinite integral. In the case of indefinite integrals of the fractional type, Pavelle showed they do not. However, as no equivalent theory presently exists to determine if a function has a special function as an indefinite integral, there is no way of telling whether an indefinite integral can be expressed in terms of these functions.

3.2.1 SERIES EXPANSION METHODS

By far the most widely used method for evaluating integrals such as Eqs (3.33) and (3.34) involve expanding the integrand in terms of convergent infinite series followed by term-by-term integration. Depending on the size of the parameter z this can be done in one of two ways. In the first, and by far most widely used form, the denominator in the integrand of Eqs (3.33) and (3.34) is expanded in terms of $\exp(-x)$ using the binomial theorem followed by integration by parts. In the second, less widely used method, the integrand is expanded in a form which depends on Bernoulli numbers. Each expansion turns out to be more suited to a different range in values for z depending on whether z is either small or large.

3.2.1.1 Large arguments

In order to handle the three fractional functions of the first kind considered in Chapter 2 as a single entity, the following generalised fractional function will be introduced

$$\Phi 1_s(z) = \frac{\displaystyle\int_z^\infty \frac{x^{s-1}}{e^x - 1} dx}{\displaystyle\int_0^\infty \frac{x^{s-1}}{e^x - 1} dx} = \frac{1}{\Gamma(s)\zeta(s)} \int_z^\infty \frac{x^{s-1}}{e^x - 1} dx, \quad s > 1. \tag{3.35}$$

Recall z is equal to the dimensionless parameter $hc/(k_B \lambda T)$ and physically must be positive while the result for the value of the integral appearing in the denominator previously found in Chapter 2 has been used (see page 65). As special cases, $s = 4$ corresponds to the radiometric case, $s = 3$ the actinometric case, while $s = 2$ the eta case.

If the integrand in Eq. (3.35) is rewritten so it takes the form

$$\Phi l_s(z) = \frac{1}{\Gamma(s)\zeta(s)} \int_z^\infty \frac{e^{-x} x^{s-1}}{1 - e^{-x}} dx, \qquad (3.36)$$

noting that $0 < e^{-x} < 1$ for all $x > 0$, one recognises the term $1/(1 - e^{-x})$ as the sum of the convergent geometric series $\sum_{k=0}^\infty e^{-kx}$. On shifting the summation index and interchanging the order of the integration and summation,[9] Eq. (3.36) becomes

$$\Phi l_s(z) = \frac{1}{\Gamma(s)\zeta(s)} \sum_{k=1}^\infty \int_z^\infty x^{s-1} e^{-kx} dx. \qquad (3.37)$$

Making the change of variable $u = kx$ the integral appearing in Eq. (3.37) can be written in terms of the upper incomplete gamma function $\Gamma(s, \alpha)$. The result is

$$\Phi l_s(z) = \frac{1}{\Gamma(s)\zeta(s)} \sum_{k=1}^\infty \frac{\Gamma(s, kz)}{k^s}, \qquad (3.38)$$

where

$$\Gamma(s, \alpha) = \int_\alpha^\infty u^{s-1} e^{-u} du. \qquad (3.39)$$

If s is a positive integer greater than one the upper incomplete gamma function can be evaluated explicitly. Performing integration by parts $(s - 1)$ times, one finds

$$\Gamma(s, kz) = e^{-kz} \sum_{i=0}^{s-1} \frac{(s - 1)!}{i!} (kz)^i, \qquad (3.40)$$

a result which can be readily confirmed by induction on s. Eq. (3.37) can therefore be expressed as

$$\Phi l_s(z) = \frac{1}{\Gamma(s)\zeta(s)} \sum_{k=1}^\infty \frac{e^{-kz}}{k^s} \sum_{i=0}^{s-1} \frac{(s - 1)!}{i!} (kz)^i. \qquad (3.41)$$

Recognising the term

$$\sum_{i=0}^{s-1} \frac{(s - 1)!}{i!} (kz)^i, \qquad (3.42)$$

is a polynomial in kz of degree $(s - 1)$ we label these polynomials *exponential E polynomials*. In general they are defined as[10]

$$\Im_n(x) = \sum_{k=0}^n \frac{n!}{k!} x^k, \quad x \geqslant 0. \qquad (3.43)$$

The first few are:

$$\Im_0(x) = 1$$
$$\Im_1(x) = 1 + x$$
$$\Im_2(x) = 2 + 2x + x^2$$
$$\Im_3(x) = 6 + 6x + 3x^2 + x^3$$
$$\Im_4(x) = 24 + 24x + 12x^2 + 4x^3 + x^4$$
$$\Im_5(x) = 120 + 120x + 60x^2 + 20x^3 + 5x^4 + x^5$$

Starting with $\Im_0(x) = 1$ and $\Im_1(x) = 1 + x$ the exponential E polynomials can be computed recursively using the following recurrence relationship

$$\Im_{n+1}(x) = (x + n + 1)\,\Im_n(x) - nx\,\Im_{n-1}(x), \quad n \geqslant 1. \tag{3.44}$$

Likewise the polynomials can be generated using the generating function

$$\frac{e^{xt}}{1-t} = \sum_{n=0}^{\infty} \Im_n(x)\frac{t^n}{n!}, \tag{3.45}$$

or alternatively can be found using the Rodrigues' formula

$$\Im_n(x) = (-1)^n \frac{x^{n+1}}{e^{-x}} \frac{d^n}{dx^n}\left(\frac{e^{-x}}{x}\right). \tag{3.46}$$

Returning to the function $\Phi l_s(z)$, in terms of the exponential E polynomials Eq. (3.41) can be expressed as

$$\Phi l_s(z) = \frac{1}{\Gamma(s)\zeta(s)} \sum_{k=1}^{\infty} \frac{e^{-kz}}{k^s}\, \Im_{s-1}(kz). \tag{3.47}$$

For the three special cases considered in the previous chapter, when $s = 4$ the fractional function of the first kind for the radiometric case can be written as

$$\mathfrak{F}_{e,0\to\lambda T} = \Phi l_4(z) = \frac{1}{\Gamma(4)\zeta(4)} \sum_{k=1}^{\infty} \frac{e^{-kz}}{k^4}\, \Im_3(kz), \tag{3.48}$$

or

$$\mathfrak{F}_{e,0\to\lambda T} = \frac{15}{\pi^4} \sum_{k=1}^{\infty} \frac{e^{-kz}}{k^4}\left[6 + 6(kz) + 3(kz)^2 + (kz)^3\right]. \tag{3.49}$$

For $s = 3$ the fractional function of the first kind for the actinometric case can be written as

$$\mathfrak{F}_{q,0\to\lambda T} = \Phi l_3(z) = \frac{1}{\Gamma(3)\zeta(3)} \sum_{k=1}^{\infty} \frac{e^{-kz}}{k}\, \Im_2(kz), \tag{3.50}$$

or

$$\mathfrak{F}_{q,0\to\lambda T} = \frac{1}{2\zeta(3)} \sum_{k=1}^{\infty} \frac{e^{-kz}}{k^3} \left[2 + 2(kz) + (kz)^2\right]. \tag{3.51}$$

While for $s = 2$, the fractional function of the first kind for the eta case can be written as

$$\mathfrak{F}_{\eta,0\to\lambda T} = \Phi l_2(z) = \frac{1}{\Gamma(2)\zeta(2)} \sum_{k=1}^{\infty} \frac{e^{-kz}}{k} \mathfrak{I}_1(kz) = \frac{6}{\pi^2} \sum_{k=1}^{\infty} \frac{e^{-kz}}{k^2} \left[1 + (kz)\right].$$
$$\tag{3.52}$$

These infinite series forms are by far the most common forms seen for the fractional functions of the first kind in the literature. We will refer to them as the exponential infinite series form.

Since the exponential infinite series form converges for all $z \geqslant 0$, each can be used for all wavelengths and temperatures. Herman Zanstra, in his work on planetary nebulae and stellar temperatures, appears to have been the first to give these forms, at least for the radiometric and actinometric cases, in 1931 [730]. However, as the work appeared in the astronomical literature, it went largely unnoticed by those working outside of the field. They were independently rediscovered a few years later, first by Zen'emon Miduno with whom we mentioned earlier, in 1938 [446] and subsequently by many others [398, 425, 501, 626, 443, 657, 388, 682, 697, 698, 213, 166, 700], and as late as 1984 results were still being claimed as new in the literature [128].

As the series appearing in Eq. (3.47) converges for all values of $z \geqslant 0$, values for the fractional function of the first kind can be found by truncating the series after N terms. In particular, when z becomes large, the series becomes dominated by the exponential factor leading to rapid convergence, meaning the number of terms that need to be summed in order to obtain reasonably high accuracy is modest.

To estimate the size of the error incurred after truncating the series after the first N terms have been summed, we begin by writing Eq. (3.47) as

$$\Phi l_s(z) = \frac{1}{\Gamma(s)\zeta(s)} \sum_{k=1}^{N} \frac{e^{-kz}}{k^s} \mathfrak{I}_{s-1}(kz) + \frac{1}{\Gamma(s)\zeta(s)} \sum_{k=N+1}^{\infty} \frac{e^{-kz}}{k^s} \mathfrak{I}_{s-1}(kz).$$
$$\tag{3.53}$$

The remainder after summing the first N terms in the series is given by the term

$$\epsilon_N(z;s) = \frac{1}{\Gamma(s)\zeta(s)} \sum_{k=N+1}^{\infty} \frac{e^{-kz}}{k^s} \mathfrak{I}_{s-1}(kz). \tag{3.54}$$

The remainder clearly depends on both the value for the argument z and the order s. We now proceed to find an upper bound for this term.

If the exponential E polynomials are written out explicitly in terms of its sum (see Eq. (3.42)), on setting $p = s - i$ the term for the remainder becomes

$$\epsilon_N(z; s) = \frac{1}{\Gamma(s)\zeta(s)} \sum_{i=0}^{s-1} \frac{(s-1)!}{i!} z^i \sum_{k=N+1}^{\infty} \frac{e^{-kz}}{k^p}. \tag{3.55}$$

An upper limit for the infinite sum containing the exponential term will now be found. This will be done using a recurrence relation for the exponential term. Define

$$A(k) = \frac{e^{-kz}}{k^p}. \tag{3.56}$$

For all $k \geqslant N > 0$ and $p > 0$ (k and p are positive integers) one has

$$\frac{A(k+1)}{A(k)} = \left(\frac{1}{1+1/k}\right)^p e^{-z} < e^{-z}. \tag{3.57}$$

Repeating the process ℓ times, by induction on ℓ it can be shown that

$$A(k + \ell) < e^{-\ell z} A(k). \tag{3.58}$$

Returning to the infinite sum term involving the exponential in the remainder, we can now write it as

$$\sum_{k=N+1}^{\infty} \frac{e^{-kz}}{k^p} = \sum_{k=N+1}^{\infty} A(k) = \sum_{\ell=0}^{\infty} A(N+1+\ell). \tag{3.59}$$

In the last sum the summation index k has been shifted to $N + 1 + \ell$. From Eq. (3.58), setting $k = N + 1$ which is clearly larger than N, one has

$$A(N + 1 + \ell) < e^{-\ell z} A(N+1). \tag{3.60}$$

A bound on Eq. (3.59) is

$$\sum_{k=N+1}^{\infty} \frac{e^{-kz}}{k^p} < A(N+1) \sum_{\ell=0}^{\infty} (e^{-z})^\ell. \tag{3.61}$$

Since $0 < e^{-z} < 1$ for all $z > 0$ the sum over ℓ which is geometric converges to $(1 - e^{-z})^{-1}$. Thus

$$\sum_{k=N+1}^{\infty} \frac{e^{-kz}}{k^p} < \frac{e^{-(N+1)z}}{(N+1)^p} \frac{1}{1 - e^{-z}}. \tag{3.62}$$

Recalling $p = s - i$, a final bound for the remainder term follows. The result is

$$\epsilon_N(z; s) < \frac{1}{\Gamma(s)\zeta(s)} \frac{e^{-(N+1)z}}{(N+1)^s} \frac{\mathfrak{I}_{s-1}[z(N+1)]}{1 - e^{-z}}. \tag{3.63}$$

Table 3.1
Number of terms needed to be summed for nine signifiance figure accuracy in the fractional function $\Phi l_s(z)$ (exponential form)

z	Number of terms		
	radiometric	actinometric	eta
0.25	59	64	69
0.50	33	35	36
0.75	24	24	24
1.00	19	19	18
1.25	15	15	15
1.50	13	13	13
1.75	11	11	11
2.0	10	10	10
2.5	8	8	8
3.0	7	7	6
3.5	6	6	5
4.0	5	5	5
5.0	4	4	4
6.0	4	3	3
7.0	3	3	3
8.0	3	3	2
9.0	3	2	2
10	2	2	1
11	2	2	1
12	2	2	1
13	2	1	1
14	2	1	1
$\geqslant 15$	1	1	1

Notice the error in truncating the series for $\Phi l_s(z)$ after N terms is no more than the size of the next term in the sum multiplied by a factor of $(1 - e^{-z})^{-1}$.

In Table 3.1 the number of terms to be summed for nine significant figure accuracy in the fractional function $\Phi l_s(z)$ for values of the argument between

0.25 and 15 for the radiometric, actinometric, and eta cases are given. Nine significance figure accuracy has been chosen as it illustrates well the number of terms that need to be summed in order to reach a given accuracy while at the same time clearly shows how this number quickly grows as the value for the argument z becomes small. It will be noticed when z is close to zero the number of terms that need to be summed becomes quite large. To deal with the increase in the number of terms needed to be summed as the argument z becomes small an alternative power series expansion for the fractional function $\Phi l_s(z)$ whose rate of convergence is much faster than the sum considered here is presented in the next section.

3.2.1.2   Small arguments

As we have seen, for small values of the argument z convergence of Eq. (3.47) is slow, meaning many terms in the series need to be used if reasonable accuracy is to be achieved. Fortunately a second power series expansion that rapidly converges when z is small can also be found. Not as widely known and consequently less frequently used compared to the former infinite series, in the limit where it applies, we shall see only the first few terms of the series need to be summed in order to achieve a relatively high degree of accuracy. As this second infinite series expansion will involve Bernoulli numbers we refer to the series found as the Bernoulli series form for the fractional function.

The Bernoulli numbers B_{2k} are defined by the power series expansion [331]

$$\frac{x}{e^x - 1} = 1 - \frac{x}{2} + \sum_{k=1}^{\infty} \frac{B_{2k}}{(2k)!} x^{2k}, \tag{3.64}$$

and converge for $|x| < 2\pi$. As a reminder, the first few non-zero Bernoulli numbers are $B_2 = 1/6$, $B_4 = -1/30$, $B_6 = 1/42$, $B_8 = -1/30$, etc. Additional details about these special numbers can be found in Appendix A.2. From properties of the definite integral, if the general fractional function is rewritten as

$$\Phi l_s(z) = 1 - \frac{1}{\Gamma(s)\zeta(s)} \int_0^z \frac{x^{s-1}}{e^x - 1} dx, \quad s > 1 \tag{3.65}$$

and the term defining the Bernoulli numbers is replaced by its power series expansion in the integrand and integrated term-by-term, one obtains

$$\Phi l_s(z) = 1 - \frac{1}{\Gamma(s)\zeta(s)} \left[\frac{z^{s-1}}{s-1} - \frac{z^s}{2s} + \sum_{k=1}^{\infty} \frac{B_{2k}}{(2k+s-1)(2k)!} z^{2k+s-1} \right], |z| < 2\pi. \tag{3.66}$$

For the radiometric case ($s = 4$) one has

$$\mathfrak{F}_{e,0 \to \lambda} = \Phi l_4(z) = 1 - \frac{15}{\pi^4} \left(\frac{z^3}{3} - \frac{z^4}{8} + \sum_{k=1}^{\infty} \frac{B_{2k}}{(2k+3)(2k)!} z^{2k+3} \right), \quad |z| < 2\pi. \tag{3.67}$$

For the actinometric case $(s = 3)$

$$\mathfrak{F}_{q,0\to\lambda} = \Phi l_3(z) = 1 - \frac{1}{2\zeta(3)} \left(\frac{z^2}{2} - \frac{z^3}{6} + \sum_{k=1}^{\infty} \frac{B_{2k}}{(2k+2)(2k)!} z^{2k+2} \right), \ |z| < 2\pi.$$

(3.68)

While for $s = 2$, the eta case, one has

$$\mathfrak{F}_{\eta,0\to\lambda} = \Phi l_2(z) = 1 - \frac{6}{\pi^2} \left(z - \frac{z^2}{4} + \sum_{k=1}^{\infty} \frac{B_{2k}}{(2k+1)(2k)!} z^{2k+1} \right), \ |z| < 2\pi.$$

(3.69)

These Bernoulli forms are less frequently found in the literature. A reason why this may be the case is possibly due to the fact that the series given for the fractional function in terms of Bernoulli numbers, despite very rapid convergence when z is small, has a finite interval of convergence and so unlike the exponential form the Bernoulli form cannot be used for all temperatures and wavelengths. A second possible reason is that authors may have simply been unfamiliar with Bernoulli numbers. Often in this limit, the first four to six terms in the infinite series expansion would be given but the general form for the nth term in terms of Bernoulli numbers would often go unrecognized [258, 683, 388, 139].

The Dutch physicist Peter Joseph William Debye (1884–1966) in his work on the specific heats of solids in 1912 was the first to represent the integral given by Eq. (3.35) as an infinite series in terms of Bernoulli numbers [184]. Following the work of Debye, Zanstra some years later represented the integral for the actinometric fractional amount as an infinite series in terms of Bernoulli numbers [730]. Both men's work however, went largely unnoticed by those working in the infrared and radiometric communities. Several years later Miduno again independently gave these two infinite series expansion forms [446]. Thereafter the forms for Eqs (3.67) and (3.68) were given and used by others [258, 501, 647, 213, 166].

If the series given by Eq. (3.66) is used to estimate the value for the general fractional function by truncating it after N terms, the size of the error can be readily estimated since the series alternates as B_{2k} alternates between positive and negative values. An estimate for the size of the error is given by the Alternating Series Estimation Theorem which states that if an alternating series is truncating after summing the first N terms the error will be no larger in magnitude than the size of the first neglected term in the sum.

For the Bernoulli form, truncating after N terms an estimate for the size of the remainder will be

$$\epsilon_N(z; s) \leqslant \frac{1}{\Gamma(s)\zeta(s)} \frac{|B_{2N+2}|}{(2N+s+1)(2N+2)!} z^{2N+s+1}.$$

(3.70)

In practice, finding Bernoulli numbers, particularly for high orders, is computationally quite intensive as they are typically calculated recursively. A significant mathematical simplification in the remainder term, resulting in a corresponding gain in speed, can be made if the Bernoulli number term is replaced with one of its known upper bounds. One such bound for the Bernoulli numbers is [468]

$$\frac{2(2n)!}{(2\pi)^{2n}}\frac{1}{1-2^{-2n}} \leqslant |B_{2n}| \leqslant \frac{2(2n)!}{(2\pi)^{2n}}\frac{1}{1-2^{\beta-2n}}, \tag{3.71}$$

where $\beta = 2 + \ln(1 - 6\pi^{-1})/\ln(2)$. From this bound the remainder term after truncating the sum after N terms becomes

$$\epsilon_N(z;s) \leqslant \frac{1}{\Gamma(s)\zeta(s)}\frac{2}{(2\pi)^{2N+2}(2N+s+1)}\frac{z^{2N+s+1}}{1-2^{\beta-2N-2}}. \tag{3.72}$$

The bound given by Eq. (3.72) is not quite as sharp as that given by Eq. (3.70) but as the latter no longer contains either a Bernoulli number or factorial term, computationally it is far simpler and quicker to calculate. In Table 3.2 the number of terms for the general fractional function in Bernoulli form needed to be summed for nine significant figure accuracy for values of the argument between 0.1 to 6.0 for the radiometric, actinometric, and the eta cases are given. We note the difference in the number of terms needed to be summed using either estimate bound are identical except in one or two cases where an additional term may be needed.

3.2.1.3   Division point

In practice it is particularly advantageous if both the small and large argument forms are used when calculating the general fractional function. As can be seen from Tables 3.1 and 3.2, for a given argument z a division point can be appropriately selected so that the number of terms needed to be summed need not be larger than the first eight to ten terms using either series. Zanstra, who was first to give both series forms for the radiometric and actinometric cases, suggested a division point at an argument of two would be suitable [730]. From such a choice four significant figure accuracy could be achieved using no more than the first five terms in either sum. Later, R. Bruce Emmons from Lockheed Missile and Space Company in the US suggested a suitable division point would be 1.75 [213]. Doing so meant only five terms in either series needed to be summed for fractional errors no larger than 1×10^{-5} while he noted that if the division point was selected at $z = 1.95$, no more than ten terms in either sum were needed for fractional errors no greater than 1×10^{-10}.

Based on the analysis presented above and summarised in Tables 3.1 and 3.2, like Zanstra and others in the astrophysical community after him [714], we suggest that an argument equal to two makes for a suitable division point. With this selection no more than ten terms in either summation need to be summed for nine significant figure accuracy in the fractional function of the first kind for the radiometric, actinometric, and eta cases. In writing a computer program to calculate the various fractional functions of the first kind,

Table 3.2
Number of terms needed to be summed for nine signifiance figure accuracy in the fractional function $\Phi1_s(z)$ (Bernoulli form).

z	Number of terms		
	radiometric	actinometric	eta
0.1	1	1	2
0.2	2	2	2
0.3	2	2	3
0.4	3	3	3
0.5	3	3	3
0.6	3	3	4
0.7	4	4	4
0.8	4	4	4
0.9	4	4	5
1.0	5	5	5
1.2	6	5	5
1.4	6	6	6
1.6	7	7	7
1.8	8	7	7
2.0	9	8	8
2.5	11	11	10
3.0	14	14	13
3.5	18	17	16
4.0	24	23	21
4.5	33	31	29
5.0	48	45	42
5.5	82	76	70
6.0	231	212	194

knowing the maximum number of terms that need to be summed in advance for a given degree of accuracy allows the first dozen or so Bernoulli numbers to be given explicitly within the program rather than having to calculate each separately on the fly. Computationally this is particularly advantageous as

having to calculate higher order Bernoulli numbers becomes quite time consuming. A program which computes values for the three fractional functions
of the first kind considered here correct to nine significant figures using GNU
OCTAVE[11] is given in Appendix B.

3.2.2 APPROXIMATION OF THE INTEGRAND FIRST

Instead of converting the integrand appearing in the expression for the fractional function into an infinite series before integrating term-by-term, as an
alternative one may approximate the integrand first using one of the approximations for Planck's law leading to a form that can be more readily integrated. For example, in the limit where Wien's law applies, namely
$x = hc/(k_B \lambda T) \gg 1$, the integrand appearing in Eq. (3.35) can be approximated as

$$\Phi \mathrm{l}_s(z) \simeq \frac{1}{\Gamma(s)\zeta(s)} \int_z^\infty x^{s-1} e^{-x} dx. \qquad (3.73)$$

The integral appearing in Eq. (3.73) corresponds to the upper incomplete
gamma function. As previously seen, when s is a positive integer it can be
evaluated in terms of the exponential E polynomials (see page 109). The result
is

$$\Phi \mathrm{l}_s(z) \simeq \frac{1}{\Gamma(s)\zeta(s)} e^{-z} \Im_{s-1}(z), \quad z \gg 1. \qquad (3.74)$$

The result dates back to at least the first decade of the twentieth century
where it was given by a number of people including the Swedish physicist
Knut Johan Ångström in 1903 [52], the English physicist Charles V. Drysdale
in 1907 [195],[12] and the German physical chemist Karl Schaum in 1908 [589].
Notice Eq. (3.74) is just the first term ($k = 1$) of Eq. (3.47).

In the limit where the Rayleigh–Jeans approximation applies ($x = hc/(k_B \lambda T) \ll 1$), we rewrite the definite integral appearing in Eq. (3.35)
as

$$\Phi \mathrm{l}_s(z) = \frac{1}{\Gamma(s)\zeta(s)} \left[\int_0^\infty \frac{x^{s-1}}{e^x - 1} dx - \int_0^z \frac{x^{s-1}}{e^x - 1} dx \right] = 1 - \frac{1}{\Gamma(s)\zeta(s)} \int_0^z \frac{x^{s-1}}{e^x - 1} dx. \qquad (3.75)$$

In this limit the integrand is approximated by using the first two terms in the
Maclaurin series expansion for the exponential function. Doing so gives

$$\Phi \mathrm{l}_s(z) \simeq 1 - \frac{1}{\Gamma(s)\zeta(s)} \int_0^z x^{s-2} dx = 1 - \frac{1}{\Gamma(s)\zeta(s)} \frac{z^{s-1}}{s-1}, \quad z \ll 1. \qquad (3.76)$$

If instead the extended Rayleigh–Jeans approximation is used, the term in
the denominator of Eq. (3.75) becomes

$$e^x - 1 = \left(1 + x + \frac{x^2}{2!} + \cdots \right) - 1 \simeq x \left(1 + \frac{x}{2} \right), \qquad (3.77)$$

giving

$$\Phi l_s(z) \simeq 1 - \frac{1}{\Gamma(s)\zeta(s)} \int_0^z x^{s-2}\left(1 + \frac{x}{2}\right)^{-1} dx. \tag{3.78}$$

As x is small the term in parentheses appearing in the integrand of Eq. (3.78) can be further approximated using the binomial expansion. Thus

$$\Phi l_s(z) \simeq 1 - \frac{1}{\Gamma(s)\zeta(s)} \int_0^z x^{s-2}\left(1 - \frac{x}{2}\right) dx, \tag{3.79}$$

giving the final result of

$$\Phi l_s(z) \simeq 1 - \frac{1}{\Gamma(s)\zeta(s)} \left[\frac{z^{s-1}}{s-1} - \frac{z^s}{2s}\right], \quad z \ll 1. \tag{3.80}$$

This result is not new. Just as the approximate expression for the fractional function of the first kind reduced to the first term appearing in the exponential series form for $\Phi l_s(z)$ (see Eq. (3.47)) when Planck's equation was approximated by Wien's law before integrating, the same is true for the Bernoulli form. Here the results obtained by first applying the Rayleigh–Jeans approximation and the extended binomial Rayleigh–Jeans approximation before integrating leads to the first and the first two terms appearing in Eq. (3.66) respectively.

3.2.3 GAUSS–LAGUERRE AND GENERALISED GAUSS–LAGUERRE QUADRATURE

Another method that can be used to evaluate the fractional function of the first kind for the various types is a quadrature method[13] known as Gauss–Laguerre quadrature. It is an extension of the well-known Gaussian quadrature method and applies to integrals on the interval $[0, \infty)$. Used to approximate integrals of the form

$$\int_0^\infty e^{-x} f(x) \, dx, \tag{3.81}$$

in such cases the integral may be expressed using [36]

$$\int_0^\infty e^{-x} f(x) \, dx = \sum_{i=1}^n w_i f(x_i) + E_n. \tag{3.82}$$

The E_n here is an error term and is given by

$$E_n = \frac{(n!)^2}{(2n)!} f^{(2n)}(\xi_n), \tag{3.83}$$

where $0 < \xi_n < \infty$.

The x_i appearing in Eq. (3.82) are known as the abscissae, nodes, or zeros. Each value for x_i corresponds to the ith root of the Laguerre polynomial $L_n(x)$

and is obtained from solving the equation $L_n(x_i) = 0$ while w_i is a weighting factor. The latter are related to the derivatives at the abscissae by[14]

$$w_i = \frac{1}{x_i[L'_n(x_i)]^2} = \frac{x_i}{(n+1)^2[L_{n+1}(x_i)]^2}. \tag{3.84}$$

In applying the quadrature rule to the generalised fractional function of the first kind, it first must be put into the form of Eq. (3.81). Substituting $x = u + z$ into the integral Eq. (3.35) can be rewritten as

$$\Phi l_s(x) = \frac{1}{\Gamma(s)\zeta(s)} \int_0^\infty e^{-u} \frac{(u+z)^{s-1}}{e^z - e^{-u}} du. \tag{3.85}$$

The generalised fractional function of the first kind can now be evaluated by Gauss–Laguerre quadrature as the sum of a finite number of terms as follows

$$\Phi l_s(x) \simeq \frac{1}{\Gamma(s)\zeta(s)} \sum_{i=1}^n w_i f(x_i), \tag{3.86}$$

where

$$f(x_i) = \frac{(x_i + z)^{s-1}}{e^z - e^{-x_i}}. \tag{3.87}$$

A calculation of this type was first performed by one of the authors and his colleague Elmer E. Branstetter in 1974 [335]. At first sight the computational effort required in using the method may seem large as values for the abscissae which require finding roots to the Laguerre polynomials and the corresponding weights are needed. To help alleviate much of the computational effort needed in applying the method meant in the past table of values for the abscissae and weight factors needed for Gauss–Laguerre quadrature up to $n = 15$ were widely available [583, 36] and used whenever this method was employed. When stored on a computer these values could be quickly retrived meaning calculations made for the fractional function using Gauss–Laguerre quadrature could be performed at great speed.[15]

In attempting to estimate the error associated with Gauss–Laguerre quadrature using Eq. (3.83), as ξ_n cannot be estimated analytically and since it can take on any positive real value, performing any form of general analysis is all but impossible. The best one can hope to do, and what Johnson and Branstetter themselves did, is to find an upper bound for the error using the absolute maximum value of the error term E_n for all positive real ξ_n. An error of less than 10^{-10} for all wavelength–temperatures products can be achieved when 15 terms in the sum of Eq. (3.86) are used. For wavelength–temperature products in the range typically considered for most terrestrial infrared systems problems, extending between 3500 to 5000 μm·K, accuracies of the order of 10^{-6} can be achieved from summing the first three to four terms. A plot of the fractional error (on a logarithmic scale) resulting from using Gauss–Laguerre quadrature when 2, 5, 10 and 15 terms are used in the sum to approximate

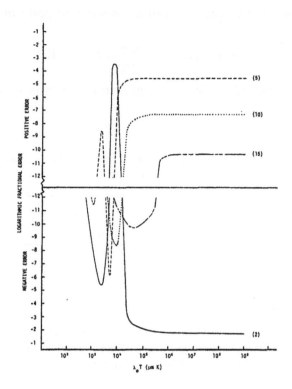

Figure 3.3 Fractional error that results when Gauss–Laguerre quadrature is used to estimate the fractional function of the first kind for the radiometric case as a function of the wavelength–temperature product. The scale used for the ordinate is a logarithmic one while the abscissa is linear. Reproduced with permission from Johnson, R. B. and Branstetter, E. E., 1974 "Integration of Planck's equation by the Laguerre–Guass quadrature method," *Journal of the Optical Society of America* **64**(11), p. 1447. Copyright 1974, OSA Publishing.

the fractional function of the first kind for the radiometric case as a function of the wavelength–temperature product (in μm·K), as given by Johnson and Branstetter, is reproduced in Fig. 3.3

A few years after Johnson and Branstetter's initial proposal, M. Janes, then with the Industrieanlagen-Betriebsgesellschaft in Ottobrun, Germany, suggested a more general method based on generalised Gauss–Laguerre quadrature [325]. The method applies to integrals of the form

$$\int_0^\infty x^\alpha e^{-x} f(x)\, dx. \tag{3.88}$$

Here α is a non-negative integer. In such cases the integral can be evaluated using [540]

$$\int_0^\infty x^\alpha e^{-x} f(x)\, dx = \sum_{i=1}^n w_i f(x_i) + E_n. \tag{3.89}$$

Known as generalised Gauss–Laguerre quadrature the error term E_n is given by

$$E_n = \frac{n!(n+\alpha)!}{(2n)!} f^{(2n)}(\xi_n), \tag{3.90}$$

where $0 < \xi_n < \infty$. Once again the x_i are known as the abscissae, but unlike the Gauss–Laguerre quadrature, correspond to the roots of the associated Laguerre polynomials obtained by solving $L_n^{(\alpha)}(x_i) = 0$. The w_i is once more a weight factor and is given by

$$w_i = \frac{\Gamma(n+\alpha+1)}{n!x_i \left(\dfrac{d}{dx}\left[L_n^{(\alpha)}(x_i) \right] \right)^2} = \frac{\Gamma(n+\alpha+1)x_i}{n!\left[(n+1)L_{n+1}^{(\alpha)}(x_i) \right]^2}. \tag{3.91}$$

For the special case of $\alpha = 0$, the generalised Gauss–Laguerre quadrature reduces to the Gauss–Laguerre quadrature.

Applying the generalised Gauss–Laguerre quadrature to the general fractional function of the first kind, from Eq. (3.84), if the term $(u+z)^{s-1}$ in the integrand is expanded using the binomial theorem

$$(u+z)^{s-1} = \sum_{k=0}^{s-1} \binom{s-1}{k} u^{s-k-1} z^k, \tag{3.92}$$

so in terms of a finite number of terms for $\Phi l_s(z)$ one has

$$\Phi l_s(z) \simeq \frac{1}{\Gamma(s)\zeta(s)} \sum_{k=0}^{s-1} \binom{s-1}{k} z^k \sum_{i=1}^n w_i f(x_i). \tag{3.93}$$

Here

$$f(x_i) = \frac{1}{e^z - e^{-x_i}}. \tag{3.94}$$

Like the Gauss–Laguerre method before it, a quadrature method such as the generalised Gauss–Laguerre method again depends on knowing values for the abscissae and weights. Once more, to help alleviate such work, a table of values for the abscissae and weights needed for generalised Gauss–Laguerre quadrature for various values of n and α existed [540]. In his work of 1984 Janes gave a table of values for the abscissae and weight factors between $n = 1$ to 5 for values of α from zero to three.

The number of calculations required if generalised Gauss–Laguerre quadrature is used to evaluate the fractional function of the first kind is considerable. To drastically reduce this number Janes suggested as the term $\exp(-x_i)$

appearing in Eq. (3.94) is small compared to the value of $\exp(z)$, if it is neglected a far simpler expression for $\Phi l_s(z)$ results. For $e^{-x_i} \ll e^z$, $f(x_i) \simeq e^{-z}$ and $f(x)$ itself can also be approximated by e^{-z}. In this limit Eq. (3.89) can be explicitly evaluated. The result is

$$\int_0^\infty e^{-x} x^\alpha \, dx = \alpha! = \sum_{i=1}^n w_i. \tag{3.95}$$

If we note the following equivalence for the binomial coefficient term

$$\binom{s-1}{k}(s-k-1)! = \frac{(s-1)!}{k!}, \tag{3.96}$$

Eq. (3.93), within this limit, can be approximated as

$$\Phi l_s(z) \simeq \frac{e^{-z}}{\Gamma(s)\zeta(s)} \sum_{k=0}^{s-1} \frac{(s-1)!}{k!} z^k = \frac{1}{\Gamma(s)\zeta(s)} e^{-z} \Im_{s-1}(z). \tag{3.97}$$

Janes seems to have been well pleased with his approximate form, stating for the radiometric case ($s = 4$) the sum in Eq. (3.97) contained only four terms compared to the n terms needed in the equation of Johnson and Branstetter (see Eq. (3.86)). Janes, however, failed to realize that by the mid-1980s, the approximate expression he had found for the fractional function of the first kind had been known in the literature for well over 80 years [52, 195, 589], being identical with the result given on page 118.

3.2.4 ASYMPTOTIC EXPANSION

Another useful tool in the computational analysis of the general fractional function $\Phi l_s(z)$ is its asymptotic expansion about a certain point. An asymptotic expansion usually takes the form of an infinite series, known as an asymptotic series, which may diverge, yet when the first few terms of the series are considered still manages to produce accurate numerical values for the value of the function in the vicinity of the point of expansion.

For small values in the argument z, starting with Eq. (3.65), integrating by parts gives

$$\Phi l_s(z) = 1 - \frac{1}{\Gamma(s)\zeta(s)} \left[\frac{z^s}{s(e^z - 1)} + \frac{1}{s} \int_0^z x^s \frac{e^x}{(e^x - 1)^2} dx \right]. \tag{3.98}$$

Continuing to integrate by parts k times leads to

$$\Phi l_s(z) = 1 - \frac{1}{\Gamma(s)\zeta(s)} \left[\frac{z^s}{s(e^z - 1)} + \frac{z^{s+1}}{s(s+1)} \frac{e^z}{(e^z - 1)^2} \right.$$

$$+ \frac{z^{s+2}}{s(s+1)(s+2)} \frac{e^z(e^z + 1)}{(e^z - 1)^3} + \cdots + \frac{z^{s+k-1}}{(s)_k} \frac{1}{(e^z - 1)^k} \sum_{n=0}^{k-1} \left\langle \begin{matrix} k-1 \\ n \end{matrix} \right\rangle e^{(k-n-1)z}$$

$$+ \frac{1}{(s)_{k+1}} \int_0^z x^{s+k} \frac{1}{(e^x - 1)^{k+1}} \sum_{n=0}^k \left\langle \begin{matrix} k \\ n \end{matrix} \right\rangle e^{(k-n)x} dx \right].$$

$$\tag{3.99}$$

Here $(s)_k = s(s+1)(s+2)\cdots(s+k-1)$ is the Pochhammer symbol [51]. Note that since s is a positive integer the Pochhammer symbol can be written in terms of factorials as

$$(s)_k = \frac{\Gamma(s+k)}{\Gamma(s)} = \frac{(s+k-1)!}{(s-1)!}. \qquad (3.100)$$

The asymptotic expansion for the general fractional function as $z \to 0$ follows from Eq. (3.99). The result is asymptotically equivalent to[16]

$$\Phi l_s(z) \sim 1 - \frac{1}{\Gamma(s)\zeta(s)} \sum_{k=1}^{\infty} \frac{z^{s+k-1}(s-1)!}{(s+k-1)!} \frac{1}{(e^z-1)^k} \sum_{n=0}^{k-1} \left\langle {k-1 \atop n} \right\rangle e^{(k-n-1)z}. \qquad (3.101)$$

In terms of the polylogarithm function of negative integer order, if x is set equal to e^z and the index n is replaced with $n-1$ in Eq. (2.5), one has

$$(-1)^k \mathrm{Li}_{1-k}(e^z) = \frac{1}{(e^z-1)^k} \sum_{n=0}^{k-1} \left\langle {k-1 \atop n} \right\rangle e^{(k-n-1)z}, \qquad (3.102)$$

leading to the following alternative expression for the asymptotic expansion

$$\Phi l_s(z) \sim 1 - \frac{1}{\Gamma(s)\zeta(s)} \sum_{k=1}^{\infty} \frac{z^{s+k-1}(s-1)!}{(s+k-1)!}(-1)^k \mathrm{Li}_{1-k}(e^z). \qquad (3.103)$$

A requirement for any asymptotic expansion is that the ratio of its error to the last term retained must tend to zero as z tends to zero. The error in an asymptotic expansion is generally taken as the difference between the function being expanded and the partial sum for the first N terms of the series considered. More formally, to show the expression for the expansion of $\Phi l_s(z)$ given by Eq. (3.101) (or Eq. (3.103)) is indeed asymptotic, for each fixed N we require

$$\left| \Phi l_s(z) - \left(1 - \frac{1}{\Gamma(s)\zeta(s)} \sum_{k=1}^{N} \phi_k(z) \right) \right| \div |\phi_k(z)| \to 0 \text{ as } z \to 0. \qquad (3.104)$$

Here $\phi_k(z)$ is the kth term of Eq. (3.101) and is equal to

$$\phi_k(z) = \frac{1}{\Gamma(s)\zeta(s)} \frac{z^{s+k-1}(s-1)!}{(s+k-1)!} \frac{1}{(e^z-1)^k} \sum_{n=0}^{k-1} \left\langle {k-1 \atop n} \right\rangle e^{(k-n-1)z}. \qquad (3.105)$$

We now show Eq. (3.101) behaves asymptotically as $z \to 0$. The difference between the function one is interested in expanding asymptotically and its partial sum is given by

$$|\Phi l_s(z) - S_N(z)| = \frac{1}{\Gamma(s)\zeta(s)} \frac{(s-1)!}{(s+k)!} \int_0^z x^{s+k} \frac{1}{(e^x-1)^{k+1}} \sum_{n=0}^{k} \left\langle {k \atop n} \right\rangle e^{(k-n)x} dx, \qquad (3.106)$$

and comes directly from Eq. (3.99). For brevity we have written the term corresponding to the Nth partial sum as $S_N(z)$. Before showing that the term given by Eq. (3.104) tends to zero as z tends to zero, we first find an upper bound for this term before returning to the general question of its asymptotic behavior.

Observing $e^{(k-n)x} \leqslant e^{kx}$ for all $k \geqslant n$, one has

$$\sum_{n=0}^{k} \left\langle {k \atop n} \right\rangle e^{(k-n)x} \leqslant e^{kx} \sum_{n=0}^{k} \left\langle {k \atop n} \right\rangle = k!\, e^{kx}. \tag{3.107}$$

Here the use of the sum identity for Eulerian numbers has been made.[17] Thus

$$|\Phi 1_s(z) - S_N(z)| \leqslant \frac{1}{\Gamma(s)\zeta(s)} \frac{k!(s-1)!}{(s+k)!} \int_0^z \frac{e^{kx} x^{s+k}}{(e^x - 1)^{k+1}} dx. \tag{3.108}$$

The value of the integral appearing in Eq. (3.108) is increased if $1/(e^x - 1)$ is replaced by $1/x$ due to the well-known result $e^x \geqslant 1 + x$ valid for all x. Thus

$$|\Phi 1_s(z) - S_N(z)| \leqslant \frac{1}{\Gamma(s)\zeta(s)} \frac{k!(s-1)!}{(s+k)!} \int_0^z e^{kx} x^{s-1} dx. \tag{3.109}$$

Next, as $e^{kx} \leqslant e^{kz}$ for $z \geqslant x$ the value of the integral appearing in Eq. (3.109) is once more increased if the e^{kx} term is replaced with e^{kz} giving an integral that can now be readily evaluated. The result is

$$|\Phi 1_s(z) - S_N(z)| \leqslant \frac{e^{kz}}{\Gamma(s)\zeta(s)} \frac{k!(s-1)!}{(s+k)!} \int_0^z x^{s-1} dx = \frac{e^{kz}}{\Gamma(s)\zeta(s)} \frac{k!(s-1)!}{(s+k)!} \frac{z^s}{s}. \tag{3.110}$$

Returning to the question of asymptoticity, combining Eq. (3.110) with Eq. (3.105), after some algebraic manipulation an upper bound for Eq. (3.104) can be written as

$$\frac{|\Phi 1_s(z) - S_N(z)|}{|\phi_N(z)|} \leqslant \frac{k!}{s(s+k)} z \left(\frac{e^z(e^z - 1)}{z} \right)^k \left[\sum_{n=0}^{k-1} \left\langle {k-1 \atop n} \right\rangle e^{(k-n-1)z} \right]^{-1}. \tag{3.111}$$

To show that this expression tends to zero as z goes to zero, using properties of limits, we split the evaluation of the limit up into the evaluation of the product of three separate limits. For the last term, the limit is

$$\lim_{z \to 0} \sum_{n=0}^{k-1} \left\langle {k-1 \atop n} \right\rangle e^{(k-n-1)z} = \sum_{n=0}^{k-1} \left\langle {k-1 \atop n} \right\rangle = (k-1)!. \tag{3.112}$$

For the middle term

$$\lim_{z \to 0} \left(\frac{e^z(e^z - 1)}{z} \right)^k = \left[\lim_{z \to 0} \left(\frac{e^z(e^z - 1)}{z} \right) \right]^k = 1, \qquad (3.113)$$

while for the first term $\lim_{z \to 0} z = 0$, thus confirming the asymptotic behavior of Eq. (3.101) as

$$\frac{|\Phi l_s(z) - S_N(z)|}{|\phi_N(z)|} \to 0 \text{ as } z \to 0. \qquad (3.114)$$

As an example, if the first two terms in the asymptotic expansion for the radiometric case are taken, one has for the fractional function

$$\Phi l_4(z) \sim 1 - \frac{15}{\pi^4} \left[\frac{z^4}{4(e^z - 1)} + \frac{z^5 e^z}{20(e^z - 1)^2} \right]. \qquad (3.115)$$

Values for the above asymptotic expression taken at five different values for arguments close to zero are given in Table 3.3. By way of comparison, exact values for the function $\Phi l_4(z)$ to nine significant figures are also given. While it is clear for values of z very close to zero the two term asymptotic expansion is able to achieve at least four significant figure accuracy, compared to the nine significant figure accuracy obtainable from summing the first one or two terms in the Bernoulli series form, there is really no comparison as to which is to be preferred. The Bernoulli series form also manages to achieve its accuracy using an expression that is far simpler in form compared to the asymptotic expression for the fractional function and probably explains why the latter is rarely, if ever, employed in practice.

3.2.5   OTHER METHODS

Besides the two infinite series expansions, Gauss–Laguerre methods, and an asymptotic method presented so far, several other methods that ultimately proved to be less popular have from time to time been put forward as alternatives to evaluating the fractional function of the first kind. When Zanstra first encountered the integral for the actinometric case in his work on planetary nebulae in 1927 he evaluated it using standard quadrature methods [729]. Accuracy obtained was only between one and two per cent. A few years later he realized that integrals of this type could be evaluated far more quickly and to far greater accuracy using infinite series expansions [730, 731]. Forrest R. Gilmore in 1956 calculated the integral numerically using Simpson's rule [258]. Surprisingly, several of the values he found were checked against those found using an infinite series. For all values checked in this way, agreement to five significant figures was found. A third method, where Chebyshev[18] polynomials were first fitted to Planck's function before being integrated, was used by Marcus Hatch [283]. Values found using this approach were checked by comparing them against those obtained using Weddle's rule for numerical

Table 3.3
Table of values for the radiometric fractional function comparing those obtained exactly with those obtained using the first two terms from its asymptotic expansion. All calculated values are given correct to nine significant figures.

z	$\Phi l_4(z)$	
	asymptotic	exact
0.1	0.999 955 702	0.999 950 569
0.25	0.999 350 858	0.999 270 656
0.5	0.995 348 394	0.994 706 840
1.0	0.970 506 676	0.965 382 309
2.0	0.858 992 334	0.818 855 317

integration [186]. As noted by Johnson and Branstetter [335], for wavelength–temperature products less than a few hundred micrometer–Kelvin, the errors found from using such an approximation are large, only becoming small when $\lambda T \lesssim 5000\,\mu\text{m}\cdot\text{K}$. A final method, used by Gonzalo Páez and Marija Strojnik Scholl, was similar to that used by Hatch in that it consisted of first approximating Planck's equation before being integrated [472, 473, 474]. They applied a truncated series approximation to Planck's equation, termed a *compensated* approximation. Their approximation to Planck's equation was compensated in the sense the number of terms used in the truncated series was adjusted according to the size of an error term with the error term being included in the series summation.

Notes

[1] Recall, as the fractional function of the second kind can be expressed in terms of the fractional function of the first kind there is no need to consider each function separately. For the relation betwen the two fractional functions see page 75.

[2] Only the spectral radiant exitance will be considered as identical approximations to the spectral photon exitance can be made.

[3] For derivations of each law along lines which closely follow those of the original, good accounts can be found in [194, 707, 559, 529, 478, 580, 288].

[4] The theorem of equipartition of energy states that molecules in thermal equilibrium have the same average energy associated with each independent degree of freedom of their motion. It is a classical result of statistical mechanics and begins to break down once quantum effects become important.

[5] The quantity Rayleigh, and later Jeans, actually calculated was the spectral radiant energy per unit volume for blackbody radiation. For a blackbody, this so-called spectral

radiant energy density, denoted by $w_{e,\lambda}^b(\lambda, T)$, is related to the spectral radiant exitance by

$$M_{e,\lambda}^b(\lambda, T) = \frac{c}{4} w_{e,\lambda}^b(\lambda, T). \tag{3.116}$$

[6] In Paul Ehrenfest's paper "Welche Züge der Lichtquantenhypothese spielen in der Theorie der Wärmestrahlung eine wesentliche Rolle?," of 1911, part four of section one is titled "Die Vermeidung der Rayleigh–Jeans-Katastrophe im Ultravioletten" (Avoiding the Rayleigh-Jeans catastrophe in the ultraviolet) and is where the term "ultraviolet catastrophe" originates from.

[7] One can see the first and second radiation constants c_1 and c_2 are equal to $2\pi ch^2$ and hc/k_B respectively. The 2014 adjusted values recommended by CODATA for international use for these two constants are $3.741\,771\,790 \times 10^{-16}$ W·m^2 and $1.438\,777\,36 \times 10^{-2}$ m·K respectively.

[8] A function is said to be *elementary* if it is a real function (of one variable) built up using a finite combination of elementary operations (addition, subtraction, multiplication, division, and root extraction) and compositions on the following class of functions: constant, algebraic, the exponential, the logarithm, the six basic trigonometric functions and their inverses, and the six basic hyperbolic functions and their inverses.

[9] As the sequence of functions $f_n(x) = x^{s-1}e^{-nx} > 0$ for all $x > 0, n \geqslant 1$, by Tonelli's theorem the interchange of the order of the integration and summation is permissible. For further discussion see end note 21 of Chapter 2.

[10] Rather than use another single Latin or Greek letter for these polynomials, to better distinguish them from the countless other named sets of polynomials which are denoted using a single letter the Cyrillic letter "e oborotnoye" or "backwards e" has been used. The reason for the use of the adjective exponential in their name is as follows. From their definition the exponential E polynomials can be seen to be related to what in the past have been called the *exponential sum function* $\exp_n(x)$ [678]. Here

$$Э_n(x) = \sum_{k=0}^{n} \frac{n!}{k!} x^k = n! \sum_{k=0}^{n} \frac{x^k}{k!} = n! \exp_n(x). \tag{3.117}$$

It should be obvious that the exponential sum function is equal to a truncated Maclaurin series for the exponential function after n terms.

[11] GNU OCTAVE is a high-level programming language intended primarily for numerical computations [200, 201]. It is one of two open-source alternatives to MATLAB, MATLAB being perhaps the most widely used commercial software package of this type.

[12] In Drysdale's paper from 1907, the result appears as Eq. (2) on page 573. There is however a small misprint in the equation given. A factor of $1/\lambda$ where λ corresponds to wavelength is missing from the third term appearing inside the curly bracket term.

[13] In mathematics a quadrature rule is a method used for calculating a definite integral numerically.

[14] In obtaining the second, derivative, free expression in Eq. (3.84) for the weights from the first, the following two recurrence relations satisfied by the Laguerre polynomials can be used [51]

$$(n+1)L_{n+1}(x) = (2n+1-x)L_n(x) - nL_{n-1}(x), \tag{3.118}$$

and

$$x\frac{d}{dx}[L_n(x)] = nL_n(x) - nL_{n-1}(x). \tag{3.119}$$

[15] On the simplicity of the Gauss–Laguerre method, shortly after it was first described the second author recalls trying to explain it to William L. Wolfe. As principal editor of the forthcoming text *The infrared handbook* which would become the sucessor to the widely regarded infrared text *Handbook of military infrared technology* from 1965, Wolfe made many of the decisions concerning what material to include in the updated edition and what to leave out. Despite repeated attempts by the second author to convince Wolfe for its inclusion in the forthcoming *Handbook*, even though very accurate results from minimal computational effort could be obtained, it never was.

[16] Asymptotic equivalence is a notion of functions eventually becoming essentially equal as their arguments become either very small or very big. More formally, let f and g be functions of a real variable. We say that f and g are asymptotically equivalent if either of the limits $\lim_{x \to 0} \frac{f(x)}{g(x)}$ or $\lim_{x \to \infty} \frac{f(x)}{g(x)}$ exist and are equal to unity with the asymptotic equivalence being denoted by $f \sim g$.

[17] The sum identity for Eulerian numbers is

$$\sum_{n=0}^{k} \left\langle {k \atop n} \right\rangle = k! \tag{3.120}$$

[18] Alternative transliterations of Chebyshev's name are: Chebychev, Chebysheff, Chebyshov; or Tchebychev, Tchebycheff (mainly French transcriptions); or Tschebyschev, Tschebyschef, Tschebyscheff (mainly German transcriptions).

4 Blackbody sources and basic radiometry

This chapter provides a brief discussion on the realization of blackbody sources and the basic elements of radiometric computations. A more in-depth treatment will not be given here, but can be found elsewhere [112, 476, 115, 360, 716, 185, 702, 364, 455, 715, 668, 95]. Our purpose of including methods of radiometric computations is to provide basic guidance to the reader in properly determining the propagation of optical radiation from its source to the receptor. Influences affecting propagation of the radiation such as the atmosphere and the calculation of optical transmittance for various materials are not covered. Instead, their affect will be included in calculations using a simple transmittance factor.

4.1 BLACKBODY SOURCES

Blackbody sources are primarily used as calibration sources and have and continue to be the subject of intense research [699]. Most real bodies encountered in practice are not perfectly black. To account for the departure of a body from blackness, a parameter known as the emissivity is introduced. Denoted by ε, it is defined as the ratio of the total radiant exitance of a body to that of a blackbody at the same temperature. It is a dimensionless quantity between zero and one, and is a measure of a body's radiating (and absorbing) efficiency. A spectral emissivity $\varepsilon(\lambda, T)$ at each temperature of the body as a function of wavelength can also be defined. In the *greybody* approximation the spectral emissivity is independent of the wavelength and temperature of the object. In such cases all radiometric and actinometric quantities corresponding to a greybody will be reduced by a constant factor equal to the body's emissivity compared to those for a blackbody. For example, the total radiant exitance and the total photon exitance for a greybody are given by

$$M_e(T) = \varepsilon\sigma T^4 \quad \text{and} \quad M_q(T) = \varepsilon\sigma_q T^3, \tag{4.1}$$

respectively.

The two basic types of experimental blackbody sources are the cavity type and the flat plate type. The fundamental design objectives for each type is the same. Each needs to be designed to achieve: (i) temperatures that are accurate, remain uniform, and are stable, (ii) high emissivity with minimal spectral variation, and (iii) Lambertian behavior over a specified field-of-view which we describe shortly. In 1898 the first description of a cavity radiation source was published. It comprised an aperture in the wall of a platinum box, divided by diaphragms, with its interior blackened with iron oxide [408]. As already mentioned in Chapter 1 (see page 13), the experimental realization of blackbody sources played a significant role in helping to develop our understanding of the radiation emitted from blackbody sources that ultimately led Planck to formulating his now famous law. Cavity sources are often designed for specialized applications requiring radiometric sources for cryogenic temperatures [137], near-ambient temperatures [346], and very-high temperatures [48]. The mechanical and electrical construction of these sources is often quite complex, particularly so when one is striving to achieve performance of the highest quality.

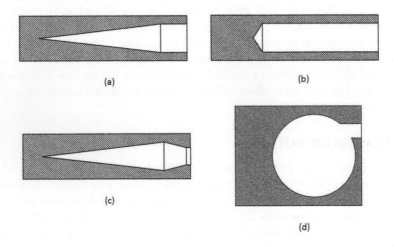

Figure 4.1 Representative cavity geometries for blackbody sources. Each is known as: (a) conical cavity, (b) cylindrical cavity, (c) reverse conical cavity, and (d) off-axis spherical cavity.

Cavity type blackbody sources are very common and provide excellent performance. A large variety of cavity configurations have been utilized in an attempt to achieve high emissivity and blackbody behavior. In general, cavities are reasonably easy to fabricate. Figure 4.1 depicts the cross-sectional view of four representative cavity geometries. Major challenges in constructing a blackbody source are in obtaining high emissivity cavity surfaces and in achieving uniform surface temperatures. Cavity geometry can also significantly impact the radiometric behavior of a blackbody source, as can the cavity surface "roughness."

Cavity type blackbody sources are limited by the physical size of the source and output aperture diameter (less than a few centimeters), which make them impractical to use for applications needing large area sources. Although producing somewhat less spatially-uniform radiance, a term to be defined shortly, than a cavity source, flat panel type sources can generally meet the large-area source requirements for many applications. Typically the plate is made from aluminium with heating elements affixed to the underside of the plate with one or more temperature sensors embedded in the plate. The front side of the plate is machined, chemically etched, or bead blasted to create a very rough surface which results in the surface emissivity being significantly increased compared to the bare aluminium surface. Machine methods include cutting relatively fine circular, parallel, or grid grooves into the surface, followed by coating the surface with a material that appears black in the spectral region of interest. A variety of coatings can be used on flat or machined surfaces. Lamp black, a black pigment made from soot, is one of the oldest used black coatings, but suffers certain durability problems. Other more modern black coatings with significantly better performance compared to traditional lamp black include *Martin Black* or *Avian DS Black* which have been shown to provide excellent low reflectance, durability, and low outgassing [525]. Other ultra-black coatings have been developed such as the *Acktar Nano Black* optical coating that incorporates carbon nanotubes. Surfaces which are almost Lambertian with emissivities of the order of 0.98 [336] in the infrared part of the spectrum can be made by bead blasting an aluminium surface with a variety of bead sizes and then applying a spray coating to the surface with a titanium-dioxide rich paint.

4.2 GONIOMETRIC CHARACTERISTICS OF SURFACES

In discussing the emission of radiation from a source, measurements related to angles play an important role. The *goniometric*[1] characteristics of surfaces therefore play an important role in radiometry and it is these concepts related to the idea of the measure of angle we wish to briefly introduce here.

In order to utilize Plank's equation in any practical application, an understanding of the concept of *solid angle* is necessary. As illustrated in Fig. 4.2 the solid angle, denoted by Ω, is defined to be equal to the surface area S of a section of the sphere having radius R, half-cone angle θ, and center C, divided by R^2. The solid angle can therefore be thought of as the three-dimensional analogue of the familiar two-dimensional angle in the plane. A method for calculating the solid angle can be developed by referring to Fig. 4.3 which shows the cross-section of Fig. 4.2 and the parameters used to determine the solid angle. Since the surface area of a sphere is given by $4\pi R^2$, the area of a hemisphere is $2\pi R^2$. Now

$$\frac{S}{2\pi R^2} = \frac{Q}{P+Q}, \tag{4.2}$$

and

$$\overline{CA} = P + Q = R. \tag{4.3}$$

Here P and Q are distances (lengths) as indicated in Fig. 4.3 while $\overline{CA}$ is the distance from the center C of the circle to the point A on its circumference.

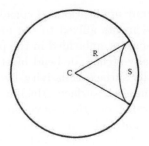

Figure 4.2 Geometric definition of solid angle.

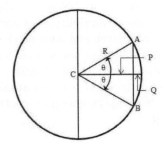

Figure 4.3 Parameters to determine solid angle.

Thus

$$S = 2\pi RQ = 2\pi R(R - P). \tag{4.4}$$

From Fig. 4.3, as $P = R\cos\theta$ the surface area of the section of the sphere S can be given by $S = 2\pi R^2(1 - \cos\theta)$. Recalling the trigonometric identity $(1-\cos\theta) = 2\sin^2(\theta/2)$, the *unweighted* solid angle can therefore be expressed as

$$\Omega = 4\pi \sin^2\left(\frac{\theta}{2}\right), \tag{4.5}$$

and on differentiating this expression with respect to θ, the differential solid angle is given by

$$d\Omega = 2\pi \sin\theta \, d\theta. \tag{4.6}$$

The radiation pattern for a point or isotropic source is illustrated in Fig. 4.4. The flux per unit solid angle, or *radiant intensity*, I_e is a constant value in any direction. If the source is a flat extended body, as shown in Fig. 4.5, the radiant exitance M_e describes the radiant flux per unit area leaving the surface. However, it gives no information about the directional dependence of the emitted radiation. For this, one introduces the concept of radiance L_e. For a perfectly diffuse or Lambertian radiator such as a blackbody, when the

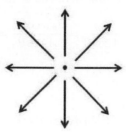

Figure 4.4   Radiation pattern for a point source.

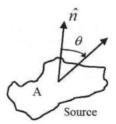

Figure 4.5   Illustration of Lambert's Law.

total radiance is summed over all possible space the radiation can be emitted into, that is over a hemispherical envelope in space above the surface of the body, the relationship between the total radiant exitance M_e and the total radiance L_e is found to be

$$L_e(T) = \frac{1}{\pi} M_e(T). \tag{4.7}$$

As a directional quantity it represents the part of the total radiant exitance falling into a certain section in space (unit area per unit solid angle) and has the units of watt per square meter per steradian [W·m^{-2}·sr^{-1}]. Johann Heinrich Lambert was the first to realize the radiance from such an ideal surface would be the same regardless of the view angle. For this to be true, the radiant intensity must behave as

$$I_{e,\theta} = I_{e,0} \cos \theta. \tag{4.8}$$

Here the $\cos \theta$ term accounts for the projected area of the radiator when viewed from a direction θ as illustrated in Fig. 4.5, and is known as Lambert's cosine law after Lambert, who first deduced it in 1760. Thus

$$L_{e,\theta} = \frac{I_{e,\theta}}{A\cos\theta} = \frac{I_{e,0}\cos\theta}{A\cos\theta} = \frac{I_{e,0}}{A}, \qquad (4.9)$$

and is constant. For a Lambertian source of area A, the differential solid angle is weighted by $\cos\theta$ so that

$$\Omega = \int_0^\theta 2\pi\sin\theta\cos\theta\,d\theta = \pi\sin^2\theta. \qquad (4.10)$$

It should be noted that the solid angle for a hemisphere for a Lambert surface is π, not 2π, as in the case for a point source [465, 67].[2] A special case occurs when the source is spherical. In this instance the projected area is constant and one has $\Omega = 4\pi\sin^2(\theta/2)$ which, for the hemisphere (here $\theta = \pi/2$), gives 2π.

Now let an element of source area of a blackbody radiator having absolute temperature T be designated dA_s as shown in Fig. 4.6. To determine the total radiant flux or power radiated into the hemisphere shown, the first step is to compute the differential solid angle $d\Omega_C$, which is given by

$$d\Omega_C = \frac{dA_C}{R^2} = \sin\theta\,d\theta\,d\phi. \qquad (4.11)$$

The spectral radiance of a blackbody $L_{e,\lambda}^b(\lambda, T)$ is given by Planck's law. The total radiant exitance of the source over the hemisphere is found by multiplying the spectral radiance by the cosine-weighted differential solid angle $\cos\theta\,d\Omega_C$ and integrating over the range of allowable angles of $\theta \in [0, \pi/2]$

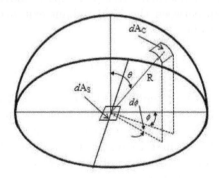

Figure 4.6 Determination of total power radiated into a hemisphere.

and $\phi \in [0, 2\pi]$. Doing so yields

$$M_e^b(T) = \int_0^\infty \int_0^{2\pi} \int_0^{\pi/2} L_{e,\lambda}^b(\lambda, T) \cos\theta \sin\theta \, d\theta \, d\phi \, d\lambda = \pi \int_0^\infty L_{e,\lambda}^b(\lambda, T) \, d\lambda.$$
(4.12)

Notice here the factor of π instead of 2π. The total radiance for a blackbody is given by

$$L_e^b(T) = \int_0^\infty L_{e,\lambda}^b(\lambda, T) \, d\lambda.$$
(4.13)

Recalling the Stefan-Boltzmann law, which gives the total power radiated per unit area from a blackbody at a specific temperature over all wavelengths, and is given by $M_e^b(T) = \sigma T^4$, from Eq. (4.12) one sees the total radiance of a blackbody is given by

$$L_e^b(T) = \frac{\sigma T^4}{\pi}.$$
(4.14)

If the source is non-Lambertian and the spectral emissivity is not constant then an appropriate dimensionless weighting function $w(\theta, \phi)$ needs to be applied to the differential solid angle. In this case the total radiant exitance becomes

$$M_e(T) = \int_0^\infty \int_0^{2\pi} \int_0^{\pi/2} \varepsilon(\lambda, T) L_\lambda^b(\lambda, T) w(\theta, \phi) \sin\theta \, d\theta \, d\phi \, d\lambda.$$
(4.15)

4.3 INVERSE SQUARE LAW

The inverse square law is of fundamental importance in electromagnetic propagation and plays an important role in radiometry, radar, lidar, and antennas. A basic statement of the inverse square law for radiometry is that the total irradiance E_e incident upon the surface of a sphere centered on a point source having radiant intensity I_e is given by the famous inverse square law of

$$E_e = \frac{I_e}{R^2}.$$
(4.16)

Irradiance is therefore a measure of the radiant flux received by a surface per unit area.

Assuming the element of area dA is small compared to the radius squared, dA can be thought of as being essentially a plane surface tangent to the sphere about the point source and the normal to the surface of dA passes through the point source. From Eq. (4.16) the irradiance at dA on-axis is

$$E_{\text{axis}} = \frac{I_e}{R^2}.$$
(4.17)

Letting this tangent plane be the observation plane, as shown in Fig. 4.7, the irradiance $E_{e,\theta}$ at dA when located at an angle θ, is equal to the radiant

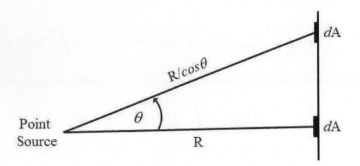

Figure 4.7 Irradiance at a plane due to a point source.

intensity multiplied by the solid angle subtended by dA when viewed from the source divided by the projected area of dA. This off-axis irradiance is therefore given by

$$E_{e,\theta} = I(dA\cos\theta) \cdot \frac{\cos^2\theta}{R^2} \cdot \frac{1}{dA} = \frac{I}{R^2}\cos^3\theta. \qquad (4.18)$$

If the source is a small extended object of area A_S and has a radiance of L_S, the irradiance at the observation plane, depicted in Fig. 4.8, is the radiance of the source multiplied by the solid angle subtended by the source when viewed from area A_0 in the observation plane, that is

$$E_{e,0} = L_{e,S}\left(\frac{A_S}{R^2}\right). \qquad (4.19)$$

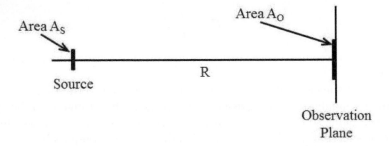

Figure 4.8 Irradiance at an observational plane due to a finite size source parallel to the observation plane.

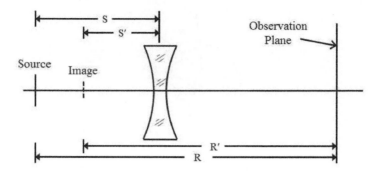

Figure 4.9 Effect of a lens on the inverse square law.

Now consider the effect on the inverse square law if a lens having a constant optical transmittance τ_0 is placed between the source and the observation plane as illustrated in Fig. 4.9. With the source located at distance S from the source, the lens forms an image of the source at a distance S' which can be readily found using a standard imaging equation such as the Newtonian imaging equation [353]. The area of the image is $A_I = A_S(S'/S)^2$. Without the lens, the source is located at distance R from the observation plane while with the lens the image of the source appears to be at a distance $R' = R + S' - S$ from the observation plane. The irradiance at the observation plane will therefore be the product between the source radiance, the optical transmittance of the lens, and the solid angle subtended by the source image. Consequently, the change in irradiance at the observation plane with and without the lens is expressed by

$$\frac{E_{e,\text{with lens}}}{E_{e,\text{without lens}}} = \tau_0 \frac{\left(\dfrac{A_I}{R'^2}\right)}{\left(\dfrac{A_S}{R^2}\right)} = \tau_0 \frac{\left(\dfrac{A_I}{R'^2}\right)}{\left(\dfrac{A_I S^2}{R^2 S'^2}\right)} = \tau_0 \left(\frac{RS'}{R'S}\right)^2. \tag{4.20}$$

4.4 EXTENDED SOURCE RADIOMETRY

Consider an extended planar source with radiance L_e having a surface spatially described by S with $\hat{n}$ as a unit normal to the surface. From Fig. 4.10 the apparent radiant intensity in the direction θ due to an element of area dA is given by

$$dI_{e,\theta} = L_e \cos \theta \, dA, \tag{4.21}$$

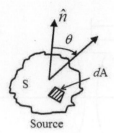

Figure 4.10 Geometry for computing apparent radiant intensity for an extended source.

while the total apparent radiant intensity from this source will be

$$I_{e,\theta} = \int_S L_e \cos\theta \, dA. \tag{4.22}$$

Figure 4.11 shows a small receptor and a small source with radiance $L_{e,1}$ at arbitrary angles with respect to one another. An element of radiant flux incident at the receptor is the source radiance multiplied by the projected area of the source viewed by the receptor, multiplied by the solid angle subtended by the receptor when viewed from the source. Thus

$$d\Phi_2 = L_{e,1} \cdot dA_1 \cos\theta \cdot \frac{dA_2 \cos\phi}{R^2}, \tag{4.23}$$

while an element of irradiance at the receptor will be

$$dE_{e,2} = \frac{d\Phi_2}{dA_2} = \frac{L_{e,1} dA_1 \cos\theta \cos\phi}{R^2}. \tag{4.24}$$

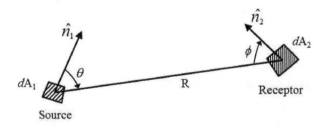

Figure 4.11 Irradiance at a small receptor due to a small source at arbitrary angles with respect to one another.

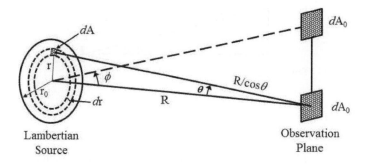

Figure 4.12 Extended source radiometry.

An important case in radiometry is the determination of irradiance at the observation plane due to a large circular Lambertian source having radiance L_e, as illustrated in Fig. 4.12. First consider an element of area of the receptor dA_0 which is parallel to the source plane and centered on the source normal vector. Recall that the source radiant intensity for an area dA in the direction θ is $I_{e,\theta} = I_{e,0} \cos \theta = L_e \cos \theta \, dA$. The differential irradiance at dA_0 due to the source element dA is

$$dE_e = I_{e,0} \cos \theta \left(\frac{\cos^3 \theta}{R^2} \right) = \frac{L_e dA \cos^4 \theta}{R^2}. \tag{4.25}$$

Notice the fourth-order power dependency on the cosine term. One of the cosine terms is associated with the projected area of dA, another with the projected area of dA_0, while the remaining two cosine terms are related to the distance between dA and dA_0, that is $R/\cos \theta$. An element of irradiance at dA_0 due to a ring on the source of width dr at a radius of r is given by

$$dE_{e,r} = \frac{2\pi r L_e \cos^4 \theta \, dr}{R^2}. \tag{4.26}$$

Recognizing $r = R \tan \theta$ and $dr = R \sec^2 \theta \, d\theta$, this can be rewritten as

$$dE_{e,r} = \frac{2\pi R \tan \theta \, R \sec^2 \theta \, d\theta L_e \cos^4 \theta}{R^2} = 2\pi L_e \sin \theta \cos \theta \, d\theta. \tag{4.27}$$

The total on-axis irradiance at dA_0 due to the large circular Lambertian source is found after summing (integrating) over all angles θ, namely

$$E_{e,0} = \int_0^{\theta_S} 2\pi L_e \sin \theta \cos \theta \, d\theta = 2\pi L_e \left[\frac{\sin^2 \theta}{2} \right]_0^{\theta_S} = \pi L_e \sin^2 \theta_S, \tag{4.28}$$

where θ_S is the angular radius of the source viewed from the center of the observation plane. The off-axis irradiance at the receptor located at an angle

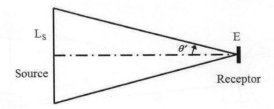

Figure 4.13 Irradiance at receptor due to a circular source of angular radius θ'.

ϕ viewed from the source is

$$E_{e,\phi} = E_{e,0} \cos^4 \phi. \tag{4.29}$$

Here two of the cosine terms in this last expression are associated with the projected areas of the source and receptor, and the two other cosine terms result from the distance between the source and the receptor.

The irradiance from a circular disk Lambertian source has been shown to be expressed by Eq. (4.28). Figure 4.13 provides a simplistic view of this arrangement. The irradiance at the receptor is

$$E_e = \pi L_{e,S} \sin^2 \theta'. \tag{4.30}$$

Consider now the circular hole is place at some arbitrary distance in front of a large uniform extended source having radiance $L_{e,S}$ where the source is much larger than the hole when viewed from the receptor as illustrated in Fig. 4.14. In this case, the solid angle subtended by the source is determined by the size of the circular hole and its distance from the receptor. The angular radius of the circular hole is θ', so the irradiance at the receptor will be

$$E_e = \pi L_{e,S} \sin^2 \theta', \tag{4.31}$$

which is the same as the case shown in Fig. 4.13. The distance between the source and the circular hole is immaterial as long as the angular radius of the source is greater than θ'. This is the principle utilized by classical light meters since the output of its detector will be directly proportional to the source radiance (see Eq. (4.44)).

If an optical component, window or lens, is placed between the source and the circular hole, as depicted in Fig. 4.15, and the image of the source maintains an angular radius greater than θ' when viewed from the receptor, the only effect upon the irradiance at the receptor is to reduce it by a factor equal to the optical transmittance of the optical component, that is

$$E_e = \tau_0 \pi L_{e,S} \sin^2 \theta'. \tag{4.32}$$

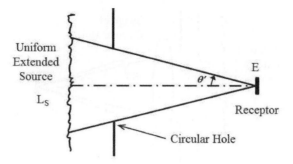

Figure 4.14 Irradiance at receptor due to an extended source where the solid angle is determined by a circular hole.

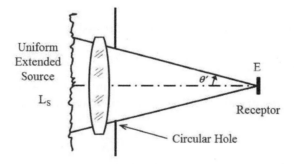

Figure 4.15 Irradiance at receptor when a lens is located between extended source and circular hole.

Should the optical component, window or lens, be placed between the the circular hole and the receptor, the situation becomes somewhat different. The lens forms an image of the circular hole as illustrated in Fig. 4.16. The angular subtense of the image of the circular hole θ'' may be larger or smaller than θ'. The irradiance at the receptor is determined by θ'' rather than θ' and is again reduced by the optical transmittance of the optical component, namely

$$E_e = \tau_0 \pi L_{e,S} \sin^2 \theta''. \tag{4.33}$$

4.5   RADIOMETRY OF IMAGES

Most radiometric systems incorporate imaging optics. Two related parameters of a lens or the imaging optics that are often used in describing radiometers are the *F-number* (focal ratio) and *numerical aperture* NA. As depicted in

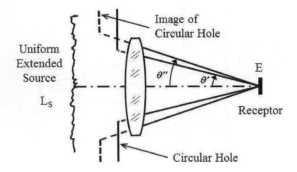

Figure 4.16 Irradiance at receptor when a lens is located between a circular hole and receptor.

Fig. 4.17, the numerical aperture can be expressed as

$$\text{NA} = n \sin \theta. \tag{4.34}$$

Here n is the refractive index of the medium in the image or detector space. If the lens is *aplanatic*, that is free of spherical aberration and linear coma, then

$$\text{F-number} = \frac{1}{2\text{NA}}. \tag{4.35}$$

When a lens is used at finite magnification m, the effective F-number is modified by a factor $(1 - m)$, where m equals the image height divided by the objective height. For example, a positive lens having an F-number of 2 for an object at infinity when used to form an inverted image of an object such that the image is one-quarter the object height. The magnification is therefore $-\frac{1}{4}$ where the minus sign indicates an inverted image. Consequently, the effective F-number will be $2(1 + \frac{1}{4}) = 2.5$.

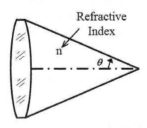

Figure 4.17 Numerical aperture of a lens, $\text{NA} = n \sin \theta$.

One of the most widely used applications of radiometric principles is determining the radiometric values at the image of a source formed by an optical system such as, for example, a telescope, microscope, camera, or infrared sensor, regardless of whether the receptor is film, an electronic detector, or the eye. Figure 4.18 shows the optical behavior of an aplanatic lens system. Carl Friedrich Gauss (1777–1855) showed in 1841 that a complex optical system can be represented by its principal surfaces and focal points [659, 334].[3] When the lens is well corrected for aberrations, it is generally aplanatic. Since the lens is aplanatic, the principal surfaces are spherical and centered on the source and image at source S and image S', respectively, and its magnification will be

$$m = -\frac{S'}{S} = -\sqrt{\frac{A'}{A}}.$$ (4.36)

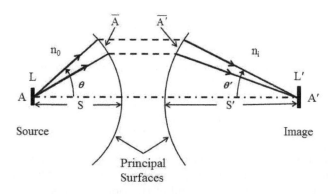

Figure 4.18   Radiometry of an aplanatic lens.

The radiant flux incident at $\overline{A}$ is $L_e A \cos\theta (\overline{A}/S^2)$ where $\overline{A}/S^2$ is the solid angle subtended by $\overline{A}$ on the anterior or front principal surface. Since the principal surfaces are images of each other with unit lateral magnification, $\overline{A}\cos\theta = \overline{A}'\cos\theta'$. The radiant flux incident at $\overline{A}$ that reaches the image is reduced only by the optical transmittance of the optical system. Hence the radiance of the image is

$$L_e' = \tau_0 \cdot \frac{L_e \overline{A} \cos\theta A}{S^2} \cdot \frac{S'^2}{\overline{A}' A' \cos\theta'}.$$ (4.37)

Now, $A = m^2 A' = S'^2 A'/S^2$, or alternatively, $AS^2 = A'S'^2$ so Eq. (4.37) becomes

$$L_e' = \tau_0 L_e,$$ (4.38)

for $n_0 = n_1$. This is a very important result: the radiance of the image and the source are the same if one ignores the transmittance factor. In cases where

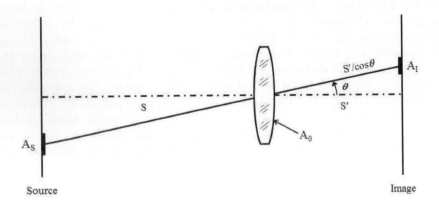

Figure 4.19 Off-axis irradiance for imaging systems.

the refractive index in which the source and image lie are not the same, the radiance of the image is given by

$$L'_e = \tau_0 \left(\frac{n_i}{n_0}\right)^2 L_e. \tag{4.39}$$

For the typical case where the source and the image lie in the same medium, the irradiance on-axis at the image plane is $E_e = \tau_0 L_e \pi \sin^2 \theta'_0$ where θ'_0 is the half-angle subtended by the exit pupil of the optical system.

Figure 4.19 shows the geometry for determining the off-axis irradiance for a generic optical system represented by a lens [554, 606]. The power incident at a lens having area A_0 due to a source having area A_S and radiance $L_{e,S}$ is

$$\Phi_0(\theta) = L_{e,S} \left(\frac{A_S \cos^3 \theta}{S^2}\right) A_0 \cos \theta = \frac{L_{e,S} A_S A_0}{S^2} \cos^4 \theta. \tag{4.40}$$

The radiant flux reaching the image of area A_I is given by

$$\Phi_{e,I}(\theta) = \frac{\tau_0 L_{e,S} A_I \cos \theta A_0 \cos^3 \theta}{S'^2}. \tag{4.41}$$

The first cosine term is from the projected area of the image A_I, the second cosine term is associated with the projected area of the lens, while the other two cosine terms arise from the distance between the lens and the off-axis image. The irradiance at the image is therefore

$$E_{e,I}(\theta) = \frac{\Phi_{e,I}(\theta)}{A_I} = \tau_0 L_{e,S} \left(\frac{A_0 \cos^4 \theta}{S'^2}\right), \tag{4.42}$$

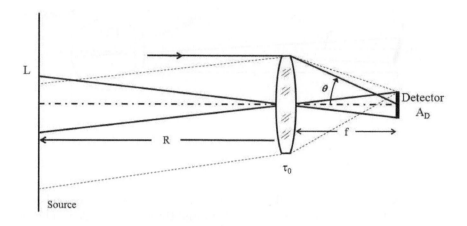

Figure 4.20 Imaging extended sources larger than the field-of-view of the detector.

where the term in brackets is the effective solid angle of the lens viewed from the image. Notice that the irradiance in Eq. (4.42) is the radiance of the image (see Eq. (4.38)) multiplied by the aforementioned effective solid angle.

Figure 4.20 shows an optical system imaging an extended source which is larger than the field-of-view of the detector. The detector is located at the focal point of the lens and the field-of-view is considered to be determined by the principal rays passing through the center of the lens and the top and bottom of the detector. The total footprint of the detector projected onto the source is determined by the size of the lens, focal length, and detector size. The dashed lines in the figure are the marginal rays that define the outer boundaries of the detector footprint. Given that the source, having radiance L_e, has a spatial extent that exceeds the field-of-view of the detector, it is evident that

$$L_{e,\text{detector}} = \tau_0 L_{e,\text{source}} \quad \text{and} \quad \Phi_{e,\text{detector}} = \tau_0 L_{e,\text{source}} \pi \sin^2 \theta A_0. \quad (4.43)$$

Since the source overfills the detector for all source-lens separations R, the radiometric power at the detector remains constant. Thus the radiance can be measured directly as

$$L_{e,\text{source}} = \frac{\Phi_{e,\text{detector}}}{\tau_0 \pi \sin^2 \theta A_D}. \quad (4.44)$$

The radiant flux measured is directly proportional to the source radiance. This system is the basic configuration for light meters and radiometers. It should be noted that if the source is Lambertian, then the measured radiance remains independent of the tilt of the source with respect to the optical system.

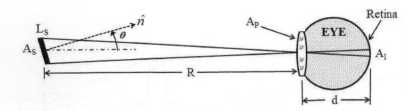

Figure 4.21   Visual radiometry.

A simple model of an eye, illustrated in Fig. 4.21, is viewing a Lambertian source of radiance $L_{e,S}$ having an area of A_S and tilted at an angle θ. The area of the image A_I is the projected area of the source times the magnification squared, namely

$$A_I = A_S \cos\theta \frac{d^2}{R^2}. \tag{4.45}$$

The irradiance at the image is the flux collected by the pupil of the eye divided by A_I. The flux at the pupil is

$$\Phi_{e,\text{pupil}} = L_{e,S} A_P \left(\frac{A_S \cos\theta}{R^2} \right), \tag{4.46}$$

where the term in the parentheses is the effective solid angle of the source viewed from the pupil of the eye. It follows that the image irradiance is given by

$$E_{e,I} = \frac{\Phi_{e,\text{pupil}}}{A_I} = L_{e,S} A_P \left(\frac{A_S \cos\theta}{R^2} \right) \left(\frac{R^2}{A_S d^2 \cos\theta} \right) = \frac{L_{e,S} A_P}{d^2}. \tag{4.47}$$

Therefore the image irradiance is the source radiance times the solid angle of the lens pupil viewed from the image. The optical transmittance τ_0 is assumed to be unity, otherwise $E_{e,I}$ should be multiplied by τ_0.

A very common optical instrument is a telescope. A Keplerian telescope, invented in 1611 by Johannes Kepler (1571–1630), is shown in Fig. 4.22. The lenses are arranged such that the focal points of the objective lens, having focal length of f_1, and the eye lens, having focal length of f_2, are coincident whereby the objective lens brings collimated light incident upon it to a focus and then the eye lens recollimates the light. The diameter of the eye lens is $D_e = f_2 D_0 / f_1$. Consider now a person looking at an effective point source, such as an unresolvable object at a distance R, with an apparent radiant intensity of $I_{e,S}$. With the unaided eye, that is no telescope, the irradiance at the retina is

$$_{\text{unaided}} E_{e,\text{retina}} = I_{e,S} \left(\frac{\pi D_P^2}{4R^2} \right) \left(\frac{1}{A_r} \right), \tag{4.48}$$

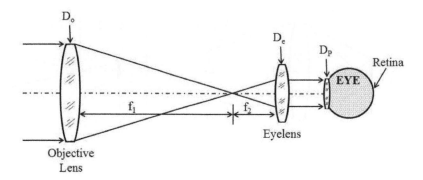

Figure 4.22 Radiometry of a telescope.

where A_r is the retinal area covered by the image of the point source and D_P is the eye pupil diameter. The term in brackets is the solid angle subtended by the eye viewed from the point source.

With the aid of the telescope and $D_P \geqslant D_e$, the irradiance at the retina is

$$D_P \geqslant D_e\, E_{\text{e,retina}} = I_{e,S}\left(\frac{\pi D_0^2}{4R^2}\right)\left(\frac{\tau_0}{A_r}\right), \qquad (4.49)$$

hence, the retinal irradiance is $(D_0/D_P)^2$ greater. When $D_P < D_e$, the eye pupil diameter is magnified by a factor of f_1/f_2 and the irradiance at the retina becomes

$$D_P < D_e\, E_{\text{e,retina}} = I_{e,S}\left(\frac{\pi D_P^2}{4R^2}\right)\left(\frac{f_1}{f_2}\right)^2\left(\frac{\tau_0}{A_r}\right), \qquad (4.50)$$

which implies that the retinal irradiance for the case with $D_P < D_e$ will be less than for the case with $D_P \geqslant D_e$. For a resolved or extended source, it should now be evident that the effect of using a telescope is to reduce the observed radiance by a factor of τ_0!

Consider an imaging system where the detector is not necessarily located at the image of the source. Figure 4.23 depicts such a situation where the detector is located at point P on the optical axis. The source is assumed to be Lambertian and, although the lens is shown to be positive, no restrictions are imposed on the sign of the lens power. Irradiance at the image has been shown to be

$$E_{\text{e,image}} = \tau_0 L_{\text{e,source}} \pi \sin^2 \theta_P, \qquad (4.51)$$

where τ_0 is the lens transmittance and θ_P is the half-angle subtended by the lens exit pupil as viewed from the image. By tracing the four marginal rays from the top and bottom of the source, as shown in Fig. 4.23, the boundaries

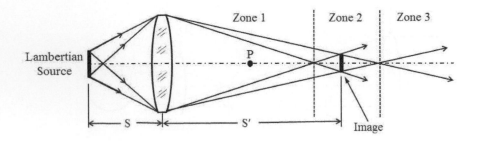

Figure 4.23 Radiometry of a lens for any axial observation point.

between regions of image space denoted as Zones 1–3 are defined where these rays cross the optical axis. The irradiance at point P is in general

$$E_{e,P} = \tau_0 L_{e,\text{source}} \pi \sin^2 \theta_P. \tag{4.52}$$

In Zones 1 and 3, θ_P is the half-angle subtended by the image when viewed from P. Likewise, in Zone 2, θ_P is the half-angle subtended by the lens exit pupil when viewed from P. As a rule, one selects the smaller θ_P of the image and exit pupil viewed from P.

An interesting and important application of the optical configuration shown in Fig. 4.23 occurs when the source is located at the focal point of the lens as illustrated in Fig. 4.24. The image is located at infinity. Since the image is at infinity, Zone 3 does not exist. Let the distance point P located from the lens be denoted as d. Following the general rule given above, it is obvious in Zone 1 that θ_P of the image is smaller than the angle subtended by the exit pupil. Consequently, θ_P is constant throughout Zone 1. In Zone 2, θ_P is determined by the angular subtense of the lens exit pupil. The irradiance at P varies inversely proportional to the square of d since the solid angle is the lens exit pupil area divided by d^2.

Now consider that a very distant object is imaged by a lens as illustrated in Fig. 4.25. When the principal ray from the object forms an angle θ with the optical axis, the image formed in the focal plane is at a height h'. An ideal lens follows the relationship $h' = f \tan \theta$, where f is the focal length of the lens. As has been shown previously, the irradiance decreases off-axis by $\cos^4 \theta$. An interesting question is, what distortion of the lens will produce an image having uniform irradiance? If the lens is designed to exhibit a large amount of barrel distortion such that $h' = f \sin \theta$, then the irradiance will be uniform over the image plane assuming no vignetting [352, 555].

Figure 4.26 shows a telecentric lens that has the property that principal rays exit the lens parallel to the optical axis. This behavior is accomplished by placing the aperture stop at the front focal plane. Consequently, the entrance

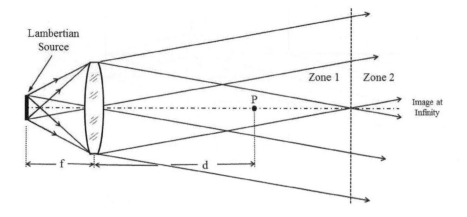

Figure 4.24 Radiometry of a searchlight or beacon for any axial observation point.

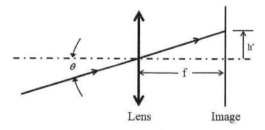

Figure 4.25 Relationship of image height and object angular height.

pupil is located at the aperture stop and the exit pupil is located to the right of the lens at infinity. Such a lens is often used in metrology for measuring the height of object features, since the center of a defocused image feature remains at the same height. A common *misconception* about telecentric lenses having zero distortion is that the irradiance at the image plane is the same regardless of the field angle [243]. In actuality, the off-axis irradiance is

$$E_{e,I}(\theta) = E_{e,I} \cos^4 \theta, \tag{4.53}$$

which is the same as with the lens shown in Fig. 4.18. In order for a telecentric lens to have uniform irradiance at the image plane, the relationship between the field angle θ and the image height must be $h' = f \sin \theta$.

It should be understood that the irradiance behavior in the image plane of a lens can be influenced by a number of factors such as distortion, image

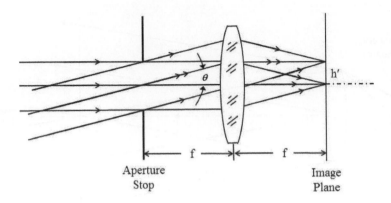

Figure 4.26   Radiometry of a telecentric lens.

aberrations, pupil aberrations, "floating" pupils often associated with very-wide angle lenses, vignetting, and fabrication errors. To account for each of these, careful modeling and analysis of the lens should be made under the conditions the lens will be used [660].

4.6   EXAMPLE PROBLEM

We now give a simple example that ties together many of the ideas presented so far. Figure 4.27 shows a system having a circular Lambertian source that irradiates a Lambert reflector with reflectance of $\rho_{\text{reflector}} = 0.7$ where the source and reflector are co-planar. A sensor comprising a lens D and detector F, which are enclosed within an optical radiation shield E, view the Lambert reflector via a mirror C. Detector F is centered on the optical axis of the lens and is located at the focal point of the lens. The following is one procedure to compute the irradiance at the detector due to a Lambertian source.

Assuming the source is a greybody at a temperature of 1000 K and emissivity of 0.8, and the detector is spectrally uniform and responsive to all wavelength (an ideal detector), the source radiance is, accounting for the emissivity, given by (see Eq. 4.14)

$$L_{\text{e,source}} = \varepsilon L_{\text{e,source}}^{\text{b}} = \frac{\varepsilon \sigma T^4}{\pi} = 1.444 \, \text{W·cm}^{-2}\text{·sr}^{-1}. \qquad (4.54)$$

The center of the source is aligned to the surface normal of the reflector at point A. The total angular subtense of the source viewed from A is 60°. The irradiance at A is $E_{\text{e},A} = \pi L_{\text{e,source}} \sin^2(30°)$ (see Eq. (4.28)). The sensor is viewing the reflector 100 cm above the center of the source. With the source being 100 cm distant from A, the view angle at point B is 45°. Following Eq.

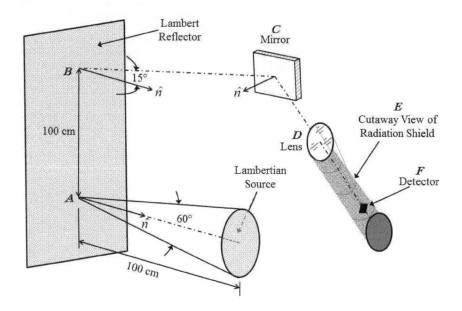

Figure 4.27 Determination of irradiance at detector due to a Lambert source. Adaptation of Figure 8.5 from Warren J. Smith, *Modern Optical Engineering*, McGraw-Hill, New York (1966).

(4.29), the irradiance at B is $E_{e,B} = E_{e,A} \cos^4(45°) = 0.283\,\text{W}\cdot\text{cm}^{-2}$ and the apparent radiance at B $L_{e,B} = \rho_{\text{reflector}} E_{e,B}/\pi = 0.063\,\text{W}\cdot\text{cm}^{-2}\cdot\text{sr}^{-1}$.

Since the reflector appears as an extended source to the detector, the reflector-to-sensor and reflector-to-detector distance is immaterial. Mirror C, having $\rho_{\text{mirror}} = 0.8$, is positioned such that the optical axis of the sensor intersects the reflector at B at an angle of 15° with respect to the surface normal $\hat{n}$. The parameters of the lens are that it has a focal length of $f = 100\,\text{mm}$, an aperture area of $A_0 = 1\,\text{cm}^2$ and $\eta_{\text{lens}} = 0.9$. As was seen in Fig. 4.13, the distance between the reflector and the lens is not important and the radiance at the detector is the same as at the reflector multiplied by the factor $\rho_{\text{mirror}}\eta_{\text{lens}}$. The solid angle subtended by the lens aperture viewed from the detector is $\Omega_{\text{lens}} = A_0/f^2 = 0.01\,\text{sr}$. Since the radiance is constant regardless of the view angle, the irradiance at the detector is

$$E_{e,\text{detector}} = \rho_{\text{mirror}}\eta_{\text{lens}}L_{e,B}\Omega_{\text{lens}} = 454\,\mu\text{·W·cm}^{-2} \qquad (4.55)$$

Notes

[1] The term *goniometry* is derived from the two Greeks words, *gonia* meaning angle and *metron*, meaning measure.

[2]Inclusion of the factor of $\cos\theta$ in the solid angle calculation results in the *projected* or *weighted* solid-angle in contrast to the unweighted solid angle as given by Eq. (4.5). See pages 62–77 of [465] for an in-depth discussion.

[3]The principal points or planes and focal points constitute four of the seven cardinal points that can be used to describe an optical system of any complexity as a first-order construction comprising the cardinal points alone. The British astronomer Joseph Harris (1702–1764) was the first to correctly specify the focal length, and discovered principal and nodal points which he presented in his *A Treatise of Optics* of 1775, while it was the Irish natural philosopher William Molyneux (1656–1698) who first invented the concept of the optical center in 1692. Gauss created and used focal and principal points to specify an optical system, while the German mathematician and physicist Johann Benedict Listing (1808–1882) extended Gauss' method to include nodal points.

Section III

Computational aids

In this section of the book we turn our attention to the diverse range of computational aids that have been developed to assist in calculations involving thermal radiation. Historically the aids developed were of three main types: tables, slide rules, and nomograms and graphs. More recently a fourth type can be added to this category, in the form of computer programs and the now ubiquitous "apps" for mobile computing devices. Tables were produced in the greatest numbers and were, before the digital age, unsurpassed in terms of accuracy. Nomograms and graphs, while numbering far fewer than tables and only suitable for order-of-magnitude estimates, managed to display a certain degree of originality and ingenuity in their design. Slides rules were not only functional devices which for a time commanded more attention from users than any of the other aids put together, many were also objects of high aesthetic value. Beautifully constructed mechanical devices of unsurpassed craftsmanship, many were capable of providing estimates with accuracies often better than one per cent. Here we tell their story and hope their history will not be forgotten.

5 Nomograms and graphs used for thermal radiation calculations

Nomography is a method of calculation based on graphical representation. Nomographic methods relate three (or more) variables of an equation graphically by means of a scale for each variable which may be either straight or curved. The functional scales are graduated and arranged so that a straight line will be made to intersect all three scales at points whose values satisfy the equation relating each of the variables. If the values for any two of the variables are given, the value for the third can be immediately found as the point of intersection of a line, referred to as an *isopleth*, formed by joining the points on the other two scales. Essentially a printed chart and requiring nothing more than a straightedge to use, nomograms are cheap and extremely easy to use. Facilitating instant answers to calculations ranging from the very simple to the highly specialized with equal ease, nomograms were a common calculational aid used in many diverse areas of science and engineering from the 1890s up until the 1960s [39]. Compared to other computational aids such as tables or slide rules, nomograms always had the advantage whenever speed was more important than precision. Not only did nomograms serve as quick calculational aids, they also provided a very succinct visual representation of how the various variables for a given quantity are interrelated. Given the tedious and often time-consuming task calculations relating to thermal radiation could take, a small number of nomograms were devised to help aid many of the most commonly encountered calculations in the field.

Graphs, too, played an important part in the estimation of quantities relating to thermal radiation. While graphs may have several perceived shortcomings compared to nomograms, such as (i) their inability to represent a change in more than one variable without the use of multiple curves, and (ii) the time it takes one to follow a line from the abscissa, to the curve, to the ordinate, and the corresponding loss in accuracy resulting from readings made in this way, graphs were none-the-less by far the most dominant graphical form of the two to be found. And while the use of nomograms in modern times has all but disappeared, graphs continue to be found in large numbers.

5.1 NOMOGRAMS

Perhaps the earliest nomogram devised for blackbody radiation was a 1926 three-scale nomogram for the calculation of spectral radiant exitance. It appeared in Appendix V of the first edition of John W. T. Walsh's influential text *Photometry* [667]. At the time, Walsh was a Senior Assistant in the Photometry Division of the National Physical Laboratory in England. He went on to become one of England's foremost experts in the field of photometry in a career which spanned more than forty years and included presidencies of both the *International Commission on Illumination* and the *Illuminating Engineering Society* (now the *Society of Light and Lighting*). In the early 1920s, few tables for the spectral radiant exitance of a blackbody existed and those which did were not very extensive. The large interval sizes found in these early tables made them almost useless for the purposes of interpolation, particularly in finding the rate of change in energy with temperature emitted from a body, especially when it occurred rapidly, as Walsh was interested in finding.

Walsh gave a surprising amount of detail concerning how he calculated his nomogram; details unfortunately lacking in many later produced nomograms. His nomogram, which we show in Fig. 5.1, consisted of two curved scales located between two vertical parallel scales. The two curved scales were for wavelength. Each covered the visible portion of the spectrum from 400 to 750 nm (nanometers, though Walsh used the equivalent unit of millimicrons [mμ]). The reason two scales for wavelength were drawn will be explained shortly. The top wavelength scale was labelled A while the scale beneath it was labelled B. Since Walsh was only interested in the visible portion of the spectrum, for temperatures up to 4000 K he found Wien's formula to be a very good approximation to use instead of Planck's equation. Doing so results in an error that is no larger than about one per cent, a value thought to be acceptable by Walsh for all practical purposes.

In the very early days of the study of blackbody radiation, it was common to express Planck's law in terms of two constants which were directly measurable. As we have already seen, these were the so-called first and second radiation constants $c_1 = 2\pi hc^2$ and $c_2 = hc/k_B$ respectively. When expressed in terms of these two constants, Wien's approximation for the spectral radiant exitance in the linear wavelength representation can be written as

$$M_{e,\lambda}^{b}(\lambda, T) = \frac{c_1}{\lambda^5} \exp\left(-\frac{c_2}{\lambda T}\right). \tag{5.1}$$

After taking the logarithm of both sides of Eq. (5.1) and rearranging algebraically one can write

$$\ln(M_{e,\lambda}^{b}/c_1) + 5\ln\lambda + c_2/(\lambda T) = 0. \tag{5.2}$$

The form of Eq. (5.2) immediately lends itself to a simple three-scale nomogram. The vertical scale to the left was a reciprocal scale for the temperature, in Kelvin [K], from 1800 to 4000 K. To the right was a vertical logarithmic scale for the spectral radiant exitance. Between each vertical scale, in order to obtain more accurate readings for the spectral radiant exitance, two separate scales for the wavelength were drawn instead of one. In use, if the top wavelength scale labelled A was selected, an estimate for the spectral radiant exitance at some given temperature could be made using values located on the left of the spectral radiant exitance scale labelled A. Similarly for the scale labelled B. The double-sided scale for the spectral radiant exitance was a way to reduce the overall size of the nomogram while retaining a reasonable level of accuracy.

Another useful feature of Walsh's nomogram was in its ability to take into account any changes in the values used for the fundamental constants. For much of the first part of the twentieth century large experimental uncertainties in the values for the two radiation constants led to a lack of general agreement in the international community as to their accepted values. Each new revision meant a nomogram could potentially be out of date before it had even been published. To overcome this, Walsh drew his nomogram using $c_2 = 1.4330$ cm·K and $c_1 = 1$. Any revised value in the second radiation constant c_2' would mean the point $1.4330T/c_2'$ on the temperature scale should be used instead of T. Similarly, with the first radiation constant set to unity the value estimated for the spectral radiant exitance was found simply by multiplying the value read from the nomogram by any desired value for c_1.

Another early nomogram devised for blackbody radiation was a very simple single scale nomogram which appeared in a 1929 text by Ellen Lax and M. Pirani [383]. It was used to find the value for the total radiant exitance at a given temperature setting (or vise versa) and was based on a simple calculation of the Stefan–Boltzmann law. A temperature range from 300 to 3000 K in steps of 100 K was covered, and to minimize its overall length the single vertical scale was divided into three shorter vertical sections. About the vertical line the scale for temperature, in Kelvin [K], and the scale for total radiant exitance, in watts per square centimeter [W·cm^{-2}], appeared back to back; temperature to the left, total radiant exitance to the right. This simple nomogram is shown in Fig. 5.2. Some years later the nomogram was reproduced in Jean D'Ans and Ellen Lax's *Taschenbuch für Chemiker und Physiker* [167, 168].

In 1938, a second early three-scale nomogram prepared for estimating the spectral exitance of a blackbody appeared in the Japanese journal *Proceedings of the Physico-Mathematical Society of Japan* [446]. Authored by Zen'emon Miduno, who has appeared on several occasions in prior chapters already, his paper presented details relating to the calculation of a number of radiometric and actinometric quantities. Infinite power series forms, described in Chapter 3 for the fractional function of the first kind for both the radiometric and actinometric cases are presented. These he used to tabulate a number of quantities relating to blackbody radiation and to produce a nomogram useful for estimating the spectral exitance of a blackbody at a given temperature and wavelength in both energetic and photonic units. The nomogram, which took up a whole journal page, consisted of two vertical parallel scales; temperature located to the left, spectral radiant exitance to the right; while a curved scale for the wavelength was found between the two parallel scales. In design and

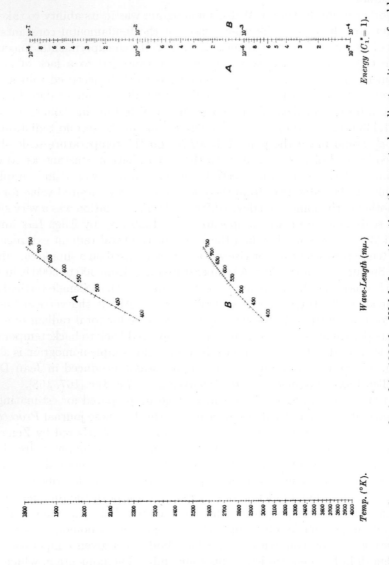

Figure 5.1 A very early nomogram devised in 1926 by Walsh for estimating the spectral radiant exitance of a blackbody. *Source:* Walsh, J. W. T., 1926. *Photometry*, London, Constable & Company Ltd. Figure 301 on page 473.

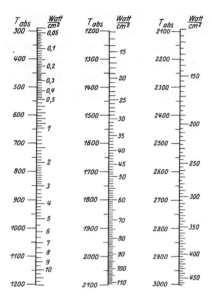

Figure 5.2 A simple early nomogram based on the calculation of the Stefan–Boltzmann law given by Lax and Pirani in 1929. With kind permission from Springer Science and Business Media: *Temperaturstrahlung fester Körper*, 1929, E. Lax and M. Pirani, Figure 8 on page 200. Copyright 1929, Verlag von Julius Springer.

appearance it was similar to Walsh's nomogram with one important difference. Miduno's nomogram could estimate both the spectral radiant and photon exitance. Running back-to-back down the right vertical parallel scale quantities to the left corresponded to the actinometric case in units of total number of photons per square centimeter per second per micron [photon·cm^{-2}·s^{-1}·μm^{-1}] while those to the right corresponded to the radiometric case in units of ergon per square centimeter per second per micron [erg·cm^{3}·s^{-1}·μm^{-1}].[1] Temperatures from 1000 to 2500 K for the radiometric case and 1250 to 3000 K for the actinometric case with wavelengths ranging from 0.15 to 2.00 μm were covered. The power given for the indices on the scale for the spectral exitance for either case where initially in error but were quickly corrected in an erratum published a few months later [447]. Interestingly, the nomogram devised by Miduno was the only one ever produced for the actinometric case; all others before and after dealt exclusively with the radiometric case.

Within a year, Kazuhiko Terada, who was working in Japan at the Meteorological Observatory in Naha, extended Miduno's radiometric nomogram by applying it to situations more suited to meteorological calculations [643]. Miduno's nomogram for the spectral radiant exitance had been for high temperatures and short wavelengths. Terada, on the other hand, considered temperatures more appropriate to terrestrial settings (from -100 to $100°C$) and extended the scale for the wavelength to far longer wavelengths (from 0.9 to $400\,\mu m$). The unit found on the spectral radiant exitance scale was also converted to the gram calorie per square centimeter per minute per micron [gcal·cm^{-2}·min^{-1}·μm^{-1}] in keeping with the unit of radiation in general meteorological use at the time.[2] This nomogram is reproduced in Fig. 5.3.

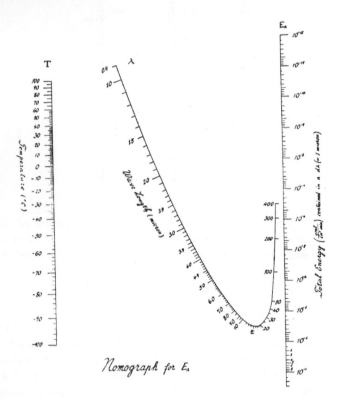

Figure 5.3 An extension of Miduno's nomogram made by Terada in 1939 to estimate the spectral radiant exitance of a blackbody. Reproduced with permission from Terada, K., 1939 "Tables and graphs of the black body radiation for meteorological use," *The Geophysical Magazine* (Tokyo) **13**, p. 140. Copyright 1939, Japan Meteorological Agency.

Terada's paper also contained a second nomogram for two quantities which had not previously been given a graphical form. It consisted of three vertical parallel scales from which two quite different quantities could be found at a given wavelength and temperature setting. These were the ratio of the spectral radiant exitance normalized relative to the value of the spectral radiant exitance at its peak (which Terada denoted by E_λ/E_m) and the integrated fractional amount (which Terada denoted by ϕ); which today we would identify as the fractional function of the first kind for the radiometric case. Each of the three parallel scales were arranged so the scale for temperature appeared to the left, the scale for wavelength to the right, while the third combined parallel scale for the spectral ratio (on the left) and the fractional amount (on the right) was located between the temperature and wavelength scales. It is shown in Fig. 5.4. Using this nomogram not only was it possible to estimate values for the spectral ratio and fractional amount at a particular wavelength and temperature setting but the peak wavelength $\lambda_{\max}$ as given by Wien's displacement law could also be found by allowing the isopleth to pass through unity on the middle scale for any given selected temperature.

A very compact graphical aid for calculating the value for the spectral radiant exitance over a very narrow wavelength range centered about 0.65

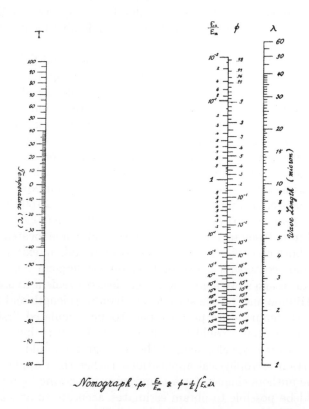

Figure 5.4 A second nomogram devised by Terada to estimate the ratio of the spectral radiant exitance of a blackbody at the spectral peak and the integrated fractional amount. Reproduced with permission from Terada, K., 1939 "Tables and graphs of the black body radiation for meteorological use," *The Geophysical Magazine* (Tokyo) **13**, p. 142. Copyright 1939, Japan Meteorological Agency.

micrometers was devised by the industrial pyrometrist Manoël Felix Béhar to help assist those working in optical pyrometry. More alignment chart than nomogram, it first appeared in 1940 [72] and was reprinted eleven years later in the *The Handbook of Measurement and Control* which Béhar edited [73]. Wavelength appeared along the abscissa. Values within the very narrow interval from 0.63 to 0.67 micrometers at steps of 0.01 micrometers were given. Running vertically were values for the spectral radiant exitance. Béhar labelled this axis the "blackbody radiant power or radiancy" and indicated the scale given was in arbitrary units. We have determined the units on the axis are actually watts per square meter per meter multiplied by a factor of 10^4. In appearance it formed a grid. Curves for different values of temperature between 800 to 6000 K were then marked off. Using a straightedge positioned horizontally, at a selected wavelength and temperature, its intersection with the ordinate allowed the corresponding value for the spectral radiant exitance to be estimated. A reproduction of Béhar's simple alignment chart is reproduced in Fig. 5.5.

In 1953 Standford S. Penner, Ralph W. Kavanagh, and Egil K. Björnerud, all of whom were working at the Guggenheim Jet Propulsion Center at the California Institute of Technology in Pasadena, California, devised two very detailed compound nomograms for the analysis of radiation spectra obtained from low pressure combustion flames as part of their work sponsored by the Office of Naval Research, U.S. Navy [494]. Each nomogram was subsequently published in two separate papers [341, 342]. As a by-product of this work, three of the ten and fourteen parallel scales found on each nomogram could be used to estimate the spectral radiant exitance of a blackbody. In arrangement each consisted of a temperature scale from 1000 to 20 000 K to the right, a wavelength–temperature product scale to the left from 0.050 to 2.00 cm·K in the first and from 0.050 to 100 cm·K in the second, while located between the two scales in both cases was a scale for the spectral radiant exitance in units of ergon per square centimeter per second per micron [erg·cm^{-2}·s^{-1}·µm^{-1}]. Operationally Kavanagh's two nomograms were less convenient to used compared to those of Miduno and Terada since for a given wavelength and temperature the wavelength–temperature product had to be calculated first before the spectral radiant exitance could be determined. The very high values found on the temperature scale also suggest the nomograms would have been more suited to certain astrophysical applications rather than applications in the infrared. The authors claimed for a nomogram measuring $8\frac{1}{2}$ by 11 inches in size it should be possible to obtain estimates accurate to at least an order-of-magnitude. The larger the size, the better the accuracy and copies of Kavanagh's larger sized nomograms could at the time be obtained on request from the authors.

Perhaps the handiest of all nomograms produced for thermal radiation calculations is due to Joseph P. Chernoch [134]. At the time Chernoch devised his nomogram in 1957 he was an infrared engineer working at the General

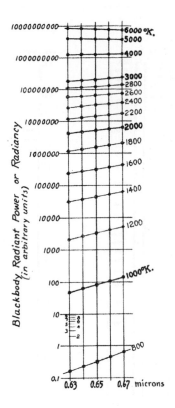

Figure 5.5 A simple alignment chart devised by Béhar to estimate the spectral radiant exitance of a blackbody over a very narrow wavelength range centered about 0.65 micrometers for temperatures between 800 to 6000 K. *Source:* Béhar, M. F., 1951. Pyrometry. In Béhar, M. F. ed. *The handbook of measurement and control*, Volume 2. Pittsburgh, PA: Instruments Publishing Company. Figure 26 on page 119.

Engineering Laboratory for the General Electric Company in Schenectady, New York. As an engineer working in the infrared in the late 1950s, the nature of his work on infrared detectors regularly called for the calculation of a number of quantities related to thermal radiation. As the work required only order-of-magnitude estimates, the speed with which computations could be made far outweighed accuracy in importance. Having found tables unsuitable as they were slow to use, Chernoch turned to developing a nomogram of his own. Published in the aerospace trade magazine *Aviation Age*, the title Chernoch chose makes his intention immediately clear. Entitled "Infrared calculations made simple," his nomogram did indeed make calculations of the type he was mainly interested in performing very simple. A copy of his nomogram is reproduced in Fig. 5.6.

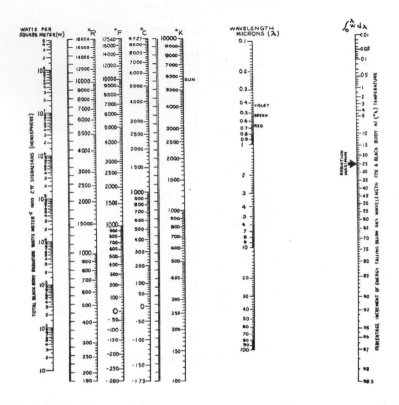

Figure 5.6 A nomogram devised by Chernoch in 1957 for the purpose of simplifying many thermal radiation calculations. General Electric Company, *Optical engineering handbook*, Section 11, page 201, J. A. Mauro, Figure 11-123, Copyright 1966. With kind permission from General Electric Company.

In layout, Chernoch's nomogram consisted of seven parallel scales. Four of the seven scales were for the conversion of temperature between Kelvin [K], Celsius [C], Fahrenheit [F], and Rankine [R]. The temperature scales for Rankine, Fahrenheit, and Celsius were enclosed on either side by two graduated but unmarked scales used for horizontal alignment. Laying a straightedge horizontally at a given temperature on one of the scales allowed the value for the temperature to be converted into equivalent values using the other three temperature scales. To the far left of the nomogram appeared a scale for the total radiant exitance in units of watts per square meter [W·m^{-2}]. For a body at a given temperature its value could also be estimated by a horizontally laid straightedge. The three scales to the right, starting with the scale for temperature in Kelvin, selecting a given temperature and wavelength allowed the fractional amount of the total radiant exitance emitted by a blackbody into a given spectral band from zero up to some arbitrary wavelength to be determined using the scale to the far right of the nomogram. The scale for wavelength ranged from 0.1 to 100 µm while the fractional scale, expressed as a percentage, extended from 0.01 to 98.5%. A gauge mark on the fractional scale labelled "RADIATION MAXIMUM" at 25% also appeared. A fractional amount from zero up to a wavelength corresponding to the spectral peak, as we saw in Chapter 2, is surprisingly close to 25% in the linear wavelength representation (see page 82). An estimate for the peak wavelength could therefore be found using the nomogram from an isopleth passing through this gauge mark and the selected temperature of interest. At the time of its publication, a limited number of reprints of the nomogram could be obtained by writing to the Reprint Department of *Aviation Age*. The actual dimensions of these reprints would have been interesting to know, given the direct correlation between the size of a nomogram and its accuracy; the bigger its size, the greater its accuracy. Unfortunately the size was not given. Some years later Chernoch's nomogram was reproduced in J. A. Mauro's text *Optical Engineering Handbook* [431].

A very simple nomogram for the calculation of the total radiant exitance and the wavelength at the spectral peak for a blackbody at a given temperature appeared in Henry L. Hackforth's text *Infrared Radiation* of 1960 [276]. He described his nomogram as "... useful for rapid calculations," and provided this was all one was interested in, it is exactly what one got. A copy of his nomogram is reproduced in Fig. 5.7. The range used for the scales is large and few to no intermediate graduation marks between major scale marks appear. The nomogram consisted of four parallel vertical scales. From left to right these were scales for temperature in Fahrenheit [F], temperature in Celsius [C], total radiant exitance in watts per square meter [W·m^{-2}] and wavelength at the spectral peak in micrometers [µm] though no units for this scale is actually given. Laying a horizontal straightedge at the temperature of interest, its points of intersection with the total radiant radiant exitance and the peak wavelength scales allowed these values to be estimated.

An all together different nomogram for estimating the spectral radiant exitance of a blackbody in the linear wavelength representation was given as late as 1978 by A. Zanker [728]. Zanker's nomogram is an example of a so-called "ladder" nomogram. A ladder nomogram is a simple connection of two function scales using curves running between matching values on each scale. In this way a ladder nomogram provides a transition between two nomograms that share a common variable. The multiple curves drawn between matching values form in appearance a so-called "ladder" and act as a guide allowing the values to be transferred from one scale over to the other. In the case of Zanker's nomogram for the spectral radiant exitance, the common variable was wavelength. The first of the wavelength scales to be used was labelled the "primary scale" while the second was labelled the "secondary scale." The

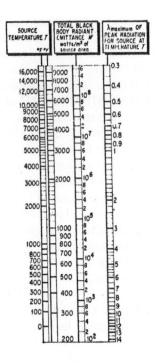

Figure 5.7 A very simple nomogram given by Hackforth in 1960. Reproduced with permission of McGraw-Hill Education. *Infrared Radiation*, 1960, H. L. Hackforth, Figure 2.5 on page 20. Copyright 1960, McGraw-Hill Book Company, Inc.

nomogram consisted of two equally sized ellipses. Each ellipse was separated and their orientation was such that their semi-major axes ran vertically.

A scale for the primary wavelength, in meters [m], ran along the circumference of the left side of the left ellipse, while running vertically down the semi-major axis was a scale for the temperature in Kelvin [K]. A scale for the secondary wavelength, in meters [m], ran around the circumference of the right side of the right ellipse. Running this time vertically down its semi-major axis was a scale for the spectral radiant exitance in units of watts per square meter per meter [$W \cdot m^{-2} \cdot m^{-1}$]. Between the two ellipses a series of oblique ladders or "tie-lines" connecting the right half of the left ellipse to the left half of the right ellipse along their respective circumferences were drawn.

In operation the spectral radiant exitance for a blackbody from Zanker's nomogram was estimated in the following manner. For a given temperature and wavelength an isopleth passing through these points was extended until it intersected the right edge of the left ellipse. Using the tie-lines as a guide, the value for the primary wavelength was transferred over to the secondary wavelength scale on the right side of the right ellipse by using a second isopleth.

Where the second isopleth intersected the central vertical scale of the right ellipse gave an estimate for the spectral radiant exitance.

Zanker's nomogram could also be used to estimate the value of the wavelength at the spectral peak. For this purpose a gauge mark labelled "MAXIMUM POWER POINT" appeared. It was represented by a small circle located on the right side of the circumference of the left ellipse. Connecting this gauge mark with a straight line and allowing it to first pass through a given temperature located on the central vertical scale and then extending the line to the primary wavelength scale on the left side of the left ellipse, the point of intersection gave the desired value for the peak wavelength. Operationally, as a two step process relying on tie-lines as guides, Zanker's nomogram was prone to inaccuracies brought about by unavoidable errors in judgment made by the user. This made it less useful compared to equivalent nomograms devised by Miduno, Terada, and Kavanagh for estimating the spectral radiant exitance of a blackbody at a particular temperature and wavelength. Zanker's unusual looking nomogram is reproduced in Fig. 5.8.

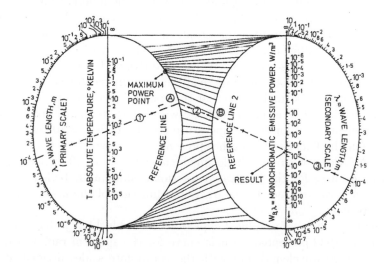

Figure 5.8 A late and unusual nomogram devised in 1978 by Zanker for estimating the spectral radiant exitance of a blackbody. This article was published in *Optik*, Vol. **49**, No. 4, Zanker, A., 1978 "Monochromatic emissive power of blackbody found quickly by nomograph," p. 411. Copyright 1978, Elsevier.

A nomogram for the calculation of the absolute and relative spectral radiant exitance or spectral radiance of a blackbody in the visible portion of the spectrum is referred to by Alwin Walther and A.-W. Kron. It is found as entry No. 54a in the extensive list of publications appended to their article "Nomographie und Rechenschieber" which appeared in the *FIAT Review of German Science* in 1948 [669].[3]

Developed by Kron in 1946 [363] and listed as being in manuscript form only, since we have not been successful in locating a copy of this document, nor is it referred to in any other secondary source, little other than its title is known about this nomogram.

5.2   GRAPHS

Graphs have always played an important part in the estimation of quantities relating to blackbody radiation. Most common amongst the graphs were those for the spectral radiant exitance as a function of wavelength for a range of different temperatures. Here the wavelength is plotted on the abscissa while the spectral radiant exitance as the ordinate using either linear scales, a semilogarithmic scale for either scale, or log-log scales as was done in Fig. 2.6 on page 44. An early example of a graph drawn with many isotherms in a double logarithmic representation is shown in Fig. 5.9. It is taken from Werner Pepperhoff's 1956 text *Temperaturstrahlung* [495]. It is typical of graphs found for the spectral radiant exitance, and makes clear the inherent limitations involved in using curves as graphical aids. Since a new curve has to be drawn for each different temperature the graph quickly fills with multiple curves, leading to it becoming considerably more difficult to work with compared to nomograms such as those given by Walsh, Miduno, or Terada where the same quantity can be estimated far more readily and with greater accuracy.

If the ordinate is normalized so that the spectral peak has a value equal to unity and is plotted as a function of the wavelength–temperature product a single universal curve for any temperature results. In Fig. 5.10 the universal curve within the linear wavelength representation is depicted. It is the curve labelled **A**. Along with the single universal curve it is often accompanied by a second curve for the integrated fractional amount. Labelled **B** in Fig. 5.10 the curve corresponds to the fractional function of the first kind for the radiometric case.

Of the more unusual graphical depictions for a blackbody that appeared, Y. Omoto in 1936 presented a single curve for the spectral radiant exitance as a function of wavelength but with the coordinate scales divided into six different scales with each being valid for a particular temperature [469]. The temperatures chosen ranged from 1000 to 6000 K in steps of 1000 K. In this way multiple curves that would have resulted from plotting a range of temperatures were avoided. Omoto's graph for the spectral radiant exitance is shown in Fig. 5.11.

A variation on Omoto's idea of drawing the spectral radiant exitance as a function of the wavelength using multiple scales was made by Aleksander Sala in 1986 [581]. In a departure from Omoto, rather than drawing a single curve, a total of seven curves were plotted. Doing so allowed for a far finer graduation in temperatures to be achieved. The first temperature interval from 350 to 650 K at steps of 50 K were drawn for wavelengths from 0 to 14 μm. The second from 700 to 1300 K at steps of 100 K using a wavelength scale that was

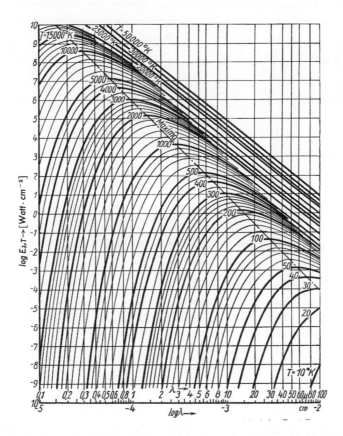

Figure 5.9 Spectral radiant exitance isotherms for blackbody radiation depicted using a double logarithmic scale. W. Pepperhoff, *Temperaturstrahlung*. Verlag von Dr. Dietrich Steinkopff, Darmstadt, 1956. Reproduced with permission of Steinkopff via Copyright Clearance Center.

half of the previous scale. The halving of the wavelength scale continued as the temperature intervals increased. The third, from 1400 to 2600 K at steps of 200 K and the fourth from 2800 to 5200 K at steps of 200 K appeared on wavelength scales from zero to 3.5 μm and zero to 1.75 μm respectively. Sala's spectral radiant exitance plots for a blackbody is reproduced in Fig. 5.12. From the seven curves drawn, estimates for the spectral radiant exitance of a blackbody for twenty-eight different temperatures could be made without any of the unnecessary clutter that would have occurred if all twenty-eight curves had been crammed into a single figure. The figure drawn by Sala should be compared with that given by Pepperhoff in Fig. 5.9 where a total of sixty-one curves are drawn, one for each different temperature considered.

A second uncommon but none-the-less compact graph for the total radiant exitance of a blackbody as a function of temperature was given by the Dutch

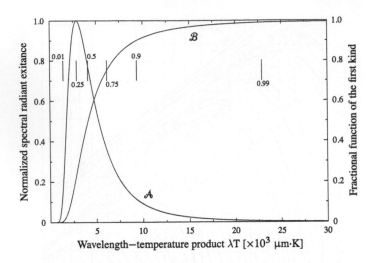

Figure 5.10 An example of two of the more common graphs seen for blackbody radiation. Curve **A**: Normalized spectral radiant exitance as a function of the wavelength–temperature product. Curve **B**: Blackbody fractional function of the first kind for the radiometric case as a function of the wavelength–temperature product. The six short vertical lines appearing across the top of the plot indicate, from left to right, the fractional amounts 0.01, 0.25, 0.5, 0.75, 0.9 and 0.99 of the total radiant exitance.

physicist Gerrit A. W. Rutgers in 1958 [579]. Six different curves labelled a through to f were drawn onto a single graph. The temperature, along the abscissa, was divided into two separate scales from 100 to 1000 K for curves a, b, and c and from 1000 to 4000 K for curves d, e, and f. The total radiant exitance for the ordinate was divided into five separate scales. To the left, scales for curve a (0.01–0.1 W·cm^{-2}), curve b (0.1–1 W·cm^{-2}) and curves c and d (1–10 W·cm^{-2}) while to the right, scales for curve e (10–100 W·cm^{-2}) and curve f (100–1000 W·cm^{-2}) appeared. Rutgers' graph is shown in Fig. 5.13. Note that Rutgers uses the symbol H for total radiant exitance rather than the modern M.

In 1957, as a result of his work in low-temperature pyrometry, T. P. Gill gave an alignment chart for the temperature derivative of the spectral radiant exitance in the linear wavelength representation as a function of wavelength [256]. The quantity seems to have been first considered by Czerny in 1944 [159]. It is given by

$$\frac{\partial M_{e,\lambda}^b}{\partial T} = c_1 c_2 \lambda^{-6} T^{-2} \frac{e^{c_2/(\lambda T)}}{(e^{c_2/(\lambda T)} - 1)^2} = \frac{c_1 c_2}{4} \lambda^{-6} T^{-2} \left[\sinh\left(\frac{c_2}{2\lambda T}\right) \right]^{-2}.$$
(5.3)

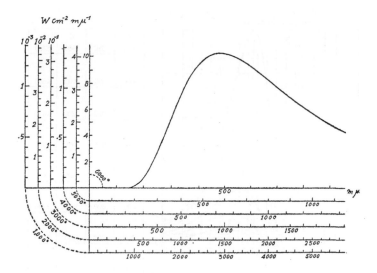

Figure 5.11 An unusual looking graph devised by Omoto in 1936 for estimating the spectral radiant exitance of a blackbody as a function of wavelength for six different temperatures. Reproduced with permission from Omoto, Y., 1936 "Table radiation function," *Journal of the Illuminating Engineering Institute of Japan* **20**(4), p. 140. Copyright 1936, IEIJ.

In shape the curve is similar to the spectral radiant exitance. At its spectral peak the maximum is given by

$$\lambda_{\max}T = \frac{c_2}{2(2.784\,705\ldots)} = \frac{c_2}{5.569\,409\ldots} \tag{5.4}$$

In this case $2.784\,705\ldots$ is the non-trivial root of the equation $x = 3\tanh x$ (compare this to Eq. (2.24)). On a log-log plot the locus of the maxima is, as with the spectral radiant exitance, a straight line. Marked out along this line were temperatures from 100 to 350 K as Gill was primarily interested in low temperatures. As all curves maintain the same shape with temperature, a single curve for $T = 300$ K is drawn. As the curve is moved along the line given by the locus of the maxima, values for the derivative of the spectral radiant exitance (in watt per cubic centimeter per Kelvin) as a function of wavelength (in centimeter $\times 10^4$) at different values for the temperature could be read off. It was an uncommon quantity to compute. As we shall see in Chapter 7, this quantity was occasionally tabulated, and other than the work of Gill, would never again be given graphic form. Gill's alignment chart for the temperature derivative of the spectral radiant exitance as a function of wavelength is reproduced in Fig. 5.14. Note the label Gill gives for the ordinate of $d\varepsilon/d\lambda$ corresponds to our $\partial M_{e,\lambda}^{\mathrm{b}}/\partial T$.

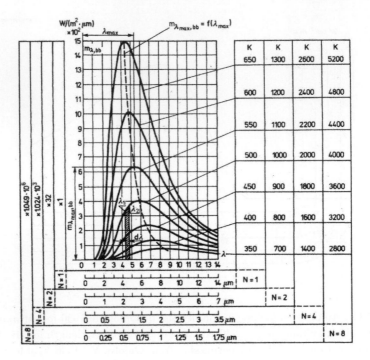

Figure 5.12 A modification on Omoto's multiscale design given by Sala in 1986 for estimating the spectral radiant exitance of a blackbody as a function of wavelength for twenty-eight different temperatures. *Source:* Sala, A., 1986. *Radiant Properties of Materials: Tables of Radiant Values for Black Body and Real Materials*, Warsaw, PWN–Polish Scientific Publishers. Figure 1 on page 5.

A final example of an unusual graph relating to blackbody radiation was given by the German astronomer Hans G. Kienle in 1941 [350]. In a paper concerned with the calculation of color temperatures for stars, Kienle showed that the derivative of the difference in magnitude $\Delta m = m_2 - m_1$ between two stars with respect to the inverse wavelength (the wavenumber), and the so-called relative gradient of a blackbody $\Delta\Phi$ over some spectral interval, are related by

$$\Delta\Phi = \frac{2}{5\log_{10} e} \frac{d(\Delta m)}{d(1/\lambda)}. \tag{5.5}$$

The absolute gradient of a blackbody $\Phi(\lambda, T)$ over some spectral interval is such that

$$\frac{d}{d(1/\lambda)} \log_{10} M(\lambda, T) = \left(5\lambda - \frac{c_2}{T}\left[1 - e^{-c_2/(\lambda t)}\right]^{-1}\right) \log_{10} e \tag{5.6}$$

$$= (5\lambda - \Phi(\lambda, T))\log_{10} e.$$

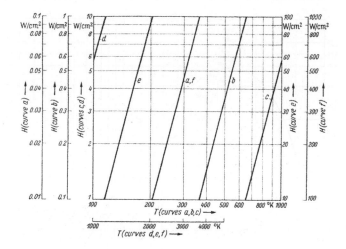

Figure 5.13 Another uncommon graph, this one devised by Rutgers in 1958, for estimating the total radiant exitance of a blackbody as a function of temperature. With kind permission from Springer Science and Business Media: *Encyclopedia of physics* Volume 5/26. Light and matter II. "Temperature radiation of solids," 1958, 129–170, G. A. W. Rutgers, Figure 5. Copyright 1958, Springer-Verlag.

In his paper Kienle presents a graph which allowed the color temperature for a star to be determined from the relative gradient for a blackbody at temperatures above 3000 K and for wavelengths within the spectral interval 2500 to 10 000. Kienle's graph is reproduced in Fig. 5.15. As relative gradients at the time could not be determined more accurately than two significant figures, using a graph for its determination represented an ideal solution to this problem. The graph gives $\Phi(\lambda, T) - c_2/T$ [$\times 10^{-4}$ cm] as a function of $\Phi(\lambda, T)$ and λ [both $\times 10^{-4}$ cm]. Within the graph itself additional scales for c_2/T [$\times 10^{-4}$ cm], $5040/T$ [K^{-1}], and T [K] are drawn, making the direct translation between absolute gradient and any of these other quantities possible. Interestingly, Kienle tells us his graph was drawn using values accurate to four significant figures before being reduced to two-fifths of its original size. His graph was reproduced thirty-three years later by Michael Golay in his text *Introduction to Astronomical Photometry*, demonstrating the usefulness it continued to find many years later with astronomers [260].

An early example of a carefully drawn series of curves for the spectral radiant exitance appeared in 1925. Prepared by M. Katherine Frehafer and Chester L. Snow, as both worked at the Bureau of Standards in the U.S. at the time, it appeared under the Bureau's *Miscellaneous Publication* series [241]. To make the plots more manageable in terms of limiting the size of the ordinate for a temperature range of between 1000 to 28 000 K, the spectral

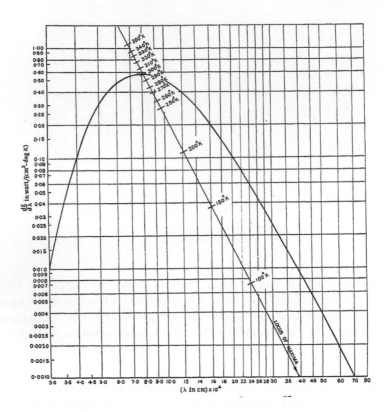

Figure 5.14 An alignment chart given by Gill in 1957 for the temperature derivative of the spectral radiant exitance as a function of wavelength. Reproduced with permission from Gill, T. P., 1957 "Some problems in low-temperature pyrometry," *Journal of the Optical Society of America* **47**(11), p. 1002. Copyright 1957, OSA Publishing.

radiant exitance was normalized relative to the value of the spectral radiant exitance at a wavelength equal to 560 μm. Not only did this make the size of the ordinates more manageable, doing so eliminated the need for knowing the value for the first radiation constant. A series of isochromatic curves for wavelengths between 400 to 720 μm at 10 μm steps were included. The curves were divided into five sections using various intervals for temperature with each graph plotted onto five separate sheets.

The sheets produced were quite large, measuring 60 cm long by 48 cm wide, while accuracies of 0.33% for a value of the second radiation constant of $c_2 = 1.4350 \times 10^{-2}$ m·K were claimed. At the time of their publication Frehafer was an Associate Physicist working at the Bureau of Standards. Snow, on the other hand, was not a scientist or engineer but a draftsman, and given the

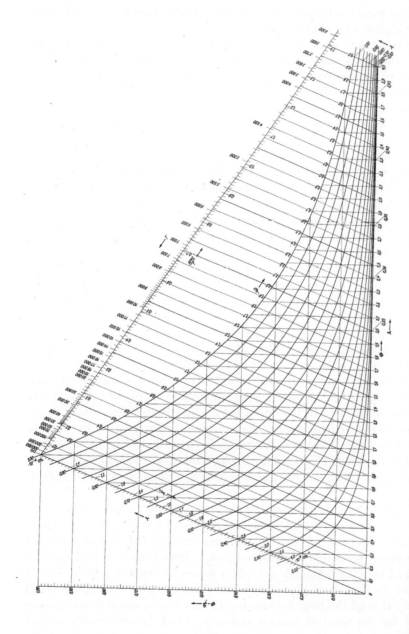

Figure 5.15 A final example of an uncommon graph, this one devised by Kienle in 1941, for obtaining the absolute gradient of a blackbody as a function of wavelength and temperature. *Source:* Kienle, H., 1941 "Zur Berechnung von Farbtemperaturen," *Zeitschrift für Astrophysik,* **20**(4) 239–245. Tafel I.

nature of the publication, consisting as it did of five graphs for the spectral radiant exitance, caused at least one reviewer at the time to remark that while the mechanical execution of the charts was of the superior quality one had come to expect from the Bureau, it was Snow's draftsmanship they had found to be of the highest order [498].

Once the technological importance of thermal radiation had been fully realized, what one finds starting to appear in the early to mid-1950s in ever greater numbers, particularly in texts where the application of the blackbody is dealt with in an essentially quantitative way, are carefully drawn graphs with finely graduated scales. Very much in the spirit of Frehafer and Snow's earlier work, they were intended to be used to provide quick, order-of-magnitude estimates for a number of the more important blackbody quantities with detailed graphs of this type having appeared right down to the present day [425, 495, 364, 696, 244, 54, 431, 300, 600, 247, 610, 697, 698, 700, 603, 453].

The most extensive set of curves for the spectral radiant exitance of a blackbody were surprisingly not produced until as late as 1980 by W. A. Feibelman for use within the astronomy community [224]. In total Feibelman gave forty spectral curves within the ultraviolet wavelength range 1150 to 3200 for temperatures ranging from 6000 to 200 000 K. Of the forty curves, twenty-one were normalized to unity at a value corresponding to a wavelength of 3200 while the remaining nineteen were normalized to unity at 1900. The curves were produced to aid astronomers working with data collected by the International Ultraviolet Explorer (IUE). IUE was an astronomical observatory satellite primarily designed to take ultraviolet spectra at high (0.2) and low (6) resolution within the 1150 to 3200 wavelength range. The mission ran for almost twenty years, from 26 January 1978 until 30 September 1996. Since the form for the spectral curves of a blackbody within the wavelength and temperature range of interest were generally only available in tabular form, Feibelman supposed users of IUE data would find having immediate access to such a set of curves more useful than constantly having to refer back to numerical tables. The curves were the product of the digital age. Values were calculated using a PDP-11/40 computer located at the Laboratory for Astronomy and Solar Physics, NASA Goddard Space Flight Center, Greenbelt, Maryland.

A useful aid for visualizing how the spectral radiant exitance of a blackbody varies with wavelength and temperature is to plot it in space using three coordinate axes. Plotting the two independent variables of temperature and wavelength along the x- and y-axes produces a three-dimensional plot of the sort given in Fig. 5.16. Although not often seen, the first attempt at such a plot was made by the French physicist André-Prosper-Paul Crova (1833–1907) in 1880 [150]. It is all the more extraordinary that Crova was able to produce such a plot for the spectral radiant exitance, which is remarkably similar to Fig. 5.16, considering the poor state measurements related to radiant energy emitted from thermal source that existed before Langley introduced the

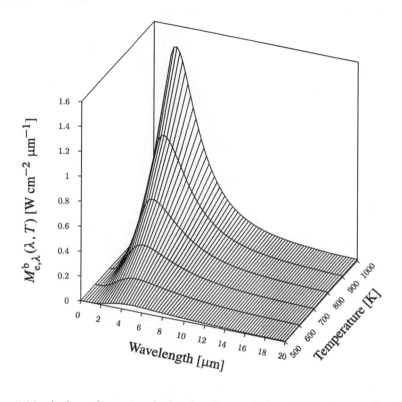

Figure 5.16 A three-dimensional plot for the spectral radiant exitance of a black-body as a function of both the wavelength and temperature.

bolometer in the early 1880s. Crova's three-dimensional plot for the radiant spectral exitance from 1880 is reproduced in Fig. 5.17. The axis marked X is temperature while Y is for wavelength. Some seventy years later a cardboard three-dimensional model for the spectral radiant exitance of a blackbody was given by the German theoretical physicist Arnold Johannes Wilhelm Sommerfeld (1868–1951) in his 1952 text *Thermodynamik und Statistik* [613].

An altogether different graph used for estimating the fractional function of the first kind for the radiometric case within finite spectral bands was given by V. I. Matveev in 1985 [430]. Concerned at the time with infrared vision devices, Matveev gave a figure that could be used to estimate the fractional amount radiated within the two principal atmospheric transmission bands in the infrared of 3.5–5 µm (midwave band) and 8–14 µm (longwave band) over terrestrial temperatures between −40°C to 100°C. Two curves were drawn; one for each spectral band. Selecting a temperature on the abscissa, to find the fractional amount a vertical line is drawn. Its intersection with the midwave band curve, the ordinate to the left is read, while for the longwave band curve, the ordinate to the right is read.

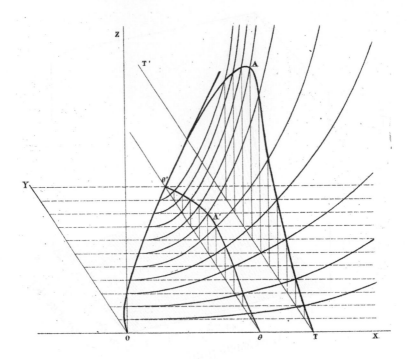

Figure 5.17 The first attempt at a three-dimensional plot for the spectral radiant exitance of a blackbody as a function of both the wavelength and temperature produced from experimental data by André-Prosper-Paul Crova in 1880 [150].
Source: Crova, A.-P.-P., 1880 "Étude des radiations émises par les corps incandescents. Mesure optique des hautes températures," *Annales de chimie et de physique*,
Série 5, **19** 472–550. Figure 1 on Plate I.

5.3 THE LEGACY OF GRAPHICAL AIDS

How widely the various nomograms and graphs were used is difficult to assess.
Now, as in the past, they were and continued to be an invaluable visual aid.
It is true that a number of authors of the more widely read infrared texts
and manuals of the day promoted their use for quick estimates of the kind
suitable for preliminary design work [324, 696, 93, 311, 82, 697]. Occasionally
one finds a particular nomogram being referred to in the literature [116],
though this is rare. Instead it is more likely nomograms and graphs were
used to provide preliminary estimates or bounds to a particular problem. If
those estimates proved encouraging, warranting further investigation, detailed
calculations made using either a radiation slide rule or tables most likely
followed and would explain the apparent paucity of references found for these
aids in the literature. It is not that nomograms and graphs were not useful,

they were. It is just that they were being used in a way they had always been designed and intended to be used — as quick and simple computational aids.

Notes

[1]The ergon is a unit of energy equal to 10^{-7} joules. Its origins lay in the centimeter–gram–second or CGS system of units, today having been replaced by the joule, the SI unit used for energy.

[2]The gram calorie or "small calorie" is the energy needed to increase the temperature of 1 gram of water by 1°C at standard atmospheric pressure and is approximately equal to 4.2 joules.

[3]The FIAT [Field Information Agency, Technical (U.S.)] Review of German Science was a comprehensive summarisation of German fundamental research results in the natural and technical sciences between the years 1939 to 1946 [347]. It was designed to disseminate many of the technical and scientific advances made by the Germans and Axis governments during the Second World War at a time when the exchange of scientific information between Germany and countries of the Allied Forces significantly decreased before effectively ceasing by the time of the war's end.

6 Slide rules used for thermal radiation calculations†

Until it was displaced by the arrival of inexpensive electronic calculators in the mid-1970s, the slide rule was an indispensable tool of the scientist and engineer. Designed to relieve the user of the drudgery associated with having to perform largely routine computations by hand, many different types and designs had been proposed and made for a bewildering variety of purposes. These ranged from relatively standard types that could be used to multiply and divide numbers, extract square and cube roots, or determine values for the trigonometric, logarithmic, and exponential functions, to those designed and used for highly specialized purposes.

Before proceeding further it is worth recalling some of the terminology associated with slide rules which may not be familiar to the modern reader. Broadly speaking, slide rules are of three general types: linear, circular, or spiral. In construction, a typical linear slide rule consisted of a central slide that slides relative to two stationary stocks and was either of the open or closed frame type. For the closed frame type the two stocks were separated by a gutter forming a body around the slide while for the open frame type the two stocks were held together at either of their ends with end braces. For the latter design, as the slide moved, an open gap would appear, hence its name, while no such gap appeared in the former. It was also usual for a slide rule to be fitted with a movable cursor with at least one hairline etched onto it. Scales were inscribed on the slide and one or both of the stocks while other fixed markings which may be found on the rule for the purpose of special calculations are known as gauge marks.

The field of thermal radiation readily lends itself to a special purpose slide rule. As noted earlier, calculations in this field are particularly tedious due to the cumbersome mathematical form the laws of thermal radiation take on. One needs little convincing of the advantages to be gained from having access to a calculational aid capable of quickly performing many of the often repetitive calculations encountered in the field. While the need was probably recognized by many, surprisingly it was not until late 1944 in war-torn Germany that the first of the many so-called radiation slide rules appeared. Prior to the availability of specialized slide rules for thermal radiation calculations, workers in the field either made use of one of a number of tables that existed for

---

†The present chapter is the outgrowth of an earlier paper we wrote in 2012 on the history of slide rules used for thermal radiation calculations as a tribute to Professor William L. Wolfe [337].

various blackbody quantities at the time,[1] gained estimates from one of the few nomograms that were available, or simply reverted to performing any needed calculation by hand.

The first of the radiation slide rules was designed by the German experiential physicist Marianus Czerny and was of the linear type. Later radiation slide rules included linear and circular forms made of plastic, aluminium, or wood coated with a white celluloid veneer, or slide charts made of either finished cardboard or more robust plastic and often given away as promotional items. All were capable of estimating various physical quantities associated with thermal radiation while the number of quantities found on each rule reflected the audience of its intended users. In what follows, each of the various radiation slide rules known to exist from the time of their initial inception up until their final demise, and on into the present day, are discussed. We record their history, their capabilities, and for whom the rule was intended.

6.1 THREE EARLY SLIDE RULES FROM GERMANY, ENGLAND, AND THE UNITED STATES

6.1.1 THE SYSTEM CZERNY RULE

By 1944, when a hand-built prototype for a radiation slide rule first appeared in the German periodical *Physikalische Zeitschrift*, [158] Marianus Czerny[2] was already well known for his work in the infrared. In his paper he writes it was due to an inability to integrate Planck's equation in closed form over an arbitrary spectral interval that prompted him to design a slide rule for the sole purpose of solving this particular problem.[3] It was the most important scale to appear on his rule. In finding the various mark locations that would make up this scale, the evaluation was performed by hand by him with the assistance of a certain Mr Kurt Schäfer who was presumably either one of Czerny's students or technical assistants who were working with him in his laboratory at the time.

Compared to the radiation slide rules that were to come, Czerny's rule was a relatively simple affair. It contained just four scales and two gauge marks. It was a single-sided linear slide rule of the closed-frame type and measured about eight inches long. Initially his slide rule came without a cursor and it seems a small number of these rules were hand built and used by his students and staff in his infrared laboratory at Frankfurt [631]. A photograph of Czerny's first hand-built radiation slide rule as it appeared in the December 1944 issue of *Physikalische Zeitschrift* is reproduced in Fig. 6.1.

A few years later Czerny's design served as a prototype for a rule produced by ARISTO, a large German manufacturer of slide rules at the time. It was designated the "System Czerny" after its inventor, was made of plastic, measured about 30 cm long, and came with a fixed cursor. Commercially it was known as the ARISTO Nr. 10048 — *Rechenschieber für Temperaturstrahlung*. It is thought the rule became available in Germany shortly after the end of

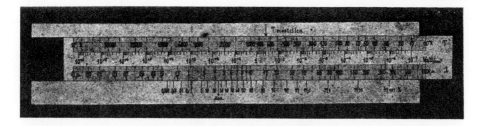

Figure 6.1   Czerny's first hand-built radiation slide rule from 1944. Reproduced with permission from Czerny, M., 1944 "Ein Hilfsmittel zur Integration des Planckschen Strahlungsgesetzes," *Physikalische Zeitschrift* **45**(9/12), p. 206. Copyright 1944, S. Hirzel Verlag.

the Second World War. By the late 1940s it was available in the U.S., being obtainable from George Haas who at the time worked at the Engineering Research and Development Laboratories at Fort Belvoir in Virginia [120]. A photograph of the front face of the ARISTO Nr. 10048 is shown in Fig. 6.2.

For the scales that appeared on the later commerically produced rule, running along the top of the bottom stock was the scale for the fractional amounts found numerically from Eq. (3.33). These were expressed as a percentage and ran from 0.001 to 99.999%. The gauge mark "Max" on the bottom stock gave the fractional amount when the interval from zero up to a wavelength corresponding to the peak wavelength in the spectral distribution curve was selected. As noted earlier, in the linear wavelength representation its value is very close to 25% (see page 82). On the slide a logarithmic temperature scale T, measured in Kelvin [K], from 10 to 10^5 K ran along the top and a reverse logarithmic scale for the wavelength λ, measured in micrometers (μm, though on the rule the unit appeared simply as μ), from 0.1 to 1000 μm ran along the bottom. Running through the middle of the slide was a scale for the total radiance. In this regard Czerny was unusual. Later radiation slide rules that appeared in England and the U.S. always used the total radiant exitance rather than the total radiance. The scale ran from 10^{-7} to 10^8 and was measured in watts per square centimeter per steradian [W·cm^2·sr^{-1}].

Labelled "T einstellen" (temperature setting) on the top stock is the second gauge mark. In operation the slide was adjusted to the desired temperature by aligning the value of the temperature with the upper gauge mark. The total radiance could then be read off from the scale appearing in the middle of the slide. At the selected temperature, the wavelength where the spectral distribution curve peaks could also be read off directly from the bottom gauge mark. Lastly, using the wavelength scale on the bottom of the slide in conjunction with the adjacent percentage scale on the bottom stock allowed the fraction of the total radiance within the wavelength range from zero up to some arbitrary value for the wavelength to be determined. A centimeter scale

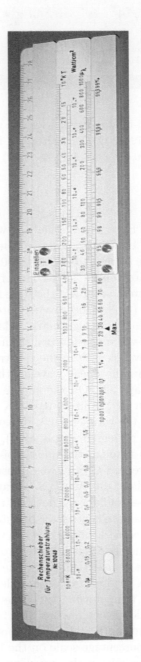

Figure 6.2 Front face of the ARISTO Nr. 10048 – Rechenschieber für Temperaturstrahlung from the late 1940s. Photograph courtesy of William L. Wolfe.

from 0 to 28 is marked along the top of the rule. The addition of this scale however, had nothing to do with the operation of the slide rule and merely allowed it to double as a simple ruler.

A specimen of what was most likely a pre-production prototype for the ARISTO Nr. 10048 is known to exist.[4] It is a large, heavy, wooden rule. Measuring 36 cm long by 4 cm wide by 1 cm thick it is of the closed-frame type. Its scales are printed on white paper and glued onto the slide and the bottom stock of the rule only, while a fixed cursor similar to the one found on the ARISTO Nr. 10048 completes the rule. The similarity in the appearance of four large screws in each of the corners of the cursor to those found on the ARISTO Nr. 10048 strongly suggests the former is indeed a prototype of the latter. Interestingly, the prototype contains all the scales found on the ARISTO Nr. 10048 together with an additional scale for the total radiant exitance. Why this scale was removed from the final commercial rule is not known. It would have made for a useful addition to the rule, at once removing the need in having to manually multiply the total radiance by a factor of π, and its removal was certainly not due to a lack of space available on the ARISTO Nr. 10048.

Four scales appear on the slide of the prototype. Labels for each of these scales run down the far left-hand end of the slide. From top to bottom they are: K (for temperature in Kelvin), σT^4 (for total radiant exitance in watts per square centimeter), $\frac{\sigma}{\pi} T^4$ (for total radiance in watts per square centimeter per steradian, though steradian is not marked), and λ (for wavelength in micrometers). The labels for the unit of each quantity appear next to each scale at the far right-hand end of the slide. In common with the ARISTO Nr. 10048, a scale for the integrated fractional amount ran along the top of the bottom stock. The hairline on the cursor seems to be quite deliberately fixed over a fractional amount equal to 25%. Doing so allowed the peak wavelength in the spectral distribution curve to be found for any given temperature setting since the fractional amount at the peak value in the linear wavelength representation is approximately equal to 25%. All graduation marks and numbers appearing on the scales were printed, while the labels for each quantity together with its associated unit appear to be handwritten. A photograph of the front face of the prototype is shown in Fig. 6.3. The rule shown was given to Professor William L. Wolfe by Dr Arthur Francis Turner [704]. Turner had worked under Czerny in Berlin in the early to mid-1930s as part of his doctoral studies [423]. After completing his *Doctor rerum naturalium* (Dr. rer. nat., the German equivalent of a PhD) in 1935 he returned to the U.S. to take up a teaching position at the Massachusetts Institute of Technology. The close personal connection between the two men suggests the rule was received by Turner from Czerny himself, though exactly when it came into his procession is not known.

An updated version of the ARISTO Nr. 10048 appeared in the late 1950s under the new model number of 922. It is likely it became available either in late 1957 or early 1958 [631] and was known as the ARISTO Nr. 922 – *Rechenstab*

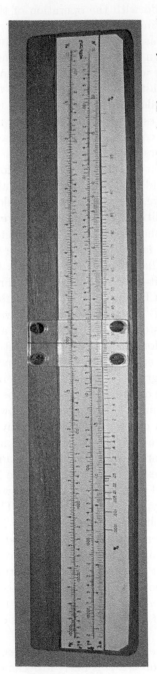

Figure 6.3 Front face of a rule believed to be a late prototype for the *System Czerny* radiation slide rule. Photograph courtesy of William L. Wolfe.

für Temperaturstrahlung. The number of scales and gauge marks remained unchanged, however, unlike the ARISTO Nr. 10048, the ARISTO Nr. 922 came with an adjustable cursor and its construction was now of the open-frame type. In operation the ARISTO Nr. 922 functioned in exactly the same manner as its predecessor. The adjustable cursor now allowed for more accurate fractional amounts to be read for any arbitrary wavelength of interest. As a further minor improvement to the rule, clear labels for each of the scales were added to the far left-hand end of the rule. Vertically aligned, reading from top to bottom these were T, Q, λ, and W(z) for temperature, total radiance, wavelength, and fractional amount. Note the ARISTO Nr. 922's symbols Q and W(z) correspond to the modern day equivalents of L and $\mathfrak{F}_{e,0\to\lambda}$ respectively. The positioning of an adjustable cursor running along the outer edges of the rule also meant the rule no longer had an uninterrupted straight edge. The former centimeter scale found on the ARISTO Nr. 10048 was therefore done away with. A photograph of the front face of the ARISTO Nr. 922 is shown in Fig. 6.4.

Outside of Germany, the two radiation slide rules made by ARISTO do not appear to have been widely available or used. Each rule is referred to in the first and second editions of Werner Brügel's important early infrared text *Physik und Technik der Ultrarotstrahlung* [108, 109]. Reference to it is also made in *Ullmanns Encyklopädie der technischen Chemie* of 1960 [227], a widely read reference work that dealt with all aspects of the science and technology of industrial chemistry. References to the rule in the English literature are somewhat more limited. It appears that the first reference to the rule in an English language publication came in 1947 and is to be found in *European Scientific Notes*, an informal, bimonthly publication of the London Branch of the U.S. Office of Naval Research (ONR) [666, 675]. Established in 1946 to survey, assess, and report on European scientific and technological activities, ONR London's best-known output would become its *European Scientific Notes*. Carrying news of noteworthy developments in European scientific research its very first issue carried a piece highlighting developments Germany had made in the infrared during the war years based on recently conducted interviews with Marianus Czerny and Gerhard Hettner. Czerny's principal areas of research were divided into five parts with the fifth entitled "Minor theoretical considerations of black-body radiation." Here one of the minor theoretical considerations is reported as being "... a useful slide rule... for obtaining the intensity (watt/cm^2) of black-body radiation in any desired finite wave length region, once the temperature is given" [635].

Some years later in a book review, John N. Howard wrote that a recent collection of tables for the fractional functions of the first and second kinds noted the radiation slide rule devised by Prof. Czerny "... was useful for quick calculations" [301]. Richard D. Hudson, Jr., in his 1969 text *Infrared System Engineering* [311] briefly mentions the ARISTO Nr. 922 in a section devoted to

Figure 6.4 Front face of the Aristo Nr. 922 – Rechenstab für Temperaturstrahlung from the late 1950s. Photograph courtesy of Nina Senger-Mertens from Arithmeum in Bonn.

a general discussion on the various slide rules available at the time for thermal radiation calculations. He manages to do this in two very short sentences and a footnote.

As late as 1978 Alan Chappell in the text *Optoelectronics: Theory and Practice* recommends a number of quantities related to the total radiance of a blackbody and the fractional amount could be most readily estimated using the ARISTO Nr. 922 "temperature slide-rule" from ARISTO [130]. So while it seems the "System Czerny" rule was used by relatively few people outside of Germany, by far its most lasting impact is to be seen in the number of radiation slide rules it helped inspire to be subsequently made.

6.1.2   GENERAL ELECTRIC RADIATION CALCULATORS

The aptly named Radiation Calculator from General Electric would come to dominate a generation of engineers working in infrared systems design and development in the United States. It was a rule first made for the General Electric Company in 1948, based on a design by the then young engineer Alfred H. Canada.[5] The design for Canada's rule appeared in the company's in-house journal *General Electric Review* [120]. Compared to the radiation slide rule made by ARISTO in Germany, Canada's rule was a very different affair. A far greater number of scales were to be found on his rule, and in the strictest sense, Canada's slide rule was not a slide rule at all but an example of what today we would call a slide chart.

In tracing the origin of Canada's rule one finds it is firmly rooted in the rule proposed, designed, and built by Czerny. In his paper of 1948 he writes that as far as he was aware the only radiation slide rule to come before his own was the rule of Czerny's. It is known that shortly after hostilities in Europe ended, Canada went to Germany as a member of one of the many scientific reconnaissance missions from the United States that entered the country at the time. His work involved assessing the developments Germany had made in infrared technology during the intervening war years and resulted in the publication of the report, *Infrared: Its Military and Peacetime Uses* two years later in December 1947 [119]. We suspect it was during one of these missions Canada first became acquainted with Czerny's rule, which in turn led to the design of his own rule.

The first radiation slide chart of Canada's was known as a "Radiation Slide Rule." It was designated GEN-15 and was made for the General Engineering and Consulting Laboratory, General Electric, Schenectady, New York by the prolific U.S. slide chart manufacturer Perrygraf Corp. Both the card and slide were made of finished cardboard; it was double sided while the various scales found on the rule were grouped into six panels, three on each side. On the front of the card were eight scales while the back contained a further six. There was no cursor. Six small metal hollow rivets located in each corner and in the middle at the top and bottom edges held the two faces of the rule together. Many of the scales provided conversions between a number of different system

of units in common use at the time. The front and back faces of the General
Electric slide chart GEN-15 are shown in Fig. 6.5.

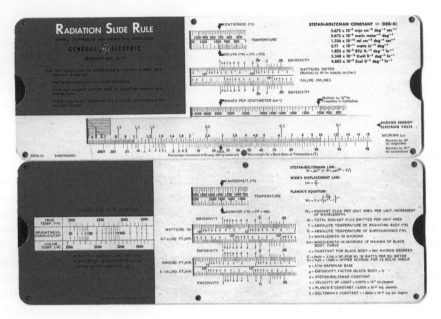

Figure 6.5 Front (top) and back (bottom) faces of General Electric's GEN-15 Radiation Slide Rule.

The top panel on the front of the slide chart was for temperature, the top
scale in units of degrees Celsius [°C] the bottom in Kelvin [K]. Two gauge
marks (heavy arrows) above and below the panel on the card were used to set
the required temperature. Reversing the slide chart, the panel at the top gave
the temperature in units of Fahrenheit [°F] and Rankine [°R]. The middle
panel on the front of the slide chart was for the total radiant exitance. The
top scale was in units of watts per square meter [W·m^{-2}] while the bottom
gave the total radiant exitance in units of calories per square centimeter per
second [cal·cm^{-2}·s^{-1}]. Aligned with the top and bottom of the second panel
and fixed on the card a scale for the emissivity running from 0.05 to 1 appears.
The slide chart could therefore be used to determine total radiant exitances
for both a blackbody ($\varepsilon = 1$) and greybodies ($0 < \varepsilon < 1$). A gauge mark at
$1/(2\pi)$ also appears on the emissivity scale. Presumably it was provided to
allow for the quick conversion between the total radiant exitance and the total
radiance of a blackbody to be made. The conversion factor given, however,
is incorrect, and instead should have been $1/\pi$ for such a purpose (see page
135). The middle panel found on the reverse side of the slide chart gave the
total radiant exitance at various emissivities in units of watts per square inch
[W·in^{-2}] and in British thermal units per square foot per hour [Btu·ft^{-2}·hr^{-1}].

Further conversions for the total radiant exitance at various emissivities in two additional systems of units appear on the bottom panel on the reverse side of the slide chart. Here the top scale gave the total radiant exitance in units of kilowatt-hours per square feet per hour [kWh·ft^{-2}·hr^{-1}] while the bottom one in units of kilocalories per square feet per hour [kcal·ft^{-2}·hr^{-1}].

The third panel located at the bottom of the front of the slide chart was the reason for the sliding part of the rule. Running along the bottom of the panel on the slide is a reverse logarithmic scale for the wavelength λ, measured in micrometers (μm, though on the rule itself it appears as microns, μ, as was custom at the time). Aligned with the wavelength scale but fixed to the card is a scale for the fractional amount of the total radiant exitance emitted by a blackbody into a given spectral band from zero up to some arbitrary wavelength for a blackbody at a temperature T. It is expressed as a percentage and runs from 0.0001% to 98%. Setting the temperature in the top panel allows the fraction of the total radiant exitance radiated by a blackbody into a given wavelength interval to be determined using the bottom panel. A gauge mark "MAX" appears on the scale running along the bottom of the card. It gives the fractional amount when the interval from zero up to a wavelength corresponding to the peak value found in the spectral radiant exitance curve is considered. Above the wavelength scale appears an additional scale in red. It gave the corresponding energy a photon has, in electron-volts [eV], for a given wavelength. Finally, running along the very top of the bottom panel is a scale for the wavenumber $\tilde{\nu}$ measured in waves per centimeter [cm^{-1}].

In the hope of further increasing its usefulness, Canada included a number of other features on the rule. On the front of the slide chart in the top right-hand corner appeared values for the Stefan–Boltzmann constant σ in no less than seven different system of units. Though not mentioned in his paper of 1948, the values Canada uses are those given in 1941 by Raymond Thayer Birge [84]. These were the most accurately known values for the fundamental physical constants at the time and were widely used. On the reverse side, one third of the right end of the slide chart is taken up with formulae for the Stefan–Boltzmann law, Wien's displacement law, and Planck's law. At the left end and taking up a further third of the slide chart is printed a nomogram showing specific characteristics of tungsten. Here color temperature (the temperature required by a blackbody to radiate light of a comparable hue to that of the light source) and brightness in candles per square centimeter as functions of the true temperature of the body in Kelvin are given. Below the equations for the three laws are defined fifteen different symbols for a number of different radiometric quantities and physical constants. Values for all physical constants appear on the rule. Formerly used symbols for the total and spectral radiant exitance, W and W$_\lambda$, and the emissivity, ρ, are given. The modern-day equivalents are M, M_λ, and ε respectively.

Shortly after Canada's paper appeared, news of the rule's availability was quickly reported [5, 3, 43, 4] and it was not long before others started referring to his rule in the technical literature [222, 359]. The rule itself could be obtained postpaid for $0.75 from the General Electric Company, 1 River Road, Schenectady, New York. A revised rule designated the GEN-15A was released in April 1952. It is almost identical to the GEN-15 except in one important regard. The gauge mark of $1/(2\pi)$ on each of the emissivity scales have been replaced with a gauge mark of $1/\pi$. Since the total radiance L for a blackbody is related to the total radiant exitance M by $M = \pi L$, the scale allowed for the determination of both total radiant exitance and total radiance and corrected the erroneous gauge mark of $1/(2\pi)$ printed on the GEN-15. An emissivity equal to one gives the total radiant exitance for a blackbody while at the gauge mark of $1/\pi$ the corresponding total radiance for a blackbody in units of watts per square centimeter per steradian [W·cm^{-2}·sr^{-1}] at the particular temperature setting chosen can be found. Another difference, though only minor, is the removal of "General Engineering and Consulting Laboratory" from just below the rule's name at the top left of the front of the rule and the shifting in the placement of the General Electric logo to a location slightly further down the front face of the rule. Values of all physical constants remained unchanged.

It is not known if the first generation GEN-15 rules came with an instruction manual. The second generation GEN-15A however, most definitely did. A reprint of Canada's December 1948 *General Electric Review* article bound between a cover and labelled as GER-121A served as the instruction manual for the rule.

Owing to the popularity and success of the first radiation slide rule, a second updated version of the slide chart was released in 1956. Its name was changed to "Radiation Calculator," the number of scales appearing on the rule compared to its predecessor was increased, and it came with a now simplified four page instruction leaflet that gave six examples showing how the slide chart could be used in practice [250]. Designated GEN-15B, its construction was also improved. It now consisted of a flexible plastic slide inside a more robust and rigid transparent vinyl sleeve that was screen printed on its inner sides. Clear panels on both the front and back of the sleeve allowed the scales on the inserted slide to be read more easily and accurately compared to the GEN-15 and GEN-15A rules.

On the front of the GEN-15B four panels with a total of eleven scales are found while on the back there appear five panels containing a further eleven scales. The top panel on the front of the slide chart contained scales for the temperature in both degrees Celsius [°C] and Kelvin [K]. Reversing the rule, the panel centered in the middle of the chart at the top had the two temperature scales in Fahrenheit [°F] and Rankine [°R]. As was the case with its predecessor, conversion of temperatures among the four different temperature units by setting the temperature on one scale and reading off values from any of the other three was once more possible. The scale immediately below the temperature panel on the front of the slide chart and to the right was a standard C/D logarithmic scale used to perform multiplication. It was a new scale for the GEN-15B not found on either the GEN-15 or GEN-15A. Presumably this scale was added in case one either misplaced or forgot to bring one's normal slide rule along for the day.

The top scale running along the middle panel of the front of the rule gave, at a particular temperature setting, the total radiant exitance in units of watts per square centimeter [W·cm^{-2}]. An emissivity scale associated with this scale running from 0.05 to 1 on the card above the panel allowed the total radiant exitance for a greybody to be found. On the emissivity scale, as with the GEN-15A, the correct gauge mark of $1/\pi$ appears. On the reverse side, the middle panel just below the temperature panel gave the total radiant exitance and corresponding emissitivies in units of watts per square inch [W·in^{-2}] along the top scale of the panel, and in British temperature units per square foot per hour [Btu·ft^{-2}·hr^{-1}] along the bottom scale of the panel. It represented a reduction of three compared to the six scales found for this radiometric quantity on the GEN-15 and GEN-15A rules. The bottom scale found on the middle panel on the front of the slide chart gave the spectral radiant exitance for a blackbody at the peak wavelength found in the spectral radiant exitance curve when plotted as a function of wavelength at a given temperature. It was given in units of watts per square centimeter per unit wavelength interval [W·cm^{-2}·µm^{-1}]. The scale did not appear on either the GEN-15 or GEN-15A but was a very useful addition since it could be used in conjunction with one of the other new scales to calculate the spectral radiant exitance, a quantity that could otherwise not be directly found using the slide chart.

The very long panel appearing on the front of the slide chart at its base on the GEN-15B was significantly extended compared to the one found on its predecessor. In common with both the GEN-15 and GEN-15A, running along the bottom of the panel on the slide was a reverse logarithmic scale for the wavelength, measured in micrometers. Aligned with this scale was a fixed scale on the card for the fractional amount of the total radiated exitance emitted into a finite spectral band from zero up to some arbitrary wavelength. Marked as $\frac{W_{0-\lambda}}{W_{0-\infty}}$ (our $\mathfrak{F}_{e,0\to\lambda}$), it was expressed as a percentage and extended from 0.0001% to 99%. Again a gauge mark labelled "MAX" appeared on the scale at 25% and gave the fractional amount when an interval from zero up to the peak wavelength was considered. A new scale ran along the card above the top of the bottom panel. It was for the ratio of the spectral radiant exitance at any given wavelength to that at the peak wavelength. The scale was marked as $\frac{W_\lambda}{W_{\lambda_{max}}}$. Aligned with this scale was a reverse logarithmic scale for the wavelength identical to that just described. Such an identical scale would not have been needed if the slide chart came with an adjustable cursor. Instead, reliance on direct alignment of each wavelength scale with these two scales was needed in order to be correctly read.

The bottom panel on the front of the rule was quite wide. Running through the middle of this panel on the slide between the two identical wavelength scales above and below it was a black curve representing the transmission spectrum for electromagnetic radiation in the atmosphere. It gave the fraction of incident radiation transmitted (the fraction being expressed as a dimensionless transmission coefficient between zero and one) through the atmosphere as a function of wavelength over a distance of one nautical mile horizontally to the surface of the Earth at approximately 80% relative humidity for an air temperature of approximately 80 °F. Finally, the curve in red found in the visible wavelength portion of the spectrum was a relative luminosity curve for the eye and was identical to that found on the GEN-15 and GEN-15A rules. On the reverse side of the rule, at the top and to the left of the temperature panel was a very narrow panel corresponding to the wavenumber $\tilde{\nu}$ in units of per centimeter $[\text{cm}^{-1}]$. It converted the peak wavelength $\lambda_{\max}$ found in the spectral distribution curve in the linear wavelength representation into wavenumbers $\tilde{\nu}_{\lambda_{\max}} = 1/\lambda_{\max}$.

The remaining two panels found on the reverse side were also new additions to the GEN-15B rule. In place of the lower panel found on the GEN-15 and GEN-15A rules for the total radiant exitance in units of kilowatt-hours per square feet per hour and kilocalaries per square feet per hour is a panel with two scales relating to actinometric quantities. The scale along the top of this panel was a reverse logarithmic scale for the total photon exitance M_q^b in units of photons per square centimeter per second $[\text{photon} \cdot \text{cm}^{-2} \cdot \text{s}^{-1}]$. The bottom scale appearing in this panel gave the energy, in electron-volts [eV], a single photon has at a wavelength corresponding to the peak wavelength. The final long narrow panel appearing on the reverse side of the slide chart at its base was used to find the irradiance normal to an area of one square centimeter at a certain distance from a blackbody source where losses in the intervening atmosphere were assumed to be negligible. The irradiance was measured in units of watts per square centimeter $[\text{W} \cdot \text{cm}^{-2}]$. The scale running along the top of the panel was a reverse logarithmic scale for the irradiance at close range. Aligned just above this scale on the card was the range in centimeters from 90 cm to 1.26×10^5 cm. The scale running along the bottom of the panel was also a reverse logarithmic scale for the irradiance at ranges beyond a kilometer. Aligned just below the scale on the card was the range in nautical miles from 0.38 to 1000 nautical miles. For convenience a gauge mark at one kilometer appeared on the nautical mile range scale.

Once again some useful information to aid the user appears on the GEN-15B. On the reverse side in the top left corner the equations for Planck's law, Wien's displacement law, and the Stefan–Boltzmann law are again given. Below these three laws the Stefan–Boltzmann constant is once more given in seven different system of units. Starting in the top right corner and listed vertically downwards are sixteen different radiometric symbols and physical constants. The values for the seven physical constants that appear have also been updated compared to those found on the GEN-15 and GEN-15A and suggest the scales found on the later Radiation Calculator should be more

accurate compared to those found on Canada's initial rules. Though not mentioned, the updated values used for all physical constants correspond with
those given in 1953 by Jesse William Monroe DuMond and E. Richard Cohen [197]. Two different panel variations found in the top left-hand corner
on the front of the GEN-15B appeared. The first of these came "courtesy of
Light Military Electronics Equipment Department, General Electric, Utica,
NY" the second was "compiled by Optics and Color Engineering Component,
General Engineering Laboratory, General Electric, Schenectady, NY." As late
as 1964 the rule was being advertised for sale for a cost of $1.50 [209, 539].

A third and final iteration of the slide chart made for General Electric first
appeared sometime in 1965 [696]. Designated the GEN-15C it was identical
in form, in both construction and in the number of scales used, to the GEN-
15B. Finding two different designations for what were essentially identical slide
charts may initially seem a little odd. However the most probable reason for
the discrete change lay in an unreported error in one of the new scales found on
the GEN-15B. When Canada introduced his rule in 1948 he characterized it as
being suitable for order-of-magnitude calculations. It turned out his rule was
usually a lot better than this, often giving estimates with an error less than
1%. The scales on the GEN-15B had been updated, having taken into account
the 1952 least-squares adjusted values for the physical constants as given by
DuMond and Cohen. Accurate readings for many of the quantities found on
the rule however not only required a certain degree of accuracy in the scales
themselves but also in the positioning of the gauge marks on the card from
where readings were made. The two gauge marks used on the actinometric
panel of the GEN-15B were incorrectly positioned. Accordingly, estimates for
both the total photon exitance and the spectral photon exitance at the peak
wavelength were in error anywhere upwards of 15%. While this may not seem
large, its significance can be better appreciated by highlighting some of its
potential consequences. For example, the "generous" nature of the GEN-15B
would yield infrared sensors that performed worse than expected and caused
some companies at the time significant financial grief and embarrassment. The
release of the GEN-15C finally corrected the ill-positioned gauge marks found
on the two actinometric scales, but not before the GEN-15B had earned the
nickname "Generous Electric." The error does not seem to have ever been
formally acknowledged, at least in the literature, though engineers working in
the infrared at the time were strongly warned to always check and be sure the
Radiation Calculator they were using was not the GEN-15B.

Like the GEN-15B, the GEN-15C is known to have come with at least three
different front panel variations. The first was the same as that found for one of
the GEN-15B's, it coming "courtesy of Light Military Electronic Equipment
Department, General Electric, Utica, NY." The second came "courtesy of
Aerospace Electronics Department, General Electric, Utica, NY," while for
the third "General Electric" simply appeared. A photograph of the front and
back faces of the GEN-15C is shown in Fig. 6.6.

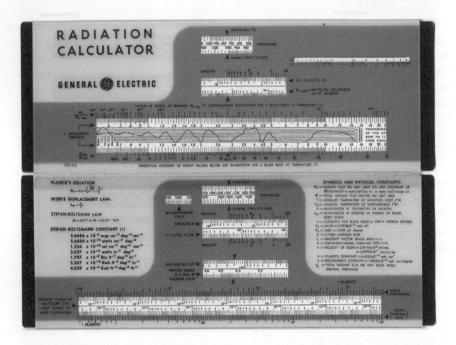

Figure 6.6 Front (top) and back (bottom) faces of the General Electric slide chart GEN-15C.

The radiation slide rule designed by Canada was by far the best known example of its kind in the U.S. By the early 1960s its adoption and use in the U.S. was very widespread. In 1962 the founding editor of the journal *Applied Optics*, John N. Howard, noted the slide chart from General Electric was "... usually found in every infrared man's briefcase" [301]. Some years later, Richard D. Hudson, Jr., in his text *Infrared System Engineering* went further. He observed Canada's rule was "... virtually a badge of the infrared fraternity," and despite Canada himself characterizing his rule as suitable for only order-of-magnitude calculations, suggested around 90 per cent of all thermal radiation calculations performed in the United States in the decade leading up to 1969 were made with one of the various slide charts from General Electric [311].[6] Descriptions of the GEN-15C slide chart were also to be found in a number of widely read infrared texts and manuals of the day suggesting their authors felt the need for those working in the field to at least be acquainted with its existence [696, 93, 311, 82, 697].

The importance of the rule is also reflected in the number of citations made to it in the literature after 1948. While it is true many who made use of the slide chart may not have referred to it explicitly, in the decades following its introduction up until the mid-1970s it was widely cited. As a relatively inexpensive device that at best was only capable of producing estimates accurate to half a per cent, its continued longevity into an age of increasingly affordable computing power was all the more remarkable. In the mid-1980s one finds Lewis J. Pinson in his text *Electro-Optics* strongly urging anyone considering to do serious work in the infrared to acquire the Radiation Calculator from General Electric [499]. The GEN-15C Radiation Calculator was still being referred to two decades later. It was mentioned in a paper outlining a sample return mission to the two Martian moons Phobos and Deimos using a proposal based on contemporary solar sails [429]. More recently still, in the text *The Art of Radiometry* written in 2010 by James M. Palmer and Barbara G. Grant [476], an image of the front face of the GEN-15C is given and is described as the "venerable" Radiation Calculator from General Electric, presumably for the benefit of a younger audience unfamiliar with calculational aids of this type. And it was not just engineers working in the infrared who found the Radiation Calculator from General Electric useful. For example, one finds the many virtues of the rule being extolled to, of all people, advanced photographic technicians [313, 314].

The success of Canada's radiation slide rule can be attributed, at least in part, to its ease of use, ready availability, and its ability to produce results accurate enough for their intended purpose. In a single setting seven radiometric and actinometric quantities could be read directly from the Radiation Calculator and all accomplished from a highly portable, relatively inexpensive device. Ultimately, Canada's rule spawned a number of successor slide charts after it ceased to be available. Each was almost identical in design and form to Canada's later Radiation Calculator and has extended the legacy of his rule right up to the present day. Made of finished cardboard and often given away as a promotional item these successor *Infrared Radiation Calculator* slide charts first appeared in the mid-1970s from firms and organizations such as Sensors, Inc. [29, 31, 592, 30, 32], the Infrared Information Analysis Center (IRIA) [316, 662], EG&G Judson [34, 35], BAE Systems, and EG&G Judson's successor Teledyne Judson Technologies, LLC.

Almost identical to the two Radiation Calculators from General Electric, the only minor differences between the earlier GEN-15B and GEN-15C rules and the Infrared Radiation Calculators which came later was in the removal of the general C/D panel found on the front of the General Electric rules and in the combining of the panel for wavenumber found at the top on the reverse side of the rule with the adjacent panel for temperature so as to form a single panel for both quantities.

The first of the Infrared Radiation Calculators to appear was made for the firm Sensors, Inc., by Perrygraf in 1976. It came with a very brief four page guide that contained seven worked examples showing how the rule could be used [595]. Single copies of the calculator could be obtained free of charge by writing to Sensors, Inc., while additional calculators and example problem guides cost $5.00 postpaid [28].

A decade later no less than three Infrared Radiation Calculators appeared. Two of these came from IRIA and EG&G Judson while a third was made specifically for the Engineering Summer Conferences program held at the University of Michigan. All three slide charts were made by Perrygraf. The Engineering Summer Conferences rule was identical to the slide chart made for IRIA except for its color; blue and white for the former while red and white for the latter. The Engineering Summer Conferences were a series of conferences, seminars, workshops, and short courses held annually at the University of Michigan. Two short courses related to the infrared were *Fundamental of Infrared Technology* and *Advanced Infrared Technology* [8, 296, 27, 736, 33]. Running back-to-back over consecutive weeks in the summer months for many years, in the early days participants had received a Radiation Calculator from General Electric as part of their general course fee [21]. After these ceased to be available commercially sometime in the early 1980s, the specially commissioned Infrared Radiation Calculator by the short course organizers was substituted in its place. Presumably these continued to be given out to participants until none remained; a time by which their general utility had long been superseded by the arrival of cheap, digital technology. Front faces for the four different Infrared Radiation Calculators made by Perrygraf in the 1970s and 1980s are shown in Fig. 6.7.

While the four Infrared Radiation Calculators were all very similar, they were by no means identical. The EG&G Judson rule was the first to introduce a simple though badly needed innovation of using labels. These appeared next to all scales found on the rule. Using capital letters running from A through to W allowed each scale to be easily referenced by its letter in the four page user guide which accompanied the rule [204]. The EG&G Judson rule also had by far the greatest longevity. Rules with copyright dates of 1991 [34, 35] and 1995 [61], in addition to the date of 1986 when the rule first appeared, are known to exist. In February 2008 EG&G Judson changed its name to Teledyne Judson Technologies, LLC, after it was acquired by Teledyne Scientific & Imaging, LLC, and it is under this latter name an Infrared Radiation Calculator with copyright dates of 2008 and 2009 are found.

Another Infrared Radiation Calculator to have come along in recent years is one made for BAE Systems by American Slide-Chart/Perrygraf. Copyrighted 2012, it is red and white in color and in layout and appearnace is very similar in style to the earlier Engineering Summer Conferences and the IRIA rules. The Teledyne Judson Technologies and BAE Systems rules currently represent the last in the generation of radiation slide rules to be made. At an incredible 64 years since Canada's Radiation Slide Rule first appeared it is testament to the extent and usefulness his initial design found throughout the infrared scientific and engineering communities. The front faces of the Teledyne Judson Technologies and BAE Systems rules are shown in Fig. 6.8.

The continued use of these slide charts well after cheap and affordable computing power had become widespread was not without its critics. Writing in a review in 1993 of Waldman and Wootton's recently published text *Electro-Optical Systems Performance Modeling*, the reviewer John Watson from the

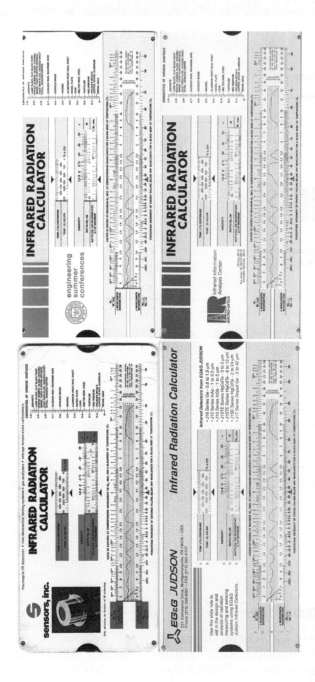

Figure 6.7 Front faces for four different Infrared Radiation Calculators made by Perrygraf for Sensors, Inc. (top left), Engineering Summer Conferences (top right), EG&G Judson (bottom left), and IRIA (bottom right).

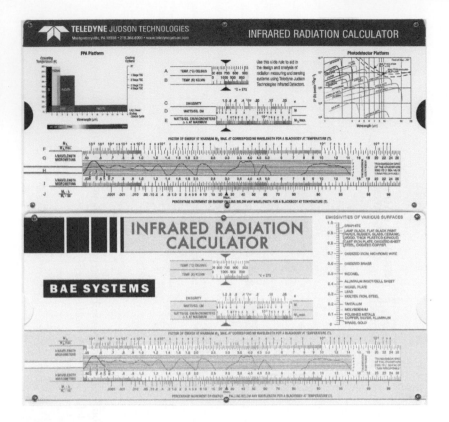

Figure 6.8 Front faces of two recent Infrared Radiation Calculators made by Perrygraf for Teledyne Judson Technologies, LLC (top) and BAE Systems (bottom).

University of Aberdeen found one of the book's "strange quirks" was in why the authors had bothered spending so much time discussing the Infrared Radiation Calculator by IRIA [676]. Seeing this "calculator" as nothing more than a flimsy piece of cardboard from the days of yore, Watson questioned whether anyone still seriously used these sliding templates in an age of powerful programmable calculators and desktop computers. And yet people still do continue to cling to their much loved "paper calculators" long after computers have become very cheap and powerful [437].

6.1.3   THE ADMIRALTY RULE

During the course of his work on thermal detection problems, having found existing tables for thermal radiation neither sufficiently comprehensive nor convenient for frequent use, the Polish émigré Mieczysław Wiktor Makowski,[7] who in early 1945 was working at the Admiralty Research Laboratory in

Teddington, England, set about to develop a series of nomograms for such work. Intended as a labor saving device they allowed Makowski to quickly approximate many of the most frequently encountered quantities in thermal radiation.

At around the time of their completion, a simple eight-inch slide rule made by Professor Czerny of Frankfurt University was received at the Admiralty. Its length suggests the rule received was one of Czerny's initial hand-built models rather than the later prototype or the Nr. 10048 production model made by ARISTO as each of the latter two rules were significantly longer than eight inches. No doubt a fortuitous arrival, to Makowski the advantages of such a simple device were immediately apparent. At once the decision was taken to extend the computations he and his co-worker, the Assistant Experimental Officer L. A. J. Verra, had already made, the intention being the development of a slide rule far superior to the one they had just received. By October 1945, development of the rule had progressed to the point where discussions between one of England's leading slide rule manufacturers, A. G. Thornton Limited of Manchester, and Makowski on behalf of the Admiralty had commenced [1]. The nature of the talks sought to understand what the most suitable form the measurements for the scales ought to take for manufacturing purposes. The issue of copyright was also raised. Here the Admiralty felt copyright should be retained by them over any future produced slide rule in view of the amount of work that was going to be involved in its preparation.

The first description of Makowski's radiation slide rule appeared in a technical report he wrote for the Admiralty in September 1947 [424]. Considerable attention is paid to explaining how the scales were calculated. He notes approximately 1000 man hours of computing time were involved in the preparation of the scales. In finding the positions of major tick marks for each scale, these were calculated using infinite power series expansions he developed in the report and correspond to Eqs (3.49) and (3.51) we already saw in Chapter 3. Intermediate marks between major tick marks were then found via interpolation using Bessel's interpolation formula up to fourth and sometime sixth differences. The majority of the calculations were made using two Brunsviga calculating machines.[8] Calculations for some of the scales were then checked using one of Britain's very early electronic computers located at the recently established Mathematics Division of the National Physical Laboratory, a close neighbor of the Admiralty's at Teddington.

In his report, Makowski mentions that three prototypes had been constructed, each being capable of calculating a great many quantities relating to thermal radiation. Experience in working with these prototypes convinced Makowski that a reduction in the total number of scales on the final commercially available rule was needed. Though not mentioned, it's likely the absence of any scale relating to radiance on the final rule, even though it was one of the scales found on Czerny's rule, was removed from the prototype. Burdened by their inclusion, Makowski saw that they could be readily done away with, since for a blackbody each could be obtained from the radiant exitance upon dividing by π. Doing so avoided any unnecessary multiplicity of scales. The technical report continued with a brief description of the slide rule, its scales, and an analysis of its accuracy. A number of examples typical of those one might encounter in practice were also included so that, as Makowski wrote, the report might serve as an instruction booklet for any future produced slide rule.

The first commercially available rule based on Makowski's design was made for the Admiralty by A. G. Thornton Limited and was designated the F5100 Radiation Slide Rule. The base of the rule was made from Honduras mahogany. White celluloid veneers containing the scales and brief instructions were then pinned and glued onto the wooden base. Its impending availability was reported in ONR's *European Scientific Notes*, in the second August issue of 1948 [2]. At the time, the cost was not known but was expected to be rather high and the initial production run was to be limited to 50 rules only, a quantity the Admiralty thought sufficient to meet the needs of other parties engaged in thermal radiation work. The Admiralty was however, willing to supply photo-printed sheets of the scales on stabilized card weight stock to any interested institution for about $2 per set and suggests the Admiralty was not expecting much of a demand for the rule. How mistaken they would turn out to be!

In 1949, a paper based on a slightly modified version of Makowski's technical report was published. It appeared in the December issue of *The Review of Scientific Instruments* [425] and was published at a time when the rule was first thought to be available in England. Makowski does not seem to have been aware of Canada's rule, as he makes no reference to it. Like Canada's rule before it, references to Makowski's rule in the literature quickly followed [279]. Summaries of his paper also appeared in a number of the abstracting journals of the day [38, 37, 6]. While the rule was at least known in the United States, being reviewed in the U.S. publication *Mathematical Tables and Other Aids to Computation* in 1950 [645], and cited by U.S.-based authors some years later [359, 567, 584, 249], it seems it could only be obtained directly from England until the early 1960s after which time it became available through a number of U.S.-based distributors [9, 11, 10, 13, 624, 696]. At a cost of around $150 in 1962, it was by no means inexpensive, and meant the rule was typically purchased for shared use amongst the employees of a company or organization involved in infrared work rather than by an individual. A reprint of Makowski's 1949 paper bounded between a lime green cardboard cover served as the instruction manual for the rule. In time, the rule would become known simply as "the Admiralty rule."

Makowski's design for the Admiralty rule was one of the most elaborate of all the radiation slide rules to be made. A total of 23 scales for 14 different quantities can be found on the rule. Like Canada before him, Makowski used Birge's 1941 values for the physical constants in the calculations made for the scales of his rule. He is quite explicit about this. Not only does Makowski refer to Birge's work in his paper of 1949, he repeats it again in print on the back of the rule as part of the brief instructions given. At the end of a list of numerical values for the constants used, one finds a reference to Birge but unfortunately the year of publication is incorrectly given as 1944 rather than 1941.

Each scale on the Admiralty rule was conveniently labelled from a̲ through to s̲ and we will use these letters when referring to them here. On the front of the stock seven scales appear, four at the bottom (labelled a̲, b̲, c̲, d̲) and three at the top (labelled e̲, f̲, g̲). Twelve scales are found on the slide which is reversible. Since in construction most of the Admiralty rules were of the closed-frame type it meant removal of the slide for reversal purposes was needed. The side of the slide labelled $\boxed{\text{ENERGY}}$ corresponding to radiometric quantities contained six scales labelled h̲ through to m̲ while a further six scales labelled n̲ through to s̲ on the reverse side of the slide labelled $\boxed{\text{PHOTONS}}$ corresponding to actinometric quantities could be found. The last three scales are located on the back of the stock running across the top of the rule but are not labelled. Relevant formulae, table of values for six physical constants, and a set of brief instructions for operation of the rule also appear on the reverse side of the rule.

Accommodating all these scales meant the rule was rather large, measuring 45 cm long by 9 cm wide by 14 mm thick. The slide was a further 4 cm longer at 49 cm. Even a sizeable cursor made of perspex came with the rule, it measuring 4.5 cm across by 9.5 cm high. A "deluxe" model of the rule was also available. In construction it was of the open-frame type. Here two large wooden end braces attached to the back of the rule held the top and bottom stocks together. These large end braces meant that the scales located on the back of the closed-frame rule had to be removed and were now found underneath the lid of the large wooden case that came with the rule. A photograph of the front face of the A. G. Thornton F5100 radiation slide rule is shown in Fig. 6.9.

The scales together with their associated label found on the Admiralty rule are given below. Note that Makowski uses the symbol H for those quantities relating to radiant exitance (possibly for irradiance which was the symbol used for this radiometric quantity at the time) and Q for those relating to photon exitance.

---

a̲ Total radiant exitance, H [W·cm^{-2}]

b̲ Spectral radiant exitance at the peak wavelength, $H_{\lambda_{\max}}$ [W·cm^{-2}·cm^{-1}]

c̲ Temperature in degrees Celsius, t [°C]

d̲ Temperature in Kelvin, T [K] (100–10 000 K)

e̲ Wavelength, λ [μm] (Black: 0.3–30 μm, Red: 30–3000 μm)

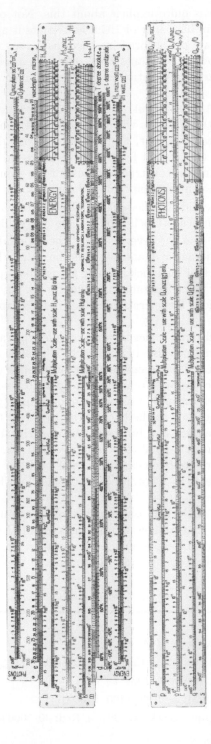

Figure 6.9 Front face of the F5100 radiation slide rule made by A. G. Thornton Ltd for the Admiralty Research Laboratory at Teddington, England. Above is the ENERGY side of the slide while below is the PHOTON side of the rule.

<u>f</u> Total photon exitance, Q [photon·s^{-1}·cm^{-2}]

<u>g</u> Spectral photon exitance at the peak wavelength, $Q_{\lambda\max}$ [photon·s^{-1} ·cm^{-2}·cm^{-1}]

<u>h</u> Spectral fractional amount in energetic units, $H_\lambda/H_{\lambda\max}$ (for $\lambda = 0.3$ to 30 µm)

<u>i</u> Spectral fractional amount in energetic units, $H_\lambda/H_{\lambda\max}$ (for $\lambda = 30$ to 3000 µm)

<u>j</u> Multiplication scale for use with scale <u>b</u> only

<u>k</u> Multiplication scale for use with scale <u>a</u> only

<u>l</u> Integrated fractional amount in energetic units, $H_{\lambda-\infty}/H = 1 - H_{0-\lambda}/H$ (for $\lambda = 30$ to 3000 µm)

<u>m</u> Integrated fractional amount in energetic units, $H_{0-\lambda}/H$ (for $\lambda = 0.3$ to 30 µm)

<u>n</u> Spectral fractional amount in photonic units, $Q_\lambda/Q_{\lambda\max}$ (for $\lambda = 0.3$ to 30 µm)

<u>o</u> Spectral fractional amount in photonic units, $Q_\lambda/Q_{\lambda\max}$ (for $\lambda = 30$ to 3000 µm)

<u>p</u> Multiplication scale for use with scale <u>g</u> only

<u>q</u> Multiplication scale for use with scale <u>f</u> only

<u>r</u> Integrated fractional amount in photonic units, $Q_{\lambda-\infty}/Q = 1 - Q_{0-\lambda}/Q$ (for $\lambda = 30$ to 3000 µm)

<u>s</u> Integrated fractional amount in photonic units, $Q_{0-\lambda}/Q$ (for $\lambda = 0.3$ to 30 µm)

In its design the temperature scale in Kelvin (scale <u>d</u>) was taken as the basis for the rule. As a logarithmic scale it meant the scales <u>a</u>, <u>b</u>, <u>c</u>, <u>f</u> and <u>g</u> were also logarithmic while the eight scales relating to the integrated fractional amounts (scales <u>h</u>, <u>i</u>, <u>l</u>, <u>m</u>, <u>n</u>, <u>o</u>, <u>r</u>, <u>s</u>) were not. Along scales <u>h</u> and <u>n</u> four gauge marks for the peak wavelengths in the spectral curves in energetic and photonic units for two different spectral representations are found. The usual peak wavelengths in the linear wavelength representation in energetic units (labelled $H_{\lambda\mathrm{mx}}(\lambda_\mathrm{m})$) and photonic units (labelled $Q_{\lambda\mathrm{mx}}(\lambda_\mathrm{m}')$) are given as are those corresponding to the peak wavelengths when a spectral representation in the linear wavenumber representation is considered. These are labelled $H_{\nu\mathrm{mx}}(\lambda_\mathrm{m}'')$ for energetic units and $Q_{\nu\mathrm{mx}}(\lambda_\mathrm{m}''')$ for photonic units and were very unusual gauge marks that would not be found on any future radiation slide rules. Gauge marks labelled "TEMPERATURE" on scale <u>m</u> on the ENERGY side and scale <u>s</u> on the PHOTONS side of the slide allowed these gauge marks to be aligned to the desired temperature on the stock.

In operation, once the temperature on either scale <u>c</u> or <u>d</u> was selected and the cursor moved over it so the two coincided, from a single setting five quantities could be immediately read from the rule. These were: (i) the peak wavelength using the gauge mark located on scale h̲ or n̲, (ii) total radiant exitance using scale a̲, (iii) spectral radiant exitance at the spectral peak wavelength using scale b̲, (iv) total photon exitance using scale f̲, and (v) spectral photon exitance at the spectral peak wavelength using scale g̲. If, on the other hand, after the desired temperature had been set, moving the cursor to a given wavelength on scale e̲ allowed the various fractional quantities located on either side of the slide to be determined. The range on the wavelength scale was divided into two parts. The first, in black, corresponded to wavelengths from $0.3 - 30 \, \mu m$ while the second, in red, corresponded to wavelengths from $30 - 3000 \, \mu m$. Color matching between either the black and red wavelength scales on the stock with identically colored scales for the fractional quantities on the slide was required when reading these scales. On the back of the closed framed model, running across the top of the stock, three unlabelled scales relating to the wavelength and wavenumber of an electromagnetic wave and the energy a corresponding photon has in electron-volts were given. As these three scales were read without the aid of a cursor, for convenience, the wavenumber scale was given twice.

It was possible to use the Admiralty rule for calculations outside the temperature limits of 100 to 10 000 K using a simple extension process. For a temperature T outside the range, a temperature T_s is selected so that it falls within the temperature range of the rule such that $T_s = T \times 10^n$. Here n is a conveniently chosen integer. Values for the various quantities are then found from the slide rule using the temperature value T_s. Values appropriate to the original temperature T can then be found using the following transformation rules:

$$H_\lambda = (H_\lambda)_s \times 10^{-5n} \quad , \quad Q_\lambda = (Q_\lambda)_s \times 10^{-4n}$$
$$H_{\lambda_1 - \lambda_2} = (H_{\lambda_1 - \lambda_2})_s \times 10^{-4n} \quad , \quad Q_{\lambda_1 - \lambda_2} = (Q_{\lambda_1 - \lambda_2})_s \times 10^{-3n}.$$

As a reminder, these transformation rules were listed on the back of the slide rule.

The accuracy of the rule varied according to the region of the scale in use. In general, calculations made at longer wavelengths (those made using the red wavelength scale) were accurate to within 1%. In the region of the respective spectral maxima, the relative error reduced to no more than a few tenths of one per cent. For shorter wavelengths (those made using the black wavelength scale) the relative error in the rule gradually increased up to about 5%. To help prevent a reduction in the accuracy at these shorter wavelengths a diagonal scale at one end of the slide for scales h̲, m̲, m̲ and s̲ (the ratio scales) were introduced and meant values for these ratios to incredibly small values could still be calculated to reasonable accuracy.

The four multiplication scales (scales j̲, k̲, p̲, and q̲) found on the rule at first appear a little mysterious. In both his technical report of 1947 and his paper of 1949 Makowski seems a little cryptic in his remarks describing how a user may find these scales useful. Nor is any clue found in the examples he provides as none make use of these scales. He writes, "[T]he multiplication scales are so arranged that it is unnecessary to move the slide, when set at a given temperature, in order to carry out the operation of multiplication" [425]. While this is true, why one would want to do such a thing is not revealed. Of course, to the infrared worker it was immediately apparent that these scales could be used to find total exitances and the spectral exitances at the peak wavelength for greybodies. Interpreting the multiplication factor, which is a number between zero and one, as a value for the object's emissivity meant the total exitance for a greybody could be found in both energetic and photonic units.

An intriguing oversight made on the part of Makowski is to be found in a short erratum [426]. Appearing four months after his paper of 1949 was published, he notes that a footnote should have been added to the paper stating that the work was carried out at the Admiralty Research Laboratory in Teddington, England, and was the organization for whom the rule was designed and where all its development took place. We find it strange Makowski should have failed to mention this. By the time of its publication in 1949 Makowski no longer worked for the Admiralty, having moved to Polish University College which was part of the Polish émigré community of post-war London. Was it merely an oversight on the part of Makowski or had his parting from the Admiralty been less than amicable? We guess this will never be known.

From its inception in the late 1940s up until the early 1960s, the rule for the Admiralty was made by A. G. Thornton Limited from Wythenshawe, Manchester. Starting in 1965, the rule continued to be produced by the firm Blundell Harling Limited, another well-known slide rule manufacturer from Weymouth in England. A notable feature of the Blundell Harling Limited produced rule was its improved construction, it now being made completely of plastic. The scales on the rule however, do not seem to have accounted for the latest values of the fundamental constants known at the time because the reverse side of the updated rule still lists the 1941 values from Birge. Nor is the error in the year of publication of 1944 for Birge's values corrected. When the rule finally ceased to be commercially available is not known, but it is more than likely it was produced by Blundell Harling Limited right up until the end of the slide rule era in the late 1970s.

Everything about the Admiralty rule was imposing. Grand in scale and ambitious in design, it was by far the most elaborate rule of its day. Until it was finally displaced by a similar though thoroughly updated rule released in the United States in 1970 by Electro Optical Industries, Inc., and shortly thereafter by affordable, hand-held programmable calculators, for a time, the Admiralty rule was unsurpassed, making it the tool of choice for very accurate thermal radiation work. Its prohibitively high cost however, limited its general availability and meant that compared to its far less expensive rivals, in particular the Radiation Calculator from General Electric, it was not as widely used as it could have been. Short descriptions of the rule are found in some of the more comprehensive infrared texts of the day [696, 311], and whenever the Admiralty rule was referred to by authors in the literature, it always came with the highest praise [293].

Even today Makowski's rule continues to draw attention from slide rule enthusiasts partly due to its size and partly due to the elaborate nature of its scales [50, 609]. Reflecting some forty years after his seminal paper with William Shockley on the thermodynamic energy conversion efficiency limit of a solar cell [598], Professor Hans Joachim Queisser recalls how in 1960, stuck in a small, cramped, rented office which was a converted old apricot barn in Mountain View, California, he was tasked with calculating the solar efficiency curve (the curve is now known as the *Shockley–Queisser limit*) by hand, using nothing more than Shockley's trusty Admiralty rule [536, 537]. At the time, in the absence of any computer, it was an exceedingly tedious calculation requiring Eq. (3.51) to be evaluated many times and each time to a reasonable level of accuracy. Queisser, who incidentally was the successor of Professor Marianus Czerny at Frankfurt University, recalls [538] that to achieve this level of accuracy required space — physical space, that is — on the grandest of all the radiation slide rules!

6.2 TWO MYSTERIOUS RULES

A semicircular slide rule designed and constructed for the rapid computation of radiance values from the output of a radiometer was described by Anthony LaRocca and George J. Zissis in May of 1958 [379]. The rule they described was designed for use with a Barnes Engineering Company model R-4B2 radiometer, though the slide rule's general design was said to be applicable to any radiometer. We have not been able to locate a copy of the report describing their rule, so any further details regarding the rule cannot be given. The information we have comes from a very brief abstract reproduced in a technical report written for the Naval Ordnance Laboratory in 1961 by Wright Instruments, Inc. [713].

In 1963 a slide chart made by Perrygraf of finished cardboard for Vahlo[9] appeared [16, 18, 19]. The rule is very different from all other radiation slide rules considered here in so far as its principle function was in the estimation of quantities relating to the performance of infrared detectors, such as detectivities, rather than with more fundamental quantities relating directly to blackbody radiation. The rule is also one of the more mysterious of the radiation slide rules as little about the rule has come down to us since its appearance in the early 1960s. Designated a Photon Detector Slide Rule, it was designed specifically to simplify calculations encountered by infrared system designers working with photon detectors. Despite this, it was still possible to find an estimate for the actinometric fractional amount using the rule.

One of the more common figures of merit used to characterize the performance of a detector is the specific or normalized detectivity D^*, pronounced "dee-star." For photon detectors such as photoconductive or photovoltaic detectors, when the detectivity is limited by the noise associated with those photons coming from background radiation rather than by noise intrinsic to the detector itself, it is referred to as the "blip" or background-limited detectivity. The blip detectivity as a function of wavelength λ for a photoconductive detector[10] is given by [311]

$$D^*(\lambda) = \frac{\lambda}{2hc}\sqrt{\frac{\eta}{M^b_{q,0\to\lambda}}}.$$

(6.1)

Here η is the quantum efficiency while $M^b_{q,0\to\lambda}$ is the photon band exitance coming from the background. As can be seen from Eq. (6.1), estimating the blip detectivity for photon detectors requires the actinometric integrated amount and presumably was the reason that this more fundamental quantity is found on Vahlo's Photon Detector Slide Rule.

In construction, the rule made for Vahlo was very similar to the first of the rules made for General Electric, the GEN-15. Made of finished cardboard and containing panels on each side of the rule, the two faces were held together by six small metal hollow rivets located in each corner and in the middle of the rule at the top and bottom edges. On the front face of the rule, three panels appeared and were used to estimate non-blip detectivities. On the reverse side a further four panels are found. Of interest to us here is the third panel from the top. Running across the stock just above the panel was a scale for wavelength in microns while on the slide a scale for temperature, in Kelvin [K], ran along the top while a scale for the actinometric fractional amount ran along the bottom. On setting the temperature to an appropriate wavelength, the value for the fractional amount was read from a gauge mark (a large black arrow) located on the stock just below the bottom of the panel. Interestingly, the top panel located on the reverse side of the rule could be used to find the wavelength–temperature product. Along the top just above the panel on the stock was a scale for the wavelength in microns. On the slide a scale for temperature, in Kelvin [K], ran along the top while a scale for the wavelength–temperature product, in units of micrometers Kelvin [μm·K], ran along the bottom. As with finding the fractional amount, on setting the temperature to an appropriate wavelength the value for the product could be found from a similar gauge mark located on the stock just below the bottom of the panel.

The Photon Detector Slide Rule was available for $2.50 by writing to Vahlo [16, 18]. Vahlo claimed their rule would eliminate the need in having to use other radiation slide rules, charts, or graphs, or in having to perform tedious calculations by hand ever again. Whether this was true or not we cannot say as no reference to the rule in the literature has been found. Front and back faces of the mysterious Vahlo Photon Detector Slide Rule are shown in Fig. 6.10.

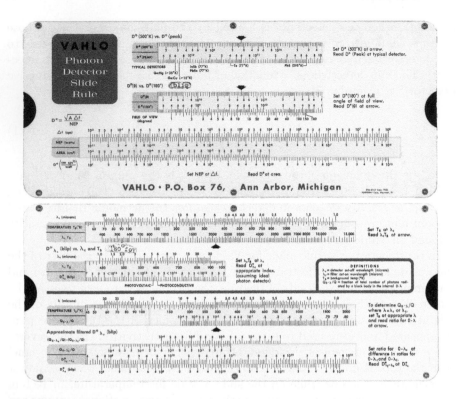

Figure 6.10 Front (top) and back (bottom) faces of the Photon Detector Slide Rule made for Vahlo. Photograph courtesy of John E. Greivenkamp from the University of Arizona.

6.3 THE DENEM NUCLEAR RADIATION CALCULATOR

Of all the radiation slide rules produced, the DENEM Nuclear Radiation Calculator is the most elusive. Undocumented in the literature and completely unknown until only very recently makes this calculator all the more intriguing. Its name together with the range in values found on its temperature scale suggest the slide chart was used for thermal radiation calculations associated with atomic explosions. In all likelihood the slide chart was probably made on contract for the U.S. military leading us to suspect it was either a restricted or classified item. What "DENEM" stood for is not known.

The slide chart we have came to us during this writing of the book. It was part of the estate of the late Martin Ravotto (1920–2007). Ravotto was a nuclear physicist for the U.S. Department of the Army who worked at the Picatinny Arsenal in New Jersey and, at times, the Los Alamos National Laboratory in New Mexico. Before his retirement in 1981, Ravotto is known to have received several awards for his outstanding contributions to nuclear development. Based on his expertise in the field of nuclear development together with the fact he had ten of these calculators in his possession leads us to conclude he was most likely responsible for its design and development.

In appearance and design, the slide chart is very similar to the General Electric GEN-15B and GEN-15C slide charts. Like the General Electric slide chart, it consists of a largely white slide that inserts into a sleeve which is predominately silver in color. Both the slide and sleeve are made of a flexible plastic while the sleeve had a number of transparent panels on either of its sides. In size it measures 24 cm long by 9.5 cm wide. The front and back faces of the calculator are shown in Fig. 6.11.

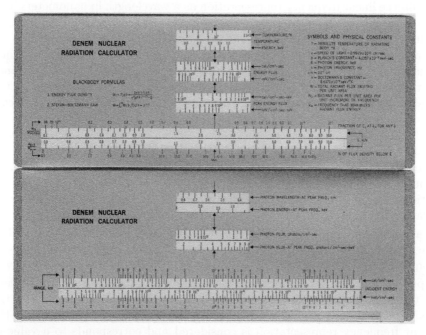

Figure 6.11 Front (top) and back (bottom) faces of the DENEM Nuclear Radiation Calculator.

The top panel on the front of the slide chart was used to set the temperature. The top scale was in Kelvin [K] while the bottom scale gave the equivalent energy content for a photon at this temperature in kiloelectron-volts [keV].[11] Two gauge marks (heavy arrows) above and below the top panel on the card were used to select the required temperature. At the selected temperature a number of quantities could be directly read off the second and third panels on the front of the rule and the first and second panels on the back of the rule.

The second panel from the top on the front of the rule, marked as "energy flux," gave the total radiant exitance in units of calories per square centimeter per second [cal·cm^{-2}·s^{-1}] on the top scale and in units of kiloelectron-volts per square centimeter per second $[keV \cdot cm^{-2} \cdot s^{-1}]$ on the bottom scale. The next panel below this, labelled the "peak energy flux," gave the value of the spectral radiant exitance in the linear frequency representation at a frequency corresponding to the peak value found in its spectral distribution curve in two different sets of units. The top scale gave the value in units of calories per square centimeter per second per unit kiloelectron-volt $[cal \cdot cm^{-2} \cdot s^{-1} \cdot keV^{-1}]$ while the bottom scale gave it in units of kiloelectron-volts per square centimeter per second per unit kiloelectron-volts $[keV \cdot cm^{-2} \cdot s^{-1} \cdot keV^{-1}]$.

On the reverse side of the slide chart, at the selected temperature, the top scale of the top panel gave the peak frequency in the spectral distribution curve within the linear frequency representation as a wavelength in nanometers [nm] while the bottom scale gave it as an energy for the corresponding photon in units of kiloelectron-volts [keV]. For the panel below, the top scale gave the total photon exitance in units of photon per square centimeter per second $[photon \cdot cm^{-2} \cdot s^{-1}]$ while the bottom scale gave the spectral photon exitance in the linear frequency representation at a frequency corresponding to the peak value in the distribution curve in units of photons per square centimeter per second per kiloelectron-volt $[photons \cdot cm^{-2} \cdot s^{-1} \cdot keV^{-1}]$.

The last two panels at the bottom of either side of the calculator were long, extending most of the way along the chart. For the bottom panel, at the front on the slide two identical scales for the equivalent energy of a photon in units of kiloelectron-volts [keV] are given. On the card, the top scale gave the ratio of the spectral photon exitance in the linear frequency representation relative to its value at its peak (maximum) and was marked as $\frac{W_\nu}{W_{\nu(\max)}}$. It is a dimensionless quantity between zero and one. Located on the card of the bottom scale of the panel the fractional amount expressed in terms of frequency. It was marked as $\frac{W_{0-\nu}}{W_{0-\infty}}$ (our $\mathfrak{F}_{e,0 \to \nu}$), is dimensionless and was expressed as a percentage. A gauge mark labelled "Max" appears on the card on the bottom scale of the bottom panel. At a given temperature it gives the fractional amount when an interval from zero up to the peak frequency in the linear frequency representation is considered and corresponds to a value of approximately 35.4%.

The final long narrow panel appearing on the reverse side of the slide chart was similar to that found on the General Electric GEN-15B and GEN-15C charts. It measured the irradiance normal to an area of one square centimeter as a function of distance from a blackbody source where losses in the intervening atmosphere were assumed to be negligible except in the case of the DENEM calculator was given in photonic units rather than in energetic units found on the General Electric rules.

Additional information can be found on the front of the slide chart. To the left, equations for the spectral radiant exitance in the linear frequency representation and the Stefan–Boltzmann law are given. To the right, starting in the top corner and running vertically downwards are nine different radiometric symbols and physical constants. From the values listed for the Planck and

Boltzmann constants it appears they correspond most closely to those given by B. N. Taylor, W. H. Parker, and D. N. Langenberg in 1969 [641]. The slide chart could therefore be said to date approximately from this period.

A unique feature of the calculator is the extraordinary high values found on the temperature scale. No other radiation slide rule ever made had temperature scales anywhere near those found on the DENEM calculator. Temperatures ranging from 10^6 to 10^8 K suggest the slide chart must have been used in calculations associated with atomic explosions, since the thermal radiation released immediately after an atomic detonation consists largely of x-rays at temperatures of the order of 100 million (10^8) degrees Kelvin, a value at the upper temperature limit found on the DENEM calculator.[12]

6.4   A CIRCULAR SLIDE RULE

Compared to other radiation slide rules the Autonetics Photon Calculator is sui generis. Circular in form it was designed with one function in mind, the estimation of the integrated spectral photon exitance $M_{q,0\to\lambda}^b$. While many other radiation slide rules readily provided estimates for the corresponding normalized fractional amount, no other rule was capable of calculating this quantity directly as an absolute amount. Designing a rule to accurately estimate an absolute amount whose value could span up to 45 orders of magnitude was no simple task, and to achieve this the circular form of the slide rule was essential. Writing $M_{q,0\to\lambda}^b$ in scientific notation it takes the form $a \times 10^b$. Here the mantissa a is such that $1 \leqslant a < 10$ while the exponent b is an integer. In doing so the problem becomes more manageable as finding the mantissa and the exponent separately is far simpler than having to find a single absolute amount. The Autonetics Photon Calculator managed to do this in a unique and rather clever way.

Almost all information we have about the Autonetics Photon Calculator comes from an article that appeared as a substitute to Frank Cooke's regular, and what turned out to be long running, column "Optical Activities in Industry" in the journal *Applied Optics*. It was written by Howard J. Eckweiler[13] and appeared in the July 1968 issue of *Applied Optics* [202]. An announcement about the rule preceding Eckweiler's paper by three months had appeared. It was briefly mentioned by John N. Howard in his April 1968 editorial of *Applied Optics* [303]. Sent a copy of the rule by Irvin Henry Swift, who at the time was the Director of the Electro-Optical Laboratory of the Autonetics Division at North American Aviation, given its broad general appeal Howard asked the author if a short description of the rule in a form suitable for inclusion in a near future Optical Activities in Industry column could be arranged. Three months later Eckweiler duly complied though perhaps misleadingly Howard in his editorial referred to the rule as an "Infrared Radiation Calculator."

As the only named author appearing on the paper, one presumes Eckweiler was responsible for the initial design and subsequent development of the calculator. Eckweiler credits Richard Ramsey, TRW, and Alfred Lawrence

Dunklee for providing contributions to the formulation and development of the rule when at Autonetics, though it would seem the bulk of the work and responsibility lay with Eckweiler. Dunklee, who took an engineering research position with the Electro-Optical Laboratory at Autonetics in August 1964, notes in the biographical statement contained in the preamble to his 1968 Masters thesis, he had co-authored a paper on an infrared radiation slide rule he helped develop while at Autonetics [199]. Subsequently published in the classified journal *Proceedings of the Infrared Information Symposia*, being a classified document we have unfortunately been unable to obtain a copy to confirm the veracity of Dunklee's claim.[14]

Autonetics started out in 1945 as a small unit in the Technical Research Laboratory located in the engineering department of the Los Angeles division of North American Aviation, Inc. As a result of winning a U.S. Army Air Corps contract to develop guided missiles, the unit rapidly expanded and led to the establishment of Autonetics as a separate division of North American Aviation in 1955. It was first located in Downey, California, before moving to Anaheim, California, in 1959. In September of 1967 Autonetics became part of North American Rockwell Corporation. Six years later saw its name change to Rockwell International. A number of divisions were included within Autonetics itself. One of these was the Electro Sensor Systems Division. The division built multi-function radar systems, armament control computers, and sensor equipment that made use of photon detectors, among other things, and it was the Electro-Optical Laboratory within the Electro Sensor Systems Division of Autonetics responsible for the development of its photon calculator. The copyright on the reverse side of the Photon Calculator is dated 1967, North American Aviation, Inc. Interestingly, Eckweiler's paper was received on 15 September 1967, the same month North American Aviation, Inc. became part of North American Rockwell Corporation, and it is the latter name Eckweiler uses as the institutional address for his paper. All this suggests the design and development of the Photon Calculator was probably completed by late 1966 and manufactured some time in early 1967 just before Rockwell-Standard Corporation acquired and merged with North American Aviation, Inc. to form North American Rockwell Corporation. Who actually made the calculator is not known.

The Autonetics Photon Calculator was a circular slide rule made of plastic. It consisted of a smaller $6\frac{3}{4}$ inch diameter inner disc that moved relative to a fixed outer 8 inch diameter disc. A movable radial cursor concentric with the two discs completed the rule. The rule came in a cardboard sleeve together with a four page instruction manual entitled "Black body photon calculator." In content the manual was very similar to Eckweiler's article that appeared in *Applied Optics* [56].

In design, four color-coded scales appeared on the front of the rule while brief operating instructions and six examples using the calculator took up the entire available space on the back of the rule. On the rim of the larger fixed outer disc were located two scales. The outermost scale was light green in color and was for the mantissa of $M^b_{q,0\to\lambda}$. It was labelled $Q^*_{0-\lambda}$ and was in units of photons per second per square centimeter [photons·s^{-1}·cm^{-2}]. It was a three-cycle ($1 \leqslant Q^*_{0-\lambda} < 10$) closed logarithmic scale. Running along side this scale on the larger fixed outer disc was a buff colored scale for temperature. It was labelled T for temperature in units of Kelvin. It was a single cycle, closed logarithmic scale. As closed scales this meant there was no limit on the values for each of these quantities. One could simply keep going round and round, mentally shifting the position of the decimal place as you went. The outer most scale on the smaller inner disc was green in color and was for the wavelength λ in micrometers. Again it was a single-cycle, closed logarithmic scale. Printed on this scale marking its start was a large black arrow with the numeral "1" appearing within it. The final white scale on the inner disc was a 33-cycle open logarithmic scale wound in eleven concentric rings for the wavelength–temperature product λT. It was measured in units of micrometer Kelvin [µm·K]. As an open scale its range was limited to: $171.6 \leqslant \lambda T < \infty$ µm·K. A photograph of the front face of the calculator is shown in Fig. 6.12.

In calculating the main scale used on the calculator, the integral appearing in $M^b_{q,0\to\lambda}$ was first converted into an infinite series of the form given by Eq. (3.51) and its logarithm (since all scales appearing on the calculator are logarithmic) was evaluated using one of the company's computers. The values Eckweiler takes for the fundamental constants are those that had been recently recommended by the National Academy of Sciences, National Research Council Committee on Fundamental Constants in 1963 [14]. In terms of the accuracy of his calculator Eckweiler says surprisingly little other than remarking he expected that values found using it would exceed those of experiment. It was a bold statement backed up with no numerical analysis.

Operating the calculator to find $M^b_{q,0\to\lambda}$ was a two part process. The first part involved the determination of the mantissa, the second required finding its corresponding exponent. The temperature setting was made by aligning the required value for the temperature on the buff scale with the large black arrow found on the green wavelength scale. The radial cursor was then moved over to the wavelength of interest from where the wavelength–temperature product could be read off the temperature scale. Note that since both the wavelength and the temperature scales were single-cycle closed logarithmic scales they functioned in exactly the same manner as the C/D scales used for multiplication on an ordinary slide rule. The preceding step of moving the cursor to find the wavelength–temperature product was only required if the product could not be performed mentally in one's head. Without moving either disc so the temperature setting remained in place, the radial cursor is swung around to the corresponding wavelength–temperature product on the inner white scale and $Q^*_{0-\lambda}$ read at the position of the cursor line on the light green scale.

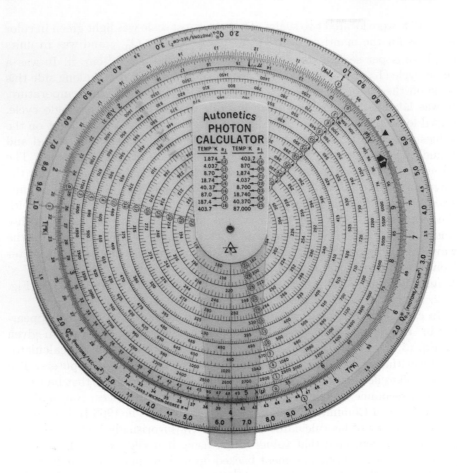

Figure 6.12 Front face of the Autonetics Photon Calculator made for North American Aviation, Inc.

With the mantissa found, finding the exponent b was broken up into three steps so that $b = a_1 + a_2 + a_3$. The three different a_i's were integers which appeared in a host of blue circles found at various places on the slide rule and its cursor and were selected by following a set of rules. Values for a_1 (> 0) were found on a cursor table at the end of the radial cursor. The positive blue circle integer lying between the temperature range shown on the cursor table for the selected temperature setting was chosen. A total of fourteen positive integer values for a_1 from $+12$ to $+25$ can be found here corresponding to temperatures ranging from 1.874 to 87 000 K. The values for a_2 (< 0) were located at 120° intervals of arc on each of the eleven concentric rings found on the white scale at the clockwise end for each of the 33 cycles. At the wavelength–temperature setting on the white scale, the first blue circle encountered by moving in a clockwise direction around the concentric ring is the value taken for a_2. There were a total of 33 negative integer values for a_2 ranging from -1 to -33. Finally, the value for a_3 (zero or unity) was found as follows. Starting at the wavelength–temperature setting, the first blue circle one meets on either the buff temperature or green wavelength scales when moving in a clockwise direction is the value taken for a_3. On the green wavelength scale there were three blue circles corresponding to the integer zero while on the buff temperature scale there were three blue circles corresponding to the integer one. Assembling the three pieces by taking their sum gave the final value for the exponent.

In addition to the large black arrow containing the numeral one printed on the green wavelength scale, three other gauge marks appeared on the rule. The first of these was a radial dashed line marked "$\lambda_m T = 3669.7$ MICRON-DEGREE K" found on the buff temperature scale. It gave the peak wavelength in the photon spectral distribution curve in the linear wavelength representation. Setting the large black arrow to some appropriate temperature followed by aligning the cursor over this gauge mark, the peak wavelength is the value read off the green wavelength scale.

The second of the gauge marks, labelled $\pi^{1/3}$, was located at 1.4646 on the green wavelength scale. It allowed the corresponding photon radiance for a blackbody to be found from the photon exitance by dividing the latter by a factor of π. To do this on the calculator, with the radial cursor held in place at the value for $Q^*_{0-\lambda}$ found on the light green outer most scale the inner disc was rotated until the $\pi^{1/3}$ gauge mark was brought exactly under the cursor line. With both discs now held fixed the cursor was moved until it aligned with the large black arrow on the green wavelength scale. The reading made from the cursor in this position at the light green scale gave the corresponding value for the mantissa of the photon radiance. The exponent remained unchanged and was equal to the value found previously for the photon exitance.

The third gauge mark was a large black triangle located on the buff temperature scale. It was used to find the fractional amount $\mathfrak{F}_{q,0\to\lambda}$ as follows. The large black arrow on the wavelength scale was firstly aligned to the large black triangle on the temperature scale. With the wavelength–temperature product calculated in advance and both discs held fixed, the radial cursor was then set to this value on the white scale. With the cursor at this setting the value for $Q^*_{0-\lambda}$ on the light green scale was read. The value for the fractional amount was then equal to $Q^*_{0-\lambda} \times 10^{a_2}$. Here the value for a_2 was found in the same manner as described previously.

Lastly, it was also possible to calculate the total photon exitance $M^b_q(T)$ at a given temperature using the calculator. With the large black arrow on the green wavelength scale set to the appropriate temperature setting on the buff scale, with both discs held fixed, moving the cursor to the "∞" setting on the white wavelength–temperature product scale and reading the value of $Q^*_{0-\infty}$ from the light green scale gave the value for the mantissa corresponding to the total photon exitance. The exponent was then found in a manner identical to that described earlier by finding integer values for the three different a_i's.

The Autonetics Photon Calculator was a specialized rule within an already specialized family of radiation slide rules. While it may not have been able to find many thermal quantities, those it could find were found quickly and efficiently. One obvious advantage to using the Autonetics Photon Calculator lay in its ability to perform many repetitive actinometric calculations in a highly efficient manner and made it particularly suited to someone working on problems relating to photon detectors. Unlike other radiation slide rules, the photon calculator came relatively unencumbered since it was not burdened with any unnecessary scales that were neither needed nor used. It was also very useful for calculating actinometric fluxes at very low temperatures.

How widely the Autonetics Photon Calculator was used is difficult to assess. Reference to it in the literature are few and far between. Hudson in his 1969 text *Infrared System Engineering* devotes two very short sentences to it [311]. Importantly, he indicates in a footnote how reproductions of the calculator could be obtained by inquiring through the Electro-Optical Laboratory of Autonetics. Here the second author recalls the calculators were difficult to acquire and if it had not been for a fortunate visit to Autonetics and being presented with one, he might never have come by the calculator in the first place. Further references to it are made by Keyes and Quist [348] who suggest some basic calculations related to the performance of photon detectors in the infrared region are best handled using a Autonetics Photon Calculator while Pinson as late as 1985 recommends the Autonetics Photon Calculator be used whenever one was required to calculate actinometric quantities [499]. When asked recently about the ten different types of radiation slide rules he had, which he used the most and which he considered to be the best of the bunch, Prof. Wolfe responded by saying that although he may have used Canada's rule the most, it was the photon calculator from Autonetics he found to be the most powerful [703].

6.5 "DO-IT-YOURSELF" SLIDE RULES AND CHARTS

Planck's equation for the spectral radiant exitance of a blackbody has an important mathematical property that readily lends itself to the design of simple sliding templates. When the logarithm of Planck's equation is taken, the general form of Wien's homologous law (i.e., the wavelength–temperature product is equal to a constant) ensures the form of the relation between the logarithm of the spectral radiant exitance to the logarithm of the wavelength is independent of the temperature. So in a log-log representation, all blackbody curves for the spectral radiant exitance have the same shape. A spectral curve plotted at any given temperature will correspond to a spectral curve at any other temperature merely by shifting the former curve along a straight line where the wavelength–temperature product remains constant. Doing so allows the value for the spectral radiant exitance of a blackbody at any wavelength for some fixed temperature to be found using a slide rule, a quantity that otherwise could not be found.

Parry H. Moon seems to have been the first to recognize this. In his text *The Scientific Basis of Illuminating Engineering* of 1936, one finds, as a folded insert, scales for the spectral radiant exitance as a function of wavelength printed on light tracing paper [455]. The straight line given by the locus of the spectral peaks is used while along this line temperatures ranging from 100 to 15 000 K are marked out. In this way, instead of sliding the spectral curve along a line where the wavelength–temperature product remains constant, Moon's insert worked in reverse. After its removal from the text the folded insert, which measured 38 cm long by 23 cm wide, was overlaid onto and moved along a fixed "constant shape" spectral curve. The fixed spectral curve appeared on a complete page of its own in the text together with the same straight line given by the locus of the spectral peaks. By aligning the two straight lines and positioning the insert so the desired temperature on the straight line coincided exactly with the peak of the spectral curve beneath it, values for the spectral radiant exitance over a range of wavelengths of interest could be read off at selected temperatures.

After Moon, sliding templates based on his idea continued to be refined and extended in the literature [149, 357] and workers in the infrared were continually encouraged by authors to make use of this simple type of sliding template for themselves [696, 244, 93, 697, 716, 662]. The idea was put into commercial form in the early 1960s in a slide chart made for Block Associates, Inc., who at the time were a small and highly unorthodox research group locate in Cambridge, Massachusetts. Their chart will be described more fully in Section 6.6.1.

A second slightly different variation on the do-it-yourself type sliding template was described in 1960 by Reid [553] and was reproduced some years later in the text by Mikaél' A. Bramson [99]. On a log-log scale if the curve for the spectral radiant exitance is plotted, not as was previously done against wavelength for constant temperature, but against temperature for constant wavelength, the shape of the curve will again be independent of the value for the wavelength chosen. The envelope formed from each constant wavelength curve as the wavelength is changed is a straight line and it is along this line, after having values for the wavelength marked off along it, any constant wavelength curve is displaced along. Sliding the constant wavelength curve to some desired wavelength, the intersection between the constant wavelength curve and the temperature scale of interest allowed the spectral radiant exitance to again be directly determined. The accuracy of the rule depended entirely on the precision with which the logarithmic scales (temperature and spectral radiant exitance) were marked onto the rule, and on the overall length of the rule. Reid claimed rules of 50 cm in length had an accuracy of about 5%. While examples of the rule of this length were constructed and used at Reid's place of work, the Eastman Kodak Company, no commercially available models of the rule are known to exist. An illustration of Reid's rule taken from his paper where it was first described is shown in Fig. 6.13.

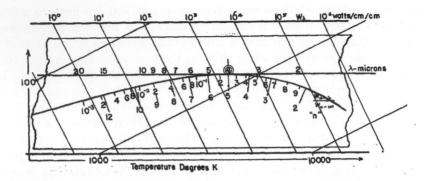

Figure 6.13 Schematic illustration of Reid's do-it-yourself sliding template. Reproduced with permission from Reid, C. D., 1960 "Nomographic-type slide rule for obtaining spectral blackbody radiation directly," *Review of Scientific Instruments* **31**(8), p. 888. Copyright 1960, AIP Publishing LLC.

6.6 RADIATION SLIDE CHARTS FROM THE 1960s

6.6.1 THE BLOCK RULE

When the availability of the Radiation Calculator slide chart made for Block Associates, Inc. was announced in the January 1962 issue of the *Journal of the Optical Society of America* (JOSA) [724], it was not done in the usual manner of a scholarly journal article, as had previously been the case with those rules developed earlier by Czerny, Canada, and Makowski, nor was it a simple paid advertisement. Instead, the text containing the announcement, while resembling a short note, is completely enclosed by a double lined border and the words "an advertisement" are found running across the top and bottom of it. While it may have read like a short note, it is clearly labelled an advertisement. As intriguing as all this may seem, it is closely tied to an interesting chapter of the Optical Society of America's decision in 1959 to start accepting paid advertising in the Society's journal, JOSA.

Block Associates, Inc. was a small company Myron Jacques Block[15] and Neils O. Young had built up in Cambridge Massachusetts. Founded in 1956, most of the so-called "associates" working at the company early on were graduate students drawn from various universities around the Boston area. Initially the company performed infrared studies on contract for the Air Force Cambridge Research Laboratories. Later it became quite a successful and prosperous optics firm with it and its offshoots being bought out in the late 1970s by Bio-Rad, Inc.

In the early days, as a research group the young associates were avant-garde. One remarkably innovative member of staff was Larry Mertz. In late 1959 he had submitted a short letter to JOSA. It contained seminal work on the electronic use of autocorrelation techniques to dig a signal out of noise. The reviewer described the letter as very clever mathematics and electronics but was not optics and suggested it probably belonged in a mathematics journal [305]. The editor at the time, Wallace Brode, accordingly rejected the letter, and a slightly dejected Mertz took his story to his boss. Block's solution was simple yet masterful. He decided to print Mertz's letter as a paid advertisement in the next issue of JOSA. It ran and, to the great surprise and shock of the editor, he could not believe how something he had turned down managed to reappear in the journal in this way. He did not see any of the ads printed in the journal ahead of time but felt as editor he should see everything that was to appear in print from cover to cover. Internally, a small battle waged at the Society before a compromise was finally reached. Block's ads were allowed to continue to be published in the journal but from the June 1960 issue onwards would now be enclosed by a prominent double lined border and would carry the consumer warning "an advertisement" [307]. Many years later Block recalled how all this was perfectly fine with him since despite having a general disdain with the peer-review process, his only concern was in gaining an audience; having the warning label attached only added to the fun [86].

Since the first few Block advertisements had been well received by readers, Block and his associates decided to build on their initial success by following these up with new advertisements every month with some, usually way-out and often patentable, idea being proposed. In time the advertisements developed quite a reputation and comments among readers of it being the first thing they looked for when each new issue of the journal arrived were not uncommon [307]. While there was nothing the editorial staff at JOSA could do, they were tolerated until the pressure of coming up with twelve truly unique or novel ideas a year finally led to their demise [270]. It was probably during one of the regular brainstorming sessions held each month to come up with new ideas for JOSA advertisements the idea for developing their rule arose.

Who was behind the idea of designing and developing the Radiation Calculator is not known. It is thought to be Block himself though the advertisement announcing its availability was written by Young who seems to have been responsible for writing many of the Block ads. Wolfe, in the *Handbook of Military Infrared Technology* of 1965 refers to it as the "Block Rule" [696] though whether he was attributing the name to Block the individual or Block the company is not clear. The rule itself was the commercial embodiment of Moon's simple do-it-yourself sliding template proposed some twenty-five years earlier. At a given temperature, it allowed the spectral radiance as a function of wavenumber in either the linear wavelength or wavenumber representations to be read directly.

Young in his advertisement writes that he and his colleagues at Block Associates had found a large 16 inch by 11 inch chart with an over sized transparent template more useful than any of the radiation slide rules or tables available at the time for estimating the spectral radiance of a blackbody when accuracies of the order of a few per cent were sufficient. The final commercially produced rule was a somewhat smaller, pocket-sized affair. Measuring $9\frac{1}{4}$ inches long by $3\frac{3}{8}$ inches wide, it was made of rigid plastic and could be obtained for \$1.98 by contacting Block Associates. The instruction manual it came with consisted of a reprint of Young's advertisement announcing the rule and was printed on a single sheet of paper.

The slide was completely transparent and on it appeared two spectral radiance curves. The curve positioned to the left gave the spectral radiance as a function of wavenumber in the linear wavenumber representation while the curve positioned to the right gave the spectral radiance as a function of wavenumber in the linear wavelength representation. Running parallel to the slide's long edge was a dashed line used for alignment as the curves were translated. On the stock two grids appeared. The grid to the left gave the spectral radiance in the wavenumber representation in units of microwatts per centimeter per steradian per unit wavenumber interval $[\mu W \cdot cm^{-1} \cdot sr^{-1}]$ as a function of wavenumber in units of per centimeter $[cm^{-1}]$ while the grid to the right gave the spectral radiance in the wavelength representation in units of microwatts per square centimeter per steradian per unit wavelength interval $[\mu W \cdot cm^{-2} \cdot sr^{-1} \, \mu m^{-1}]$ as a function of wavenumber in units of per centimeter $[cm^{-1}]$. Unfortunately the symbol I instead of L is used for the spectral radiance in either representation (I_ν in the spectral wavenumber representation and I_λ in the spectral wavelength representation) while the symbol ν instead of $\bar{\nu}$ is used for wavenumber. A solid line running parallel to the long edge of the stock ran through each grid along which temperatures from 20 to 500 K were marked. Finally, a sample curve which was dashed appeared in each grid and gave the spectral radiance of a blackbody in the wavenumber and wavelength representations at 300 K.

In operation the slide was moved until the solid spectral radiance curve on the slide intersected the solid parallel temperature line on the stock at the temperature of interest. With the dashed guide line on the slide coincident with the solid temperature line beneath it ensuring proper alignment, the value for the spectral radiance within the range from 100 to 10 000 cm^{-1} in either the wavelength or wavenumber representations could be simultaneously read from either grid on the calculator. A photograph of the front and back faces of the Block Rule is shown in Fig. 6.14.

It was possible to extend the temperature range to temperatures outside the direct range of the rule by multiplying the coordinate scales by suitable multiplies. For example, for temperatures in the range 500 to 5000 K one simply multiplied the wavenumber scale by 10, the spectral radiance in the wavenumber representation by 10^3 and the spectral radiance in the wavelength

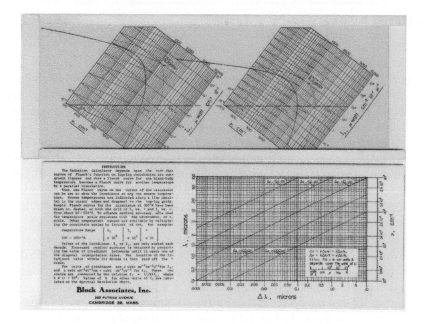

Figure 6.14 Front (top) and back (bottom) faces of the Radiation Calculator made for Block Associates, Inc.

representation by 10^5. As a useful aid to the user these multiplication factors were found printed on the reverse side as part of a brief set of instructions for the rule. Also on the reverse side of the rule was a spectral resolution chart which contained curves of $\Delta\bar{\nu}$ and $\Delta\lambda$ as functions of wavenumber ($\bar{\nu}$) and wavelength (λ). It had been the subject of an earlier Block advertisement in JOSA the previous year [723] and was presumably included on their Radiation Calculator since from the time of its announcement the chart had been found to be useful but remained unavailable for purchase. The relation between the two spectral quantities found on the spectral resolution chart is

$$\frac{\Delta\lambda}{\lambda} = -\frac{\Delta\bar{\nu}}{\bar{\nu}}.$$

Wolfe pointed out that despite the curves being correct, the text appearing as an insert to the spectral resolution chart giving the relation between these two quantities was in error [696].

While the Block rule may have been capable of estimating the spectral radiance directly, and was something no other radiation slide rule could do, unless one was only interested in estimates accurate to within a few per cent, the rule was unlikely to be used for serious computational work. References to the rule in the literature are found sparingly [301, 209, 696, 113], suggesting that it was not widely used.

6.6.2   THE INFRARED SLIDE RULE

Very little has been recorded in the formal literature about the Infrared Slide
Rule, a slide chart made for Infrared Industries, Inc. (IRI) in the early 1960s.
It seems to have been used mainly as a promotional item given away by the
company. Even so, the slide chart still managed to find its way into the hands
of many. It was often given away in bulk to individuals involved in the teaching
of infrared-related courses at universities. Gradually these would have been
handed out to a considerable number of students over a period of time, further
increasing the extent of its distribution. As a smaller, more diminutive slide
chart it was clearly overshadowed by its much bigger brother, the Radiation
Calculator from General Electric, but we do not think its intention was ever
to compete with such a rule. Instead, it was seen as a handy aid to have if
nothing more than a quick, order-of-magnitude estimate was required.

The history of the rule is intricately tied to the early history of the company
for which the rule was made. IRI was founded in the late 1950s in Waltham,
Massachusetts, by E. Douglas Reddan. Initially the company produced infrared detecting elements, instruments, and control systems for military and
civilian use. The person responsible for the design and development of the rule
is thought to have been Arthur John Cussen,[16] an employee who joined IRI in
1958. By the late 1950s Cussen already had a close connection with radiation
slide rules. At the end of the Second World War he was with the U.S. Navy,
eventually becoming the head of the Infrared Division at the Naval Ordnance
Laboratory located at Corona, California. During his time with the Navy, he
first learned of Makowski's rule made on behalf of the British Admiralty which
saw him becoming partly responsible for making the availability of the rule
better known in the United States [304]. In 1958 he left the Naval Ordnance
Laboratory to become the General Manager of the Infrared Standards Laboratory in Riverside, California (though the plant itself was actually located in
Pomona) [7]. The laboratory had recently become part of the engineering and
production division of IRI in Waltham, Massachusetts. In 1960 Reddan, as
company president, decided to establish his residence in Santa Barbara, California. In August of that year this led to the selling of additional stock for the
construction and equipping of a new factory in nearby Carpinteria (though
the future company address would always be given as Santa Barbara). On
completion in 1962, a reorganization of a number of former divisions of IRI
took place and saw Cussen along with William E. Standring, Jr., who was
the General Manager of IRI's Photoconductor Division, being appointed vice
presidents of the new IRI plant in Santa Barbara [12].

Very similar in design to Canada's General Electric rule, the IRI rule was a
slide chart rather than a slide rule despite its designation to the contrary. At
least two different types that differed mainly in their construction are known.
The first was made of finished cardboard like the GEN-15 and GEN-15A
while the second was made of plastic like the GEN-15B and GEN-15C rules.
Who the IRI rules were manufactured by is not known, but judging by their

appearance, suggest they were probably made by Perrygraf. Nor is any year marked on the rules making dating them particularly difficult. One source we have comes from a former employee of IRI. Max J. Riedl, who joined IRI at the new company plant in Carpinteria, recalls a cardboard version of the rule was already there when he arrived in early 1962 [560]. He also believes the earlier of the two rules was the one made of cardboard. A later cardboard version of the rule also appeared. It is more recent than the plastic version, since on the reverse side of each rule the noise equivalent bandwidth, Δf, is defined and its units given. The plastic rule lists the units for Δf as cycles per second (cps) while the cardboard version gives its units as hertz (Hz). In 1960 the SI unit for frequency was changed from cps to Hz and therefore suggests the former plastic rule is older than the latter cardboard rule.

In size the more recent cardboard IRI rule measured 16.2 cm long by 9.9 cm wide and came with a very simple instruction card printed on either side. Four small hollow metal rivets located in each corner held the front and back faces of the rule together. On the front of the rule were six panels containing fourteen scales for eleven different quantities. Unlike the General Electric rule whose scales placed most of their emphasis on the estimation of radiometric quantities, the scales on the IRI rule were evenly split between radiometric and actinometric quantities. The top panel located in the middle of the rule was a scale for temperature in Kelvin only. Immediately below the temperature panel, the scale running along the top of the second panel gave, at a selected temperature, the total photon exitance in units of photons per second per square centimeter [photons$\cdot$s$^{-1}\cdot$cm^{-2}] while the scale running along the bottom of the panel gave the spectral photon exitance at the corresponding spectral peak in units of photons per second per square centimeter per unit wavelength interval [photon$\cdot$s$^{-1}\cdot$cm$^{-2}\cdot\mu$m^{-1}]. The third panel was identical to the preceding panel except it was for radiometric quantities in energetic units. The two longer panels running along the bottom and second from the bottom gave spectral and integrated ratio amounts. The scale above the bottom panel running along the top of the stock was for the spectral fractional amount in photonic units while the scale running below the panel along the bottom of the stock gave the integrated fractional amount in photonic units. On the slide between these two scales was a scale for the wavelength in micrometers though the unit of measure is not given on the rule itself. The scales found on the panel second from the bottom were again identical to those found on the bottom panel but were for the corresponding radiometric quantities. The final panel, located at the top and to the left of the temperature panel, gave as a wavenumber, in units of per centimeter, the peak value for the spectral curve in the linear wavelength representation. On the reverse side of the rule, running along its base, was a temperature nomogram for converting between Fahrenheit, Celsius and Kelvin. The remainder of the reverse side of the rule was devoted to a listing of a number of useful formulae relating to sensor detectivity and a listing of eleven symbols together with their meaning. On the reverse side of the slide, approximate transmissions (for wavelengths of transmission greater than 10%) for various optical materials and their refractive index at 3 μm were given. For the cardboard version, the transmission spectrum for radiation in the atmosphere similar to the one given on the General Electric GEN-15B and GEN-15C rules was also given. To read either of these scales the slide needed to be removed from the rule first.

Sometime earlier, though exactly when is not known, IRI had made their second type of Infrared Slide Rule. Identical in size and almost similar in design to its cardboard successor, it was made from plastic. A flexible plastic slide was now housed in a rigid vinyl sleeve. On the front of the plastic rule a grey, rather than the former white, background was punctuated by five clear panels instead of the six found on the cardboard version. No panels were again found on the back of the rule. The second and third panels for the total exitance and the spectral exitance at the spectral peak in both photonic and energetic units found on the cardboard rule were combined into a single panel on the plastic rule. The temperature panel located at the top of the latter rule was also expanded in two ways resulting in two minor variations. In the first variation, which we will call P1, the temperature panel had a scale for the temperature in Celsius running along the top and a scale for Kelvin running along the bottom. In the second variation, which we will call P2, the temperature panel was expanded further to also include a scale for Fahrenheit. The remaining three scales found on the plastic rule were identical with those found on the cardboard rule. On the back of the plastic rule, a list of formulae and symbols are once more given but were oriented in portrait rather than landscape form. Other minor variations found between the two plastic rules include the symbols used on the integrated ratio scales for the total radiant exitance and total photon exitance. On rule P2, one finds W and Q being used for the total radiant exitance and the total photon exitance respectively, and were the same symbols as those used on the cardboard rule, while on rule P1 these two symbols were changed to $W_{0-\infty}$ and $Q_{0-\infty}$. As was the case with many of the other radiation slide rules discussed, older symbols of W for the radiant exitance and Q for photon exitance were used. On the reverse side of rule P1 the incident flux density H is one of the symbols defined while the symbol P for peak is not given while on rule P2 it is the exact opposite, the symbol P being given while H is not. Finally, at the bottom on the reverse side of both plastic rules the Waltham and Santa Barbara addresses for IRI are given. On rule P1 this appears over three lines while it takes up five lines on rule P2. The front faces for the cardboard (top) and one of the plastic (bottom) IRI rules are shown in Fig. 6.15.

Due to their size the IRI rules were without doubt the handiest of all the radiation slide rules made. Given away at trade shows and to students alike, they no doubt found their way into the hands of many potential users. Reserved for very rough order-of-magnitude estimates only explains why one fails to find any reference to these rules in the literature. As the precision

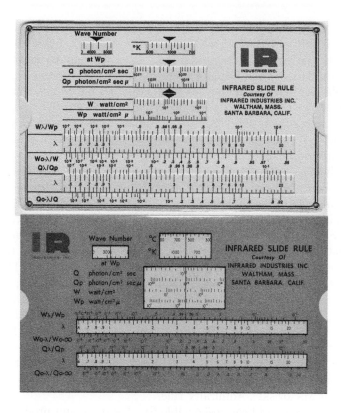

Figure 6.15 Front faces of two different Infrared Slide Rule made for Infrared Industries, Inc. On top is shown the more recent cardboard rule, while below the plastic version P1 is shown.

with which the scales on the IRI rule could be read was low, it meant any serious work would always be passed over to one of its more advanced and able rivals. It would, however, not be the last radiation slide rule Cussen would find himself closely associated with. In 1964 he would leave IRI to establish his own company, Electro Optical Industries, Inc. [20], and it would be from here he would go on to play a pivotal role in the development of the last of the true radiation slide rules, a rule that would finally rival the Admiralty rule.

6.7 THE LAST OF THE REAL SLIDE RULES

The radiation slide rule made for Electro Optical Industries (EOI), Inc. was an improved and updated version of the Admiralty rule designed by Makowski. First available in 1970, it represented the pinnacle of what was possible in a general purpose radiation slide rule. It was also the last of the true slide rules to be made for performing thermal radiation calculations. As we have seen, several Infrared Radiation Calculators came later but these were not slide rules in the strictest sense, and since each was almost identical to the Radiation Calculator from General Electric, they simply extended the legacy of Canada's rule without adding anything further to its development. The usefulness of the EOI rule to those working in the infrared would however turn out to be relatively short lived. Through no fault of its own the EOI rule would come to be surpassed in a few short years by the arrival of affordable handheld programmable electronic calculators.

Established in 1964, EOI had quickly become one of the leaders in the design and manufacturer of infrared and visible sources for testing and calibrating equipment in the United States. In particular, the company was widely regarded for the quality of the blackbody radiation sources it produced. As a company whose main line of business was intimately connected to blackbody radiation [155], it makes sense that they would have more than just a passing interest in developing a radiation slide rule of their own. The founder and president of the company was Arthur J. Cussen who we met earlier. By 1970 his involvement and connection to radiation slide rules was already well established. Responsible for the Infrared Slide Rule made for IRI during his time spent working there in the early 1960s, and earlier still in his part responsibility for making the availability of the Admiralty rule better known in the U.S. when he was with the Naval Ordnance Laboratory; it must have always been a desire of his to design an ambitious rule of his own that would one day rival Makowski's Admiralty rule.

The rule Cussen had been responsible for designing and producing for IRI was functional perhaps, but hardly inspiring, and was more toy than high-precision workhorse. The task of starting a new company and taking on the role of its founding president understandably diverted Cussen's attention for a time. While Cussen had a design for a radiation slide rule in mind, having little time himself to devote to the task of its development meant any immediate ambition he may have had in producing a rule to rival Makowski's Admiralty rule had to wait. And there his idea to produce a radiation slide rule to rival all others languished for several years as the day to day operations of running his company took precedence.

Within a few years, as company president of a by now small but prosperous enterprise, Cussen finally had at his disposal the means to allocate resources to such a task.[17] This he did in the summer of 1968 by hiring a young engineering student by the name of Raymond James Chandos[18] who was charged with seeing Cussen's project through.

Chandos started working at EOI during the late 1960s, first as a summer intern student, and then full-time after graduating. His first job at the company was as the computer for the scales that would appear on Cussen's rule. The design of Cussen's rule was similar to the Admiralty rule, but represented an improvement in several important ways. For a start, the positioning of the

scale markings had been updated using the latest known values for the fundamental constants. Secondly, in construction the rule was duplex in form. As
a double-sided rule, this meant the slide no longer had to be removed to be
reversed; one simply flipped the rule over instead.

The commercially available rule of 1970 made for EOI was made by Pickett
and designated the Model 17. At the time Pickett was one of the larger slide
rule makers in the United States who were famous for their all yellow "eye-
saver" rules made from aluminium. The Model 17 rule, while being made from
aluminium, was "traditional" white in color. It measured 10 inches long by
2 inches wide and as a duplex rule came with a nylon cursor that wrapped
around both sides of the rule. Raised end-braces at either end of the rule held
the top and bottom stocks in place and prevented the cursor from sliding off.
Hand engraved in the bottom corner of the right end brace on the side of the
rule labelled ENERGY, a serial number can usually be found. Scales printed
in both black and red appeared on the rule.[19] In its short life it would come
to be referred to as "Cussen's rule" [697].

As was the case with the Admiralty rule, one side of Cussen's rule was used
to calculate radiometric quantities while the other side calculated actinometric
quantities. As with many slide rules sold in the U.S. at the time, it came with
the customary leather holster case and an interim manual [123]. A year later
the interim manual was replaced with a two color typeset manual [124] and
some time after this a one page errata appeared. Unlike the Admiralty rule,
as a duplex rule it meant there was no available space on the rule to print
a brief set of instructions and made having access to its detailed instruction
manual all the more important.

Most of the information we have about Cussen's rule comes from the aforementioned manuals supplied with the rule. The values Chandos used for the
fundamental constants when calculating the scales were the revised values
given by the National Bureau of Standards in May of 1969 [17]. Details on
how the calculations for the scales were made are not provided. Jack R. White,
who went to work for EOI as an electronics technician in early 1965, recalls
the calculations were performed by Chandos renting time on one of the mainframe computers available at the University of California, Santa Barbara [680].
Moreover, of the two references given in the manual but not referred to in the
text, the second of these is to the extensive set of tables of blackbody radiation functions compiled by Mark Pivovonsky and Max R. Nagel in 1961 [501].
It therefore seems more than likely these were consulted by Chandos during
their preparation, probably being used to check against the accuracy of the
mainframe computed values to ensure the computer calculations were free
from error. Cussen closely oversaw all the computational work performed by
Chandos. Cussen's wife Marjorie, who at the time of the rule's development
served as Operations Manager for the company, recalls a very young Chandos
sitting at his desk quietly working away as he computed the data to be used
for each scale found on Cussen's rule [156].

As alluded to earlier, part of the motivation behind the origin of the rule stemmed from a desire by Cussen to have a radiation slide rule of his own design and making. The other was to produce a rule comparable to the Admiralty rule yet more competitively priced. As Chandos recalls [127], in the late 1960s only Cussen as company president and Chandos's brother Robert as vice-president owned copies of the Admiralty rule while everyone else employed at the company worked with the relatively inexpensive Radiation Calculator slide chart from General Electric. The latter was fine for most tasks but for more accurate work the Admiralty rules would need to be "loaned." As Cussen's rule was manufactured by Pickett, a manufacturer of slide rules conveniently located in the Santa Barbara area not far from EOI, Cussen was able to get the final retail price for the rule down quite a bit. In 1972 it sold for $50 [304]. While considerably more expensive than the Radiation Calculator slide chart from General Electric which at the time were selling for about $4 [697], at this price it was still one-third the cost which the Admiralty rule had sold for ten years earlier. The rule's availability was widely announced in the literature, typically appearing under the "new products" section of various journals and trade magazines [22, 24, 23, 25, 304, 26].

An incredible 34 scales for 13 different quantities are packed onto Cussen's rule, 18 on the front side labelled "ENERGY" relating to radiometric quantities and a further 16 on the reverse side labelled "PHOTONS" relating to actinometric quantities. Photographs of either side of the Pickett Model 17 radiation slide rule are shown in Fig. 6.16. Running vertically down the far left hand end of the rule on either side are the symbols used for the various quantities next to each scale while at the far right hand end the corresponding unit, if any, for each quantity next to the appropriate scale is given. Unlike the Admiralty rule, Cussen and Chandos unfortunately chose not to use any labels for the various scales found on the rule.

We list the scales found on Cussen's rule below. Like the Admiralty rule, the Model 17 used the symbol H for those quantities relating to radiant exitance (again possibly for irradiance or was it used in homage to its predecessor) and Q for those relating to photon exitance. The order is given as they appear on the rule running from the top down.

| Top Stock – ENERGY Side |

1. Total radiant exitance, $H_{0-\infty}$ [W·cm^{-2}]
2. Spectral radiant exitance at the peak wavelength, H_{λ_m} [W·cm^{-2}·µm^{-1}]
3. Temperature in Fahrenheit, t_f [°F]
4. Temperature in Celsius, t_c [°C]
5. Temperature in Kelvin, T [K] (100–10 000 K)

| Slide – ENERGY Side |

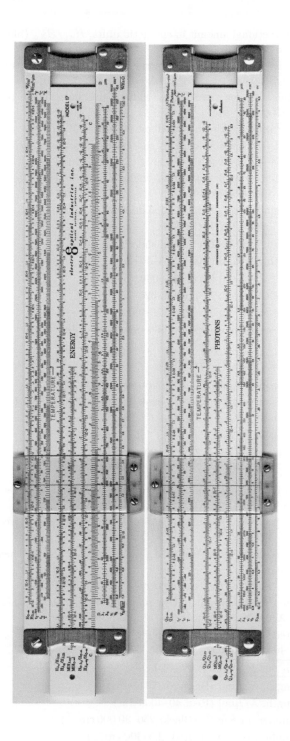

Figure 6.16 Front and back faces of the Pickett Model 17 radiation slide rule made for Electro Optical Industries, Inc.

6. Spectral fractional amount in energetic units, $H_{\lambda_1}/H_{\lambda_m}$ (black and for $\lambda = 0.3$ to $30\,\mu m$)

7. Spectral fractional amount in energetic units, $H_{\lambda_2}/H_{\lambda_m}$ (red and for $\lambda = 30$ to $3000\,\mu m$)

8. Multiplier scale for the spectral radiant exitance at the peak wavelength, $M(H_{\lambda_m})$

9. Multiplier scale for the total radiant exitance, $M(H_{0-\infty})$

10. Integrated fractional amount in energetic units, $H_{0-\lambda_1}/H_{0-\infty}$ (black and for $\lambda = 0.3$ to $30\,\mu m$)

11. Integrated fractional amount in energetic units, $H_{\lambda_2-\infty}/H_{0-\infty}$ (red and for $\lambda = 30$ to $3000\,\mu m$)

12. C scale found on a conventional slide rule

    $\boxed{\textit{Bottom Stock – ENERGY Side}}$

13. D scale found on a conventional slide rule

14. Wavelength, λ_1 [μm] (Black: 0.3–$30\,\mu m$)

15. Wavelength, λ_2 [μm] (Red: 30–$3000\,\mu m$)

16. Wavenumber, ν_1 [cm^{-1}] (Black: 320–$40\,000\,cm^{-1}$)

17. Wavenumber, ν_2 [cm^{-1}] (Red: 32–$400\,cm^{-1}$)

18. RMS Johnson noise potential per root ohm–hertz, $V_n/\sqrt{R\Delta f}$ [$V\cdot Hz^{-1/2}\cdot\Omega^{-1/2}$]

    $\boxed{\textit{Top Stock – PHOTONS Side}}$

19. Total photon exitance, $Q_{0-\infty}$ [$photons\cdot s^{-1}\cdot cm^{-2}$]

20. Spectral photon exitance at the peak wavelength, Q_{λ_m} [$photon\cdot s^{-1}\cdot cm^{-2}\cdot cm^{-1}$]

21. Temperature in Fahrenheit, t_f [$°F$]

22. Temperature in Celsius, t_c [$°C$]

23. Temperature in Kelvin, T [K] (100–$10\,000\,K$)

    $\boxed{\textit{Slide – PHOTONS Side}}$

24. Spectral fractional amount in photonic units, $Q_{\lambda_1}/Q_{\lambda_m}$ (black and for $\lambda = 0.35$ to $40\,\mu m$)

25. Spectral fractional amount in photonic units, $Q_{\lambda_2}/Q_{\lambda_m}$ (red and for $\lambda = 40$ to $4000\,\mu m$)

26. Multiplier scale for the spectral photon exitance at the peak wavelength, $M(Q_{\lambda_m})$

27. Multiplier scale for the total photon exitance, $M(Q_{0-\infty})$

28. Integrated fractional amount in photonic units, $Q_{0-\lambda_1}/Q_{0-\infty}$ (black and for $\lambda = 0.35$ to $40\,\mu m$)

29. Integrated fractional amount in photonic units, $Q_{\lambda_2-\infty}/Q_{0-\infty}$ (red and for $\lambda = 40$ to $4000\,\mu m$)

    $\boxed{\textit{Bottom Stock – PHOTONS Side}}$

30. Wavelength, λ_1 [μm] (Black: 0.35–$40\,\mu m$)

31. Wavelength, λ_2 [μm] (Red: 40–$4000\,\mu m$)

32. Wavenumber, ν_1 [cm^{-1}] (Black: 250–$30\,000\,cm^{-1}$)

33. Wavenumber, ν_2 [cm^{-1}] (Red: 2.5–$300\,cm^{-1}$)

34. Energy of a photon with a wavelength equal to the peak spectral wave-
    length, E_{λ_m} [eV]

As was the case with the Admiralty rule, the Kelvin temperature scale was
taken as the basis for the rule with all other scales on the top and bottom
stocks of the rule taking their relative positions from this scale. The range of
the wavelength (and wavenumber) scales on each side are not quite identical
and there is a very good reason for this. For temperatures between 100 to
10 000 K, the peak wavelength in the spectral curve in energetic units $\lambda_{e,max}$
lies between 0.28977 and 28.977 μm, almost the range for the wavelength scale
λ_1 used on the ENERGY side of the rule, while in photonic units $\lambda_{q,max}$ lies
between 0.36697 and 36.697 μm which is within the range for the wavelength
scale λ_1 used on the PHOTONS side of the rule. So for any temperature setting
on the rule, the corresponding value for the peak wavelength was always to
be read from the first of the wavelength scales, λ_1.

A single gauge mark labelled "TEMPERATURE" is found on the slide of
either side of the rule. In operation the temperature is selected by adjusting the
slide to the appropriate setting. After moving the cursor so as to coincide with
the selected value for the temperature, a total of twelve readings, six on each
side, could be made. On the ENERGY side these were: (i) the total radiant
exitance using the $H_{0-\infty}$ scale, (ii) spectral radiant exitance at the spectral
peak wavelength using the $H_\lambda$max scale, (iii) the peak wavelength using the
λ_1 scale, (iv and v) the corresponding temperatures in both Fahrenheit using
scale t_f and Celsius using scale t_c, and (vi) the RMS Johnson noise potential
using scale $V_n/\sqrt{R\Delta f}$.[20] On the PHOTONS side, the quantities were the same
as those given for the first five on the ENERGY side except in photonic units
while the sixth reading gave the energy of a photon in electron-volts with a
wavelength equal to $\lambda_{q,max}$. For greybodies with emissitivies less than one, the
total exitance and the spectral exitance at the peak wavelength could also be
found using the two multiplication scales that ran down the middle of the slide
on each side. With the temperature set, moving the cursor until its hairline
coincides with the value of the emissitivy for the greybody on the multiplier
scale, the corresponding exitance for the greybody could be determine from
either of the exitance scales running along the top of the stock. If, on the
other hand, after the temperature had been set, moving the cursor to a given
wavelength allowed the various fractional quantities on either side of the slide
to be calculated. Importantly, if one were interested in wavelengths found on
the black λ_1 scale, the black fractional scales on the slide needed to be read.
Similarly for the red λ_2 scale and red fractional scales.

Extension in the range of the scale, as with the Admiralty rule, was also
possible. Once again for a temperature T outside of the temperature range of
the slide rule, a temperature T_s is selected so that it falls within the temperature range of the rule such that $T_s = T \times 10^n$ where n is a conveniently chosen
integer. In addition to the extension rules already given (see page 208), for

the scales not found on the Admiralty rule one has

$$E_{\lambda_{\max}} = (E_{\lambda_{\mathrm{m}}})_s \times 10^{-n} \quad , \quad \frac{V_n}{\sqrt{R\Delta f}} = \left(\frac{V_n}{\sqrt{R\Delta f}}\right)_s \times 10^{-n/2}.$$

Chandos gives no indication to the accuracy of Cussen's rule. In one sense it was more accurate than the Admiralty rule since the scales had been prepared using the latest known values for the fundamental constants and therefore put Cussen's rule at a twenty-eight year advantage compared to the Admiralty rule. As the design of the two rules are very similar one would expect their respective accuracies to also to be very similar. However, the greatest difference between the two came in how the scales on the rules were calculated and in the precision in which the scales on each rule could be read. Firstly, regarding how the scales were calculated. All tick marks for Cussen's were calculated using a computer, while for the Admiralty rule the major tick marks were directly calculated using humans as computers while the minor tick marks were then found using interpolation. Secondly, on the Admiralty rule the Kelvin temperature scale consisted of two logarithmic cycles which was exactly 40 cm long and meant the working modulus chosen in the design of the rule was 20 cm. As all other logarithmic scales on the rule were prepared against this scale, it meant their moduli were also 20 cm. Makowski in fact explicitly mentions this and it meant that while his rule was rather long it allowed him to include many intermediate graduation marks in between the major scale tick marks. The Kelvin temperature scale on Cussen's rule on the other hand was only 10 inches long and meant Chandos only had the luxury of a working modulus of 5 inches (about 12.7 cm). This meant it was simply not possible to draw onto the rule as many intermediate graduation marks between major scale tick marks. The precision in which readings could be made using Cussen's rule compared to the Admiralty rule was therefore reduced. The duplex nature of Cussen's rule did, however, represent an improvement in design as it was possible to read both radiometric and actinometric quantities using a single temperature setting without the need of having to remove and reverse the slide, as was the case with the Admiralty rule.

To mark the fiftieth anniversary of the founding of the company in 1964, in 2014 Electro Optical Industries released a commemorative edition of Cussen's rule. Though made of plastic instead of the obligatory Pickett aluminium, the scales found on the commemorative rule were identical with those found on its 1970 predecessor. Still labelled as a Model 17, it was manufactured by American Slide-Chart/Perrygraf. In size it measured 31 cm long by 7 cm wide, making it slightly longer and significantly wider than its predecessor, and is held together by eight small metal hollow rivets. A very small, minor, modern addition of "Slide rule manual can be found at electro-optical.com" can be found printed on the ENERGY side of the rule. Photographs of the front and back of the commemorative rule are shown in Fig. 6.17.

Cussen's rule appeared right at the very end of the slide rule era. A few years later and its realization might never have come to pass. The arrival

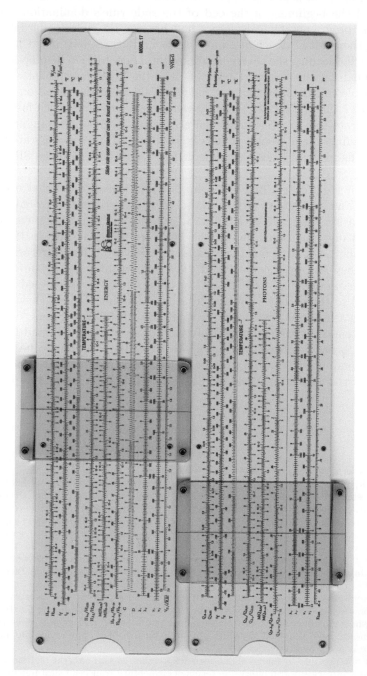

Figure 6.17 Front and back faces of the commemorative rule made for Electro Optical Industries, Inc. in 2014 as part of their fiftieth anniversary celebrations to mark the founding of the company.

of programmable hand-held electronic calculators at the beginning of the 1970s signalled the beginning of the end of the slide rule's domination. By the mid-1970s, programmable calculators had become an almost affordable item, putting them within reach of many engineers. Surprisingly, it is Chandos himself we find leading the charge into the digital age. In 1975 he prepared two short manuals for the Hewlett-Packard 65 and the Texas Instruments SR-52 programmable calculators that could calculate all quantities found on the Pickett Model 17 rule [125, 126]. Not only were these programs much faster, they were accurate to a greater number of significant figures not possible to obtain from readings made using a slide rule. Here the HP-65 was accurate to five significant figures while seven significant figure accuracy with the SR-52 was possible.

But all was not quite lost. In 1979, John N. Howard suggested that even with these fast and accurate programs available, it was still useful to have a blackbody radiation slide rule on hand as it gave one a "... better physical feeling for what one was calculating" [306]. Wolfe refers to the rule in 1978, together with the GEN-15C, but they found themselves no longer receiving the attention they had once commanded. The amount of space he devotes to discussing an "antiquated" technology, the slide rule now represented was small compared to the space he devoted to program listings for the Hewlett-Packard 25 and the Texas Instruments SR-52 and SR-56 calculators [697]. By 1993 the attention Wolfe had previously paid to slide rules and hand-held calculators had been replaced with a number of BASIC programs to be run on a personal computer [700]. And so the radiation slide rule era had come to pass. No longer the calculational aid of choice, in a few short years it all but disappeared.

Notes

[1] These tables form the subject of the chapter that follows.

[2] Professor Dr. phil. Dr. rer. nat. h. c. MARIANUS CZERNY was born on the 17 February 1896 in Breslau and died on 10 September 1985 in Munich. He received his *Doctor philosophiae* (Dr. phil.) from the Universität Berlin in 1923 for a thesis on the so-called "Reststrahlen" (residual ray) method. The sheer breadth of his scientific activities, much of it pioneering experimental work in the infrared, was the result of a sixty-year career which extended well into his retirement. For example, in 1929 he developed what he called "evaporographie," a photographic technique for making infrared radiation visible while a year later, working with A. Francis Turner, he developed their now famous *Czerny–Turner* monochromator. Five years after his retirement in 1966, he was awarded a *Doctor rerum naturalium honoris causa* (Dr. rer. nat. h. c., literally "honorary doctor of the natural things"), an honorary doctorate by his former institution the Johann Wolfgang Goethe-Universität. A short summary of his most important scientific work in English is given by Genzel, Martienssen, and Mueser [252] while a more extensive biography of his life and work in German is given by Wiesbaden [691].

[3] While it is not possible to find a closed-form expression to the integral of Planck's equation over an arbitrary spectral interval in terms of elementary transcendental functions [492, 493], as we saw in Chapter 2, it is possible to express the integral in closed form in

terms of one of the higher transcendental functions of mathematical physics known as the polylogarithm function. This however, was a later development not known to Czerny at the time he tackled the problem in 1944.

[4]The only known specimen of the rule was held in the private collection of Emeritus Professor William L. Wolfe of the University of Arizona. In 2012 he donated his collection of rules to the university where a permanent display in the College of Optical Sciences will house these together with other historical and more recent paraphernalia relating to blackbody radiation.

[5]ALFRED H. CANADA was born in Portland, Oregon, on 14 November 1918. In 1940 he graduated with a degree in electrical engineering from Oregon State University and shortly afterwards joined his country's war effort, performing military service at the Engineering Research and Development Board at Fort Belvoir in Virginia. Much of his early work focused on military applications of the infrared and was done while he worked at the General Engineering and Consulting Laboratory at General Electric in Schenectady, New York. He later moved into engineering management before retiring in 1974. He was issued eleven United States patents between the years 1946 and 1963. As part of a self-funded retirement avocation he was involved in the design and development of large-scale solar photovoltaic generation power plants. He died in 2002.

[6]A first-hand vote of confidence regarding the general utility of the General Electric rule we found appeared scrawled on the back of the cardboard envelope that came with a GEN-15B rule. Presumably given as a gift to a colleague, a very satisfied user of the rule writes "Jack Haven, Here is what you have always needed! C.".

[7]Very little is known about MIECZYSŁAW WIKTOR MAKOWSKI. He was born on 5 January 1900 in St. Petersburg, Russia to Polish parents. At the time his paper on the radiation slide rule was published in December 1949 he was working at the Polish University College in London after having spent a number of years working for the Admiralty Research Laboratory in Teddington, England. He is known to have held a Diplom-Ingenieur, the traditional engineering degree from Germany, and was an Associate Member of the Institution of Electrical Engineers in the UK. Through naturalization he had become a citizen of the United Kingdom and its Colonies in June 1949. In the mid-1950s Makowski is found working at the British Electrical and Allied Industries Research Association, or ERA as it was more generally known (ERA was short for "Electrical Research Association"). It is not known when he died.

[8]*Brunsviga* was the brand name of a popular series of mechanical calculating machines produced by the German manufacturer Grimme, Natalis & Co. from the 1890s onwards. By appropriately setting the rotors to some initial state and cranking a handle, the machine performed multiplication by repeated addition.

[9]Vahlo, PO Box 76, Ann Arbor, Michigan 48107. As an entity, Vahlo is deeply mysterious. We have not been able to ascertain what Vahlo was though it is likely it was a small firm involved in either the design, development, manufacture, and/or supply of photon detectors.

[10]For a photovoltaic detector an additional factor of $\sqrt{2}$ is needed due to the absence of recombination noise [311].

[11]For a photon at a temperature T its equivalent energy content is equal to $k_B T$.

[12]An obvious question to ask is, is the radiation emitted from a nuclear explosion blackbody-like in its behavior? The answer is yes. In a mid-air explosion, the initial thermal radiation released consists largely of x-rays. These are quickly absorbed by the surrounding air to form the characteristic fireball associated with a nuclear explosion. The assumption of blackbody behavior for the initial fireball is made since it is generally found to serve as a reasonable approximation for interpreting the primary thermal emission characteristics from a nuclear explosion.

[13]Dr HOWARD JESSE ECKWEILER was born in New York City on 11 July 1906. In 1928 he obtained a Bachelor of Science degree from New York University before joining the Electrical Testing Laboratories in New York City as an assistant to the technical director. In 1935 he joined the mathematics department at New York University. After completing a Master of Science in 1937, followed by a PhD at the Courant Institute of Mathematical Sciences in 1942, he returned to industry, joining Kollsman Instrument Corporation in Elmhurst, New

York, as Chief of their Optical Section. In 1958 he left Kollsman Instrument Corporation to found Lyle Co. Some time in the early 1960s he moved to Autonetics, a Division of North American Aviation, Inc. in Anaheim, California, where he assumed the role of a Senior Staff Scientist. He died on 20 December 1996 in Orange County, California, aged 90.

[14]The *Infrared Information Symposia* (IRIS) were held under the sponsorship of the U.S. Office of Naval Research and were a continuing series of classified symposia convened to discuss the latest developments in military infrared [311]. Attendance required a security clearance and was on a need-to-know basis with papers presented at the symposia being published in their classified proceedings.

[15]MYRON JACQUES BLOCK was a real out-of-the-box thinker. Born in New York City on 13 April 1924, after graduating with a Bachelor of Science degree from Rensselar Polytechnic Institute in New York in 1943 he went to work at what was then the National Advisory Committee for Aeronautics (now NASA) as a physicist. Shortly after the war ended he joined the Physics Department at Johns Hopkins University in 1946 as an assistant where he worked for three years before joining the Engineering Research and Development Laboratories at Fort Belvoir in Virginia. In 1953 he moved to Cambridge, Massachusetts, joining Baird Associates, Inc., as an industrial physicist to work on their infrared spectrophotometers. When the company merged with Atomic Instruments, Inc. in 1956 he, together with Neils O. Young who had also been working at Baird Associates, left and founded Block Associates, Inc., with Block taking on the role of company president. It was here Block surrounded himself with a group of young and very talented individuals. Unconventional and avant-garde, under Block's leadership the group advocated new techniques in optics which many practitioners at the time found beyond the pale. In 1968 Digilab, a wholly owned subsidiary of what had by then become Block Engineering, Inc., was formed to develop a line of commercial Fourier transform infrared spectrometers (FTIR). Block continued as president of the company until it and Digilab were acquired in 1978 by Bio-Rad, Inc. In 1980 he, together with Michael Hercher and Gerry Wyntjes, left Block Engineering; now a division under Bio-Rad, to form Optra, Inc., in Everett, Massachusetts. The company specialized in the research, development, and manufacture of state-of-the-art electro-optical systems. In his early seventies, having sold his share in Optra, Inc., Block established Optix, LP to focus on the research and development of analytical instrumentation. Block was incredibly inventive, being issued with many U.S. patents, and despite a personal disdain for the peer-review process, had two short notes published in *Nature* during the 1950s. It is not known when he died.

[16]Born in St. Louis, Missouri on 23 June 1925 ARTHUR JOHN CUSSEN spent his entire professional career working in the optical and infrared regions of the electromagnetic spectrum. Graduating with a Bachelor of Arts degree in 1947, he got his start in infrared working for the government at the Naval Ordnance Laboratory in Corona, California. After becoming the head of the Infrared Division in 1956, he left two years later to head up Infrared Standards Laboratory, a newly formed subsidiary of Infrared Industries, Inc. in Riverside, California. In 1964 he left Infrared Industries, Inc. to establish Electro Optical Industries, Inc. in Santa Barbara, California. Here he served as the founding company president for just over thirty years before stepping down on retirement in 1995. Known to all who knew him simply as "Art," after a long and productive life he died on 6 September 2013, aged 88.

[17]Some say his "dream" radiation slide rule.

[18]Professor RAYMOND JAMES CHANDOS was born in 1949 and received a Bachelor of Arts degree from the University of California, Irvine. After graduation he worked for a number of years at EOI before leaving in the early 1980s to join REC Corporation, a spin-off of EOI started by the former Vice-President and Chief Engineer, his brother Robert E. Chandos. As of June 2012 he was a Professor of Electronic Technology in the School of Physical Sciences and Technologies at Irvine Valley College, California.

[19]On at least two examples we have seen, the red scale markings are not red but a teal green in color. In provenance the first belonged to Cussen himself and is marked with a relatively high engraved serial number (No. 711). The second comes from an old-timer still working at EOI and is unmarked. Unlike rules with red scales, the latter have no markings

on it indicating it was made by Pickett. Perhaps the teal green colored scale rules were a later model or came from a second batch of rules made.

[20]The Johnson or Nyquist noise given on this scale has units of volts per square root of ohms per square root of Hertz, that is, $V \cdot \Omega^{-1/2} \cdot Hz^{-1/2}$. For example, if $T = 1000\,K$, $R = 10\,k\Omega$ and $\Delta = 5\,kHz$, then from the rule one reads $2.34 \cdot 10^{-10}$. The noise voltage is therefore $2.34 \cdot 10^{-10} \cdot \sqrt{10^4} \cdot \sqrt{5 \cdot 10^3} = 1.66\,\mu V$ (rms).

7 Tables used for thermal radiation calculations

Before the widespread availability of digital computers, of all the aids available to the user, tables were by far the most numerous and the most accurate. Many of the earliest tables were prepared to four significant figure accuracy. As the methods used to calculate these tables shifted from human computers before the Second World War to mechanical and electronic computers after it, so their accuracy increased. In the latter electronically produced era, tables with between seven to nine significant figures were not uncommon. Compared to using nomograms which were little better than providing order-of-magnitude estimates, or a slide rule where estimates to three significant figures at best was possible, the accuracy gained from using tables was unsurpassed.

Over the years many tables for various quantities related to blackbody radiation appeared. The most extensive set of tables produced were those published as entire books. Not seen until the early 1960s, a number were produced and were devoted entirely to tabulations of quantities related to blackbody radiation. A second common source of tables were those included in books consisting of either more general sets of tables or in handbooks listing physical and chemical data. These appeared shortly after Planck introduced his equation and remained a staple feature of such texts for many decades to come. Tables for many of the more important quantities relating to blackbody radiation were also to be found in many discipline specific texts where the topic of thermal radiation arose. Here the extent of the tables varied considerably from short listings found in the main body of the text to whole appendices given over to extensive tabulations, and remain one of the few places today where one still continues to encounter such tables. Various technical journals before the Second World War often carried articles devoted entirely to the calculation and presentation of tabulated values for various quantities related to blackbody radiation. One also occasionally finds during this period whole student theses, or parts thereof, devoted to the calculation and tabulation of such quantities.

The most prolific period in the creation of blackbody radiation tables coincided with the rise of the digital revolution starting in the early 1950s and lasted until the arrival of affordable computing power for all in the mid-1970s. Many of these tables were produced as either technical reports or notes, particularly in the United States, by individuals or small research groups working either for or attached to various universities, scientific organizations, governmental agencies, or companies, and closely paralleled the advances being made in the application of infrared technology to systems design and engineering.

The new and unprecedented demands in accuracy these advancements called for led to tables with far greater accuracy over far smaller interval sizes being produced. Thereafter, the production of tables devoted to thermal radiation slowed before almost completely disappearing as former users of these tables took full advantage of the growing computational abilities offered by relatively inexpensive and widely available digital computers. The era of fast computations on demand had arrived, and with it, the need for computational aids such as tables was at once consigned to history.

Diverse as the literature was where tables relating to blackbody radiation were found, so to was the number of different quantities tabulated. The two main quantities considered were the spectral radiant exitance (Planck's equation) and the fraction of energy radiated by a blackbody into a given spectral band (the fractional function of the first kind). These were given in both energetic and photonic units and for various spectral representations including linear wavelength, frequency, and wavenumber. Other quantities often tabulated included the spectral radiance, total exitance and radiance, ratio of the spectral radiant exitance or radiance at either its peak value or some other fixed value, value of the wavelength, frequency, or wavenumber at the spectral peak, to less frequently needed and used quantities such as the derivative of the spectral radiant exitance with respect to temperature, wavelength, wavelength–temperature product, spectral peak values, and so on. Entries were often tabulated as either a function of the independent variables individually, such as the temperature and wavelength (or frequency or wavenumber), or as a single combined variable such as the wavelength–temperature product. The convenience of the former was often bought at the expense of the latter in order to keep the tables down to a manageable size, which meant in use many of the intermediate interpolations between tabulated values had to be made. Selected interval sizes also varied considerably.

Of the various ways tables could be tabulated and presented, by far the most convenient and immediately useful were those that gave the particular quantity of interest directly. For the two most commonly tabulated quantities, this meant the spectral radiant exitance as a function of wavelength (or frequency or wavenumber) over a convenient range of temperatures and the corresponding integrated fractional amount from zero up to some arbitrary wavelength (or frequency or wavenumber) would be presented. Values calculated in this way were susceptible to frequent change since they depended on the values used for the fundamental constants, which themselves were constantly being revised and updated.

In the very early days of the study of blackbody radiation, it was common to express Planck's law in terms of two constants that were directly measurable. As seen already these are the so-called first and second radiation constants $c_1 = 2\pi hc^2$ and $c_2 = hc/k_B$ respectively. When expressed in terms of these two constants, Planck's equation for the spectral radiant exitance in the linear wavelength representation can be written as

$$M_{e,\lambda}^{b}(\lambda, T) = \frac{c_1}{\lambda^5 (e^{c_2/(\lambda T)} - 1)}. \tag{7.1}$$

The large experimental uncertainties in the values for the two radiation constants led to a lack of international agreement in their accepted values for much of the first part of the twentieth century. Tabulated values for the spectral radiant exitance calculated directly would quickly become obsolete with each change in value for either radiation constant. To help overcome this, values for the spectral radiant exitance normalized relative to its peak value (or often some other convenient value) as a function of wavelength–temperature product were commonly tabulated. Here

$$\frac{M_{e,\lambda}^{b}(\lambda, T)}{M_{e,\lambda}^{b}(\lambda_{max}, T)} = \frac{b^5 (e^{c_2/b} - 1)}{(\lambda T)^5 (e^{c_2/\lambda T} - 1)}. \tag{7.2}$$

Since $b = c_2/[5 + W_0(-5e^{-5})]$ (see page 43), tabulations made in this way had the advantage of avoiding the uncertainity in the value for the first radiation constant.

If longevity in the values tabulated was desired, a reduction of the spectral radiant exitance to a completely dimensionless form was necessary. However, it came at the expense of convenience as a number of intermediate calculations had to be made in order to produce a final value for the spectral radiant exitance. In the linear wavelength spectral representation, if one lets $x = \lambda T/c_2$, the spectral radiant exitance can be rewritten as

$$M_{e,\lambda}^{b}(\lambda, T) = \frac{c_1}{c_2^5} \frac{1}{x^5 (e^{1/x} - 1)} T^5 = \frac{c_1}{c_2^5} y T^5. \tag{7.3}$$

and it is the dimensionless quantities x and y one finds tabulated.

Another important factor affecting the design of a table depended on the table's intended audience. The heaviest users of blackbody radiation tables were made by those working in infrared systems design, photometry, colorimetry, and optical pyrometry. Many of the tables prepared catered to these particular communities with the range in temperatures and wavelengths (or frequencies or wavenumbers) selected accordingly. Tables for blackbody radiation were also prepared for those found working in the astrophysical and meteorological fields and are differentiated from the former principally by their choice in range of temperature. For the astrophysical case, tabulations to very high temperatures were the norm while a much narrower terrestrial temperature range with much smaller interval sizes are found in tables intended for meteorological use.

In discussing the many tables produced to aid calculations involving thermal radiation, we shall employ three broad divisions based on time of publication. While the division into these three periods is essentially arbitrary, each has been chosen to broadly correspond to related developments in those fields where the tables were principally used. The first period encompasses the very earliest of tables up until 1939. During this time the tables were produced by human computers mainly for those working in the field of either illumination or for astrophysical work. The second period from 1940 to 1954 represents a transitional period spanning the Second World War and the years immediately following it. While these tables were still the result of human computers, they are characterized by greater accuracy and by a tendency to becoming more general in nature. The third and final period is from 1955 onwards and represents the era of digitally produced tables. They are of very-high accuracy while the vast majority were produced for those working in fields related to the infrared.

7.1   NOTATIONAL CONVENTIONS USED FOR TABLES

As the modern reader may not be familiar with some of the general notational conventions used when discussing tables, we begin by giving a brief description of these. For any table the information it presents to the user is a listing of values for some given dependent quantity as a function of typically one (single entry) or two (double entry) independent variables. The independent quantities are tabulated over some predefined range of interest using interval widths of a given size. The interval width sizes may, however, not be uniform throughout their range.

Consider some quantity y is tabulated as a function of x. If lower and upper bounds for x of x_{min} and x_{max} are selected, for a tabulation with uniform interval size of Δx throughout, one would write this as

$$x_{min}(\Delta x)x_{max}.$$

If instead the interval is divided into two portions with different interval sizes of Δx_1 and Δx_2 selected, for each we would write this as

$$x_{min}(\Delta x_1)x^*(\Delta x_2)x_{max}. \tag{7.4}$$

Here $x_{min} \leqslant x \leqslant x^*$ and $x^* < x \leqslant x_{max}$. As the number of differently sized interval widths increases, we write this by extending the notation of Eq. (7.4) by adding each new interval width size in brackets over the corresponding range where it applies.

As an example, if values for a temperature T are written as

$$T = 100(0.1)200(1)1000(5)5000(20)15\,000\,\text{K}$$

tabulations at the following temperatures have been made:

- 100 to 200 K at a step of 0.1 K
- 200 to 1000 K at a step of 1 K

- 1000 to 5000 K at a step of 5 K
- 5000 to 15 000 K at a step of 20 K

This corresponds to 1000 temperature values in the first interval, 800 in the second and third, and 500 in the fourth giving a total of 3100 entries for the temperature.

7.2 FROM THE EARLIEST TABLES TO 1939

The first calculations known to have been made for the spectral radiant exitance of a blackbody using Planck's equation were performed in early October of 1900. It comes down to us as an interesting footnote in the history surrounding the initial deduction by Planck of his radiation law. In late 1900 the correct form for the spectral radiant exitance of a blackbody was still very much the subject of intense experimental and theoretical research. The distribution law proposed by Wien four years earlier [687] was thought to give the correct description of a blackbody over all wavelengths and at all temperatures. Subsequent improvements made by experimentalists in their ability to more accurately measure the energy radiated from a blackbody at longer wavelengths extending deep into the far infrared portion of the spectrum, were beginning to show serious discrepancies between what was observed compared to predictions based on Wien's law, discrepancies that could not be accounted for by experimental error alone.

Two experimentalists who had been performing very careful work in the far infrared over a wide range of temperatures were Heinrich Rubens and Ferdinand Kurlbaum. In a pivotal event, not only in the history of blackbody radiation but for the new theory of the quantum that was to come, Ruben and his wife paid Planck a visit at his home on Sunday 7 October 1900 [290, 338, 439]. During the course of the visit their discussion naturally turned to Rubens' latest findings. His colleague Kurlbaum was set to give a short talk on their most recent work at the next meeting of the *Deutsche Physikalische Gesellschaft* to be held in Berlin on 19 October 1900 [366] while a more extensive talk before the *Königlich Preussische Akademie der Wissenschaften* was to follow a week later on October 25, 1900 [572].

Rubens' visit to Planck came at a time just before he commenced writing a short note to accompany the talk to be given before the *Akademie*.[1] The most recent measurements of Rubens and Kurlbaum had shown that at very long wavelengths the distribution function proposed by Wien was totally inadequate and clearly in need of modification. On hearing this, shortly after Rubens left, Planck immediately set himself to work. By modifying Wien's law using a simple mathematical interpolation he arrived at a distribution law that seemed a better fit compared to Wien's while at the same time being consistent with both the Stefan–Boltzmann law and Wien's homologous law [508]. Planck communicated his new radiation law to Rubens by postcard that evening. Received the following morning, a day or two later Rubens returned to Planck to inform him he had checked the results given by his new radiation formula against his own measurements and had found a "satisfactory concordance" at every point between the two [290, 356]. How he performed his calculations posterity does not record, but they stand as the first in a very long succession of first human, and later electronic, computers to do so.

In the coming years after its announcement, as the predictions of Planck's new radiation law were checked and confirmed against experiment, many calculations of the spectral radiant exitance using Planck's equation had to be made. In those very early days, given each investigator's preoccupation in confirming Planck's new radiation law, how values were actually calculated from Planck's equation tended not to be given.

One of the earliest tabulations of a quantity relating to thermal radiation comes to us from the Danish astronomer Ejnar Hertzsprung (1873–1967) in 1906 [289]. Working at the time as an amateur astronomer at the Copenhagen University Observatory and the Urania Observatory in Frederiksberg, in a paper published in the somewhat obscure journal for scientific photography, *Zeitschrift für Wissenschaftliche Photographie, Photophyik und Photochemie*, a table related to the base 10 logarithm of the spectral radiant exitance appeared. In the paper, Hertzsprung was concerned with establishing a relation between the intensity of blackbody radiation in the optical spectrum to relative sensitivities of the eye. He avoided uncertainties associated with values for the two radiation constants by tabulating, for the following wavelengths $0.40(0.05)0.75\,\mu m$, values of $\log_{10}(M_{e,\lambda}^{b}/c_1)$ for various values of the ratio c_2/T ranging from 1 to $20\,\mu m$. In all, a total of 80 tabulated values correct to five significant figures were given. Shortly after their publication, they were reproduced by Karl Schaum in his 1908 German text *Photochemie und Photographie* [589].

A second early tabulation for quantities relating to the radiation of a blackbody is to be found in the Fifth Revised Edition of the *Smithsonian physical tables* of 1910 prepared by Frederick E. Fowle [234]. Given are two very short tables. The first is a tabulation for the total radiant exitance for a number of temperatures ranging between -273 to $5000\,°C$, while the second gives the spectral radiant exitance as a function of wavelength ranging between 2.0 to $100\,\mu m$ at the six temperatures of: -80, -30, 0, 15, 30, and $100\,°C$. In both tables, all values are tabulated with a number of significant figures which varies between two to four. The very-low temperatures selected for the second table suggests they were likely intended for meteorological use at terrestrial temperatures.

Values for the two radiation constants and the Stefan–Boltzmann constant in the first decade of the twentieth century were by no means known with any degree of certainty. Fowle chose values of $3.688 \times 10^4\,W{\cdot}cm^2$, $1.4550\,cm{\cdot}K$, and $5.32 \times 10^{-12}\,W{\cdot}cm^{-2}$. In 1914 for the Sixth Revised Edition, he updated these tables using values for the three constants of $3.86 \times 10^4\,W{\cdot}cm^2$, $1.4450\,cm{\cdot}K$, and $5.75 \times 10^{-12}\,W{\cdot}cm^{-2}$ [235]. By the time the Seventh Revised Edition was published in 1920, a further change in the value of the second radiation constant to 1.4350 cm·K had been made [236].

From Fowle's work one can see how the convenience gained in giving tabulations explicitly in terms of the quantity one is interested in are quickly negated by their limited expected lifetime during a period when the values for the fundamental constants were themselves still very much in a state of flux. Fowle himself recognized this. The Seventh Revised Edition also saw an extension in the tabulations made to a third and fourth table for the spectral radiant exitance as a function of wavelength ranging from (i) 1.0 to 100.0 μm at temperatures from 50 to 600 K, and (ii) 0.1 to 100.0 μm at temperatures from 800 to 20 000 K. Tabulations for the spectral radiant exitance to higher temperature were no doubt driven by the now growing needs of photometrists. Fowle also mentioned for the first time how small changes in the value for the second radiation constant could be handled while still making use of his tables. It would however, not be until 1929 when Fowle was preparing a more extensive set of tables for the total and spectral radiant exitance for the fifth volume of *International Critical Tables of Numerical Data, Physics, Chemistry and Technology* [237] that he dealt with this problem in a rather explicit fashion. An auxiliary table is presented to account for any small change in c_2. As a change in this constant had the greatest affect on values for the spectral radiant exitance, it allowed these values to be corrected.

Fowle extended all previous tabulations he had made for one last time in the Eighth Revised Edition of the *Smithsonian Physical Tables* of 1933 [238]. Relative values for the spectral radiant exitance were given for the first time, but unlike his tables of 1929, no auxiliary table that allows for any change in the value for the second radiation constant is included. In contrast to the tabulations of Fowle, in this year a very modest set of tables in dimensionless form for Planck's equation appeared in the first edition of Jahnke and Emde's *Funktionentafeln mit Formeln und Kurven* [320], a highly regarded and widely used text on mathematical tables that went through many subsequent editions during its lifetime.

Another early tabulation for the spectral radiant exitance came from a paper published by Emil G. Warburg, G. Leithäuse, E. Hupka and C. Müller in the German journal *Annalen der Physik* in 1913 [673]. In this paper an experimental determination for the second radiation constant was made. Towards the end of the paper one finds two separate tabulations for the spectral radiant exitance of a blackbody relative to the value found at the spectral peak for wavelength ratios relative to the spectral peak $\lambda_{\max}$ are given. In the first $\lambda < \lambda_{\max}$ and $\lambda/\lambda_{\max}$ runs from 0.50 to 0.99 in steps of 0.01 while in the second the direction of the inequality was reversed. The tabulations however, were not utilized until a followup paper published in 1915 under the same title by the first and fourth authors [674].

Starting in the 1920s, a concerted effort began to be made towards dedicated tables for use in finding the spectral radiant exitance of a blackbody. In

1920 William W. Coblentz, working in the United States, gave a three-page table for a number of different intermediary quantities that he used to compute the spectral radiant exitance of a blackbody [141]. From his tables the spectral radiant exitance was found indirectly. Starting with Planck's equation for the spectral radiance $L(\lambda, T)$, on taking the base 10 logarithm of both sides, Coblentz wrote

$$\log_{10} L(\lambda, T) = \log_{10} k - 5 \log_{10} \lambda. \qquad (7.5)$$

Here $\log_{10} k = \log_{10} c_{1L} - \log_{10}(e^u - 1)$, $u = c_2/(\lambda T)$ while c_{1L} is the first radiation constant for spectral radiance.[2] To simplify matters further, for his intended purpose, for arbitrary units Coblentz chose a value of exactly $\log_{10} c_{1L} = 5.00000$. When compared to the modern day value of 5.075 927, we see Coblentz's choice was more than reasonable given the large uncertainties that existed in the value for either radiation constant at the time. Large uncertainties in the values for the radiation constants was also the reason Coblentz did not prepare tables for the spectral radiance at various temperatures and wavelengths directly, because he thought it too premature to do so.

For the dimensionless parameter u, ranging from 0.10 to 14.0, the three quantities e^u, $\log_{10}(e^u - 1)$, and $\log_{10} k = 5 - \log_{10}(e^u - 1)$ were tabulated to six significant figures. In using these tables, a value for the wavelength had to be selected so that the resulting exponent u appeared in his table. In doing so the value for the wavelength rarely came out as some simple whole value, but as Coblentz noted, this was not important. Finally, after having found a value for $\log_{10} k$, additional calculations were needed before a final value for the spectral radiance at a given temperature and wavelength could be found. Coblentz recommended these additional calculations could most easily be accomplished using an ordinary slide rule. As Coblentz's tables did not depend on a value for the second radiation constant c_2, the latest accepted value for c_2 could simply be inserted into the calculations when needed.

In his report to the Standards Committee for 1919 on pyrometry [231], as part of a need to collect reference material together into one place, and in a form others working in the field of pyrometry would find useful, William E. Forsythe presented a one page table for the spectral radiant exitance of a blackbody in the visible range (0.40 to 0.76 µm) for twelve different temperatures from 1000 to 5000 K. Interestingly, all values except the last column for 5000 K were calculated using Wien's approximation with the last column being calculated using Planck's equation. A table of correction factors for five different wavelength–temperature products was also provided. These could be used to convert values obtained using Wien's approximation to those obtained using Planck's equation for a given wavelength–temperature product. Tabulating values for the spectral radiant exitance using Wien's form, which is far simpler, and correcting the resultant values found, was an idea returned to later by others [216, 41, 668]. His table was reproduced a few years later as part of Leonard T. Troland's report to the Committee on Colorimetry for 1920–21 [649].

The growing demand for the need to compute the spectral radiant exitance of a blackbody in the visible portion of the spectrum at some particular temperature prompted M. Katherine Frehafer and Chester L. Snow to produce a number of tables and graphs relating to this quantity. Initially Frehafer presented her tables and graphs as a contributed paper at the *7th Annual Meeting of the Optical Society of America* held at the National Bureau of Standards in Washington between October 25–28, 1922 [240]. Given the amount of time needed to compute values for the spectral radiant exitance by hand, Frehafer thought it worthwhile to publish these aids for the benefit of others.

Three years later they were published as part of the Bureau of Standard's miscellaneous publications and sold for a cost of 35 cents [241]. Comprising seven large, loose, sheets measuring 48 cm wide by 60 cm long, the first sheet consisted of text, the next five were of graphs, while the last contained a number of tables. The first sheet of Frehafer's 1925 publication was devoted to a discussion of the various laws of thermal radiation and gave values of the constants used in the computations ($c_2 = 1.4350 \times 10^{-2}$ m·K and $b = 2890$ μm·K). It also provided some details on the graphs and tables that followed. Sheets two to six contained various plots for the relative spectral radiant exitance (these were discussed on page 175) while sheet seven contained a number of tables. The first table given was for the spectral radiant exitance relative to the spectral radiant exitance at the spectral peak as a function of the wavelength–temperature produced to four significant figures. The next two sets of tables gave the relative spectral radiant exitance at values equal to the spectral radiant exitance at wavelengths of 590 nm and 560 nm, respectively, as a function of both wavelength and temperature. Wavelengths between 400 to 720 nm at intervals of 10 nm were given for 38 different temperatures between 1000 to 28 000 K. Values were again tabulated to four significance figures with a claimed accuracy of 0.15%.

In 1931, the tables of Frehafer and Snow for the relative spectral radiant exitance were extended by Raymond Davis and Kasson S. Gibson [172]. Resulting from their work on various filters developed for use in photographic sensitometry, colorimetry, and photometry, additional values for the relative spectral radiant exitance for the photographically important wavelength region between 350 and 390 nm were given. Forty-eight different temperatures between 2000 and 20 000 K for wavelengths ranging from 350 to 720 nm at 10 nm intervals can be found. The value used for the second radiation constant was identical with that used by Frehafer and Snow. The tabulated values for the relative spectral radiant exitance were given to either three or four significant figures and were claimed to be accurate to within 0.29%. The values in Davis and Gibson's tabulations had either been (i) calculated from the first of Frehafer and Snow's tables which gave the spectral radiant exitance in terms of the wavelength–temperature product, taken directly from their second set of tables, or (ii) read from one of their graphs they had produced for values that were not tabulated.

The early tables were by and large designed to aid calculations encountered in the field of photometry. With wavelengths confined principally to the visible region and temperatures ranging from one to several thousand Kelvin, most were relatively small and compact in form with large interval sizes usually being employed, particularly in values for the temperature. A growing need for more accurate calculations would ultimately lead to more extensive tables with greater accuracy being produced. In 1929 James F. Skogland, who like Frehafer was an Associate Physicist working at the Bureau of Standards in the United States, was the first to produce one of these more extensive and accurate set of tabulations [605]. For wavelengths ranging from just beyond violet to the upper red end of the visible spectrum (0.32–$0.76\,\mu$m) values for the spectral radiant exitance relative to the spectral radiant exitance at a wavelength corresponding to $0.59\,\mu$m were tabulated for temperatures ranging from 2000 to 3120 K at intervals of 20 K. Values given were accurate up to five significant figures. By using a relatively small interval size, direct linear interpolation would yield values for the relative spectral radiant exitance that retained their accuracy to the last digit. Since relative values for the spectral radiant exitance are calculated, Skogland's tabulations did not depend on the value for the first radiation constant except at $\lambda = 0.59\,\mu$m where actual values for $M_{e,\lambda}^{b}(0.59\,\mu\text{m}, T)/c_1$ are given. The value Skogland used for the second radiation constant was 1.4330×10^{-2} m·K. To accommodate a change in value for this constant, Skogland also gave a correction table which could be used to adjust values in the relative spectral radiant exitance for a change in c_2 between 1.4300 and 1.4360×10^{-2} m·K. The small interval step sizes used coupled with the five significance figure accuracy meant Skogland's tables were very popular at the time. Despite covering only a limited range of temperatures and being confined to the visible part of the spectrum, his tables continued to be used for many years.

Without tables, the calculation of the spectral radiant exitance by hand already required the expenditure of much time and effort. And if this was not bad enough, finding the proportion of the total energy radiated by a blackbody into a given spectral band saw the time needed to perform such calculations move into a whole new realm as Planck's equation needed to be integrated between finite spectral limits. The first attempt to tabulate such a quantity was made by the French physicist M. P. A. Charles Fabry. They appeared in his influential text *Introduction générale a la photométrie* of 1927 [221]. In addition to tabulating the spectral radiant exitance in dimensionless form to three and four signifiance figures as a function of the dimensionless wavelength parameter $x = \lambda/\lambda_{\max} = \lambda T/b$, the fractional amount radiated into the spectral band between zero and x is given with accuracies varying between two to five significant figures.

A year later L. L. Holladay in the United States also tabulated the fractional amount except this time as a function of the wavelength–temperature product [294]. From Eq. (7.2) for the normalized spectral radiant exitance, the form for the fractional function of the first kind for the radiometric case can be rewritten as

$$\mathfrak{F}_{e,0 \to \lambda} = \frac{15 c_2^4}{\pi^4 b^2 (e^{c_2/b} - 1)} \int_0^{\lambda T} \frac{M_{e,\lambda}^b(\lambda, T)}{M_{e,\lambda}^b(\lambda_{\max}, T)} d(\lambda T), \qquad (7.6)$$

where the integral can be approximated by summing the products

$$\frac{M_{e,\lambda}^b(\lambda, T)}{M_{e,\lambda}^b(\lambda_{\max}, T)} \Delta(\lambda T), \qquad (7.7)$$

between the limits of integration and dividing the sum by the constant factor appearing in front of the integral. In doing so, Holladay was able to take advantage of Frehafer and Snow's previously tabulated values for the spectral radiant exitance ratios by using these values in the calculation for the fractional amount. The wavelength–temperature product range considered extended from 400 to 21 000 µm·K while all values for the fractional amount appeared to four significant figures. A value of $c_2 = 1.4350 \times 10^{-2}$ m·K was used for the second radiation constant.

During the earliest period considered here, tables also came from individuals working in fields quite unconnected to illumination. In 1927, as part of his work on planetary nebulae and stellar temperatures, Herman Zanstra first encountered a need for the evaluation of the fractional function of the first kind for the actinometric case [729]. The integral was calculated for eleven different values of the dimensionless parameter $z = h\nu/(k_B T)$ ranging from 0.2 to 10. The range in values for the argument given meant temperatures from 15 000 to 200 000 K were achieved within the visible portion of the spectrum. Values of the integral for $z = 0.2$ to $z = 5$ were found using quadrature with an accuracy of at least one per cent for $z = 0.6, 1, 2, 3, 4, 5$ and an accuracy of two per cent for the first two non-zero values of z of 0.2 and 0.3. For $z \geqslant 6$ the negative one term appearing in the denominator of the integral was dropped allowing the integration to be performed analytically.[3] Values obtained within this approximation were accurate to within one quarter of a per cent. A few years later Zanstra realized that integrals involving the fractional function of the first type could be carried out quickly and to far greater accuracy using series methods, as was discussed in Section 3.2.1. A re-tabulation of the same integral for exactly the same values of the dimensionless parameter z up to 11 were given in his later work of 1931 [730, 731].

A year later, A. Brill in the German astrophysical text *Handbuch der Astrophysik* presented a number of tables for various quantities relating to thermal radiation [103]. In the first of his tables, for fifteen different temperatures ranging from 1497 to 23 750 K, he tabulated the corresponding peak wavelength $\lambda_{\max}$ in the spectral distribution curve, the maximum spectral radiance at the spectral peak and its base 10 logarithm, and the total radiance and its base 10 logarithm. The first column of the table gave values for the parameter c_2/T.

These were to be used in all subsequent tables in place of the temperature T. In the second of Brill's tables, using the same fifteen values for the parameter c_2/T, values for the spectral radiance and its base 10 logarithm are given as a function of wavelength ranging from 0.30 to 1.20×10^{-6} m. In his third table the correction factor $\log_{10}(1 - e^{-c_2/(\lambda T)})^{-1}$ as a function of the previously considered values for the wavelength and the parameter c_2/T are tabulated. This factor corresponded to the amount one needed to add to the logarithmic amount of the spectral radiance as given by Wien's approximation, to give the correct amount as found using Planck's equation. Brill's fourth and final table, while a little unusual, turns out to be rather useful for astrophysical calculations. It gave, for the previously considered values for the wavelength and the parameter c_2/T, the gradient of the logarithm of the spectral radiance curve and could be used in determining the color temperature of a star; a quantity we already saw in Chapter 5 was put into graphical form some years later by Kienle [350] (see Fig. 5.15 on page 177).

A rather different approach to the tabulation of the spectral radiance was undertaken by the Dutch physicist Willem de Groot in 1931 [180]. By converting Planck's equation for the spectral radiance into its base 10 logarithmic form, the spectral radiance could be quickly found using an auxiliary table first, where values for the wavelength and temperature in units of Ångström and in Kelvin respectively were combined as a single value. Dimensionless numbers corresponding to the selected values of the temperature and wavelength were then read from the table and summed. The sum of these two numbers gave an auxiliary parameter that could be used in the main table to read off corresponding values for the spectral radiance.

In rewriting Planck's equation for the spectral radiance in base 10 form, de Groot wrote

$$L(\lambda, T) = AT^5 \phi(x) = AT^5 x^5/(10^x - 1), \qquad (7.8)$$

where

$$A = \frac{c_1}{(c_2 M)^5} \quad \text{and} \quad x = \frac{Mc_2}{\lambda T}. \qquad (7.9)$$

Here M is a mathematical constant equal to $\log_{10} e$. For any combination of wavelength or temperature in units of either Ångström or Kelvin respectively, the value for the dimensionless parameter $\log_{10} x$ could be found from an auxiliary table. To begin with, values for the wavelength and temperature at values numerically equal to each other and

$$\log_{10} x = \log_{10} M + \log_{10} c_2 - \log_{10} \lambda - \log_{10} T = 0, \qquad (7.10)$$

were found. Calling these values λ^* and T^*, de Groot then set $\log_{10} \lambda = \log_{10} \lambda^* - a_\lambda$ and $\log_{10} T = \log_{10} T^* - a_T$. After substituting these into the expression for $\log_{10} x$ one finds

$$\log_{10} x = a_\lambda + a_T. \qquad (7.11)$$

Tabulating a (for either a_λ or a_T) between zero and 0.99 corresponded to a wavelength and temperature range between 807.1 and 7887 Ångström or Kelvin respectively. By selecting a particular wavelength and temperature value found in the auxiliary table, the value for $a_{\{\lambda,T\}}$ could be read off and summed to give the value for $\log_{10} x$. The corresponding value for the function $\phi(x)$ located in the primary table could then immediately be read off and the resulting spectral radiance calculated. Use of the auxiliary table in this way meant it was totally impractical for wavelength and temperature values not found in the table. Instead, de Groot's tables were found to be of use for those wishing to compare data measured experimentally with theoretical plots for the spectral radiance of a blackbody. Of course de Groot's primary table could have been used directly, as it is presented in dimensionless form. This however, would have required considerable calculational effort since finding the wavelength or temperature from $\log_{10} x$ was required. The values de Groot used for the two radiation constants were $1.4325 \times 10^{-2}\,\mathrm{m\cdot K}$ and $\frac{3.703}{\pi} \times 10^{-12}\,\mathrm{W\cdot m^2\cdot sr^{-1}}$ for the first and second radiation constants respectively.

A surge in the tabulation of various quantities relating to blackbody radiation came from the Japanese in the second half of the 1930s. The impetus behind these tables stemmed primarily from work in the field of illumination. Over a four year period, no less than four different sets of tables were published by various authors. The first of these tables was published in 1936 by Y. Omoto [469]. The dimensionless function $y(x)$ relating to the spectral radiant exitance of a blackbody as given by Eq. (7.3) as a function of the logarithm to base 10 of the dimensionless quantity x related to the wavelength–temperature product are tabulated. Four to five significant figure accuracy is given for y while its logarithm is given to six significance figures. In the same year, Z. Yamauti and M. Okamatu, both of whom were working at the time at the Electrotechnical Laboratory (Denki Sikenzyo) in Tokyo, published first in concise form [719] and then in more extended form in the laboratory's *Researches of the Electrotechnical Laboratory* [720] the first of their multiple set of tables. In a similar manner to Omoto, Yamauti and Okamatu chose to tabulate a dimensionless quantity related to the spectral radiant exitance as a function of the dimensionless parameter $x = \lambda T/c_2$. They did this to an accuracy of five significance figures. In their case, by writing the spectral radiant exitance[4] as

$$E(\lambda, T) = E_m(T)E(x), \qquad (7.12)$$

it is $\log_{10} E(x)$ as a function of $\log_{10} x$ one finds tabulated. Here

$$E_m(T) = c_1 c_2^{-5}\beta^5(e^\beta - 1)^{-1}T^5, \qquad (7.13)$$

corresponds to the spectral radiant exitance at a wavelength where the spectral curve is a maximum while $\beta = 4.9651$ was a mathematical constant arising from the non-trivial root of a transcendental equation,[5] $E(x) = Kx^{-5}(e^{1/x} - 1)^{-1}$, while K is a second mathematical constant which appears when the spectral radiant exitance at the spectral peak is considered and is equal to $\beta^{-5}(e^{\beta}-1) = 0.047\,167$.[6] In their first more concise form [719] the following range and interval sizes for $\log_{10} x$ were chosen: $-2.700(0.002)-1.500$, $-1.500(0.005)-0.950$, and $-0.950(0.01)0.25$. In practice however such a choice often meant interpolation was required. In the more extensive set of tables which followed [720], the range was considerably extended and the interval sizes greatly reduced.

It was not long before Yamauti and Okamatu realized their initially produced tables required considerable effort to use. From the tabulated values, many subsequent calculations were required to extract a value for the spectral radiant exitance at a given wavelength and temperature making them extraordinarily tedious and time consuming to use. A year later they went some of the way towards remedying the situation by producing a second slightly more user friendly set of tables [721]. This time the logarithm to base 10 of the spectral radiant exitance normalized relative to the maximum spectral radiant exitance at the spectral peak was given to four significant figures for a range of temperatures within the visible wavelength range of 0.30 to 0.76 µm at intervals of 0.01 µm. What they gained in convenience they lost in permanency as the tabulations now depended explicitly on the value for the second radiation constant. They chose $c_2 = 1.4320 \times 10^{-2}$ m·K. The range in temperatures from 2000 to 3200 K at 20 K intervals suggest the tables were most suited to illumination work with incandescent filaments. Both authors claimed that the small interval sizes in temperature allowed direct linear interpolation to be used for temperature values not appearing in their tables without any significant loss in accuracy.

As a way of mitigating against obsoleteness brought about by any future change in the accepted value for the second radiation constant, a more extensive but slightly modified set of tables produced by both Yamauti and Okamatu appeared in the *Researches* in the same year [722]. Again, the normalized base 10 logarithm for the spectral radiant exitance as a function of wavelength at $0.30(0.01)0.76$ µm, but to five significant figures, was now given. However this time, in place of temperature, the normalized amount is given as a function of the quantity $\log_{10}(T/c_2)$. Values for $\log_{10}(T/c_2)$ were taken from -1.140 to -1.350 at intervals of 0.001 (which roughly corresponds to temperatures between 1977 to 3206 K when a value of $c_2 = 1.4320 \times 10^{-2}$ m·K is used). The change from temperature to $\log_{10}(T/c_2)$ was made to increase the longevity of their tables. Any future change in the value of c_2 could be readily incorporated into their tables using only minor adjustments compared to tables where the temperature appeared explicitly.

In 1938 Zen'emon Miduno, with whom we previously met in our discussion on nomograms and infinite series expansions for the fractional functions, produced what could be considered the first set of tabulations suitable for radiometric work in the infrared [446]. After developing both infinite exponential and Bernoulli series forms for the fractional amount in both energetic and photonic units, Miduno used these results to tabulate both fractional quantities to five significant figures as a function of the wavelength–temperature product which ranged from 300 to 100 000 μm·K. The spectral radiant exitance normalized relative to the value of the maximum spectral radiant exitance at the spectral peak in both energetic and photonic units as a function of the same wavelength–temperature product to five significant figures are also given. All values presented in Miduno's table were computed using a value for the second radiation constant equal to exactly 1.4490297×10^{-2} m·K. Miduno was particularly anxious that his tables not become obsolete as a result of any future change in the accepted value for this constant. To this end, listed next to each entry was a value that could be used to correct the tabulated value for any future change in c_2, which was small.

Prompted by Miduno's work, Kazuhiko Terada modified Miduno's tables in 1939 to produce tables of thermal quantities relating to blackbody radiation suitable for meteorological use [643]. The three simple tables he gave were for the peak wavelength in the spectral curve, the maximum spectral radiant exitance at the spectral peak, and the total spectral exitance, as functions of temperature ranging from −100 to 100 °C at one degree intervals.

Starting in 1936, the electrical engineer Parry H. Moon working at the Massachusetts Institute of Technology embarked on what would ultimately become a twelve year table making odyssey. Interrupted by other work and the intervening war, by its conclusion Moon and his human computers had produced a comprehensive set of tables for the spectral radiant exitance of a blackbody accurate to five significant figures, covering temperatures from 2000 to 8000 K. At first Moon simply extended the tables Skogland had given a few years earlier for the relative spectral radiant exitance. Here he considered the same range in wavelength from 0.32 to 0.76 μm but instead chose temperatures that differed from the temperatures Skogland used by a degree or two. So for example, Skogland's temperature range started at 2000 K while Moon's started his at 1999 K. The table is presented as Appendix C to his 1936 text *The Scientific Basis of Illuminating Engineering* [455]. The temperature range considered by Moon was however, an extension on the range considered by Skogland. In Moon's table, values for the relative spectral radiant exitance beyond Skogland's upper temperature of 3120 K were given. On a slightly reduced wavelength scale ranging from 0.40 to 0.72 μm, tabulations for temperatures up to 23 950 K appeared.

A year later saw a renewal in Moon's table making efforts. Motivated by the growing use of Planck's equation in fields such as photometry, radiometry, and colorimetry to describe the radiation emitted by a blackbody, together with his own personal dissatisfaction with many of the tables that existed at the time, Moon embarked more fully on his own tabulation of Planck's equation. His tabulations gave the spectral radiant exitance directly as a function of temperature and wavelength over a wide range for each independent variable. By the time of Moon, many tabulations for the spectral radiant exitance had been given in dimensionless form in terms of a single combined independent variable for the wavelength–temperature product. Moon noted that while such tables may have the advantage of compactness, allowing an extensive range in wavelengths and temperatures of interest to be covered, in use they almost always required one to interpolate, not to mention the many additional steps still needed to find the value of the spectral radiant exitance from the dimensionless parameter x and the given values for the wavelength and temperature.

Moon was particularly scathing of the tables produced by Yamauti and Okamatu (1936, 1937). He writes that he had found it quicker to calculate the spectral radiant exitance directly from Planck's equation using a table of logarithms and an analogue calculating machine, compared to using one of their tables. Of the so-called double entry tables which contained tabulations of the spectral radiant exitance directly as a function of the two independent variables λ and T, Moon was similarly unsatisfied. He despaired that these tables had not been made using international standard values for the two radiation constants. This was the case with the tables prepared by Forsythe (1920), Fowle (1910, 1914, 1920, and 1929), Frehafer and Snow (1925), and Skogland (1929). While a change in the value of c_2 was easily accommodated within these existing tables, it only further diminished their value in the eyes of Moon. Finally, many of these later tables were limited in the range of temperatures and wavelengths covered, and those which were not, such as those produced by Fowle (1929), and Frehafer and Snow (1925), were only calculated to three significant figures, a number barely sufficient for calculations requiring greater accuracy as had started to emerge in the 1930s.

Moon planned to put straight many of the shortcomings he saw in all previous attempts at tabulating values for the spectral radiant exitance. He started by choosing the current, internationally accepted values for the two radiation constants of $c_1 = 3.6970 \times 10^{-16}\,\mathrm{W \cdot m^2}$ and $c_2 = 1.4320 \times 10^{-2}\,\mathrm{m \cdot K}$. Next he considered temperatures from 2000 to 8000 K selecting 10 K intervals from 2000 to 3500 K and 100 K intervals from 3500 to 8000 K. For what he considered the more useful temperature range between 2000 to 3500 K the following wavelength ranges and interval sizes were employed: 0.26(0.01)0.75 µm, 0.75(0.05)1.00 µm, and 1.00(0.10)3.00 µm. The range in wavelengths considered took the tabulation of Planck's equation into the mid-infrared portion of the spectrum. The wavelength values were sufficiently close so as in most cases interpolation was not needed. At the higher temperatures of 3500 to 8000 K the wavelength range considered was restricted to the visible portion of the spectral only (0.38 to 0.76 µm) at 0.01 µm intervals.

Unlike many previous authors, Moon was very particular in the exact details of how he and his human computers went about finding values for the spectral radiant exitance. To take advantage of the many logarithmic tables that existed at the time Moon, like de Groot before him, started by converting Planck's equation to base 10 form. After values for the various physical and mathematical constants had been substituted into Planck's equation, actual computations were made using the equation

$$M_{e,\lambda}^{b}(\lambda, T) = \frac{36970}{\lambda^5} \left(10^{6219.096981/(\lambda T)} - 1\right)^{-1}. \qquad (7.14)$$

Here $c_2' = c_2 \log_{10} e = 6219.096981\ \mu m \cdot K$ while the spectral radiant exitance was given in units of watts per square centimeter per micrometer [W·cm^{-2} · μm^{-1}]. Using an auxiliary table of values for c_2'/T at those temperatures of interest, on dividing these values by λ the value for the power of 10 in Eq. (7.14) could be read directly from a standard table of logarithms. Next the necessary multiplications and divisions were made on an electrically driven mechanical calculating machine. Results were tabulated to seven significant figures and were checked using fifth differences. After re-checking, the results were rounded off to five significant figures. Values for the spectral radiant exitance at the higher temperature range of 3500 to 8000 K were made in this way and published in 1937 [456] followed with a reprinting in 1938 [457].

For temperatures below 3500 K, it became apparent that values for the spectral radiant exitance could not be checked by means of differences, causing Moon's table making efforts to be split into two parts. The first, for temperatures from 3500 to 8000 K, as we saw, was completed in 1937. Other work and the intervening war however, interrupted progress, and it would not be until 1947 that the project was finally completed. It was published in complete form as a *Contribution from the Department of Electrical Engineering, Publications from the Massachusetts Institute of Technology* [458] and in a highly condensed form a year later [459]. Copies of the complete tables could be obtained gratis by writing to the Electrical Engineering Department at Massachusetts Institute of Technology for as long as the supply lasted.

In the same manner as values had been obtained for the spectral radiant exitance at the higher temperature range in 1937, this was again done for the lower temperature range. Using once more electrically driven mechanical computing machines, values were tabulated to 10 significant figures before being rounded off to 8 significant figures after being checked. All tabulated results found in the table were believed to be correct to the last digit, though as a result of the rounding process there may have been an error in a few instances of as much as 5 in the sixth figure. Moon used the same value for the second radiation constant for his 1947 tabulations as he had used for his tabulations completed nine years earlier.

The hiatus between the two tables proved problematic for Moon. In 1929 the best value for c_2 was $1.43174 \pm 0.0006 \times 10^{-2}\ m \cdot K$ [83], but based on new experimental data, by 1941 the value had been revised to $1.43848 \pm 0.00034 \times 10^{-2}\ m \cdot K$ [84], an amount which differed from the previous value considerably more than the probable error of 1929. Moon stuck with his older value for c_2 and reasoned it was conceivable that further research might produce a still higher value, or quite possibly a value of $1.4320 \times 10^{-2}\ m \cdot K$ would eventually prove to be approximately correct. He was rather optimistic in his predication. Today the best estimate for the second radiation constant is $1.438\,777\,36 \times$

10^{-2} m·K [454], a value just within the probable error of the 1941 result. Recognizing this could lead to a possible deficiency in his tables, a recurring problem for any table maker who chose to work with one or both radiation constants, Moon devotes a section of his 1947 tables showing how his values could be adjusted to accommodate other values for either of the radiation constants.

The 1947 set of tables was thought to be particularly suited to photometrists and colorimetrists working with incandescent lamps made of tungsten, as the lower temperatures covered the region of filament temperatures found in such lamps. The earlier set of tables from 1937 at higher temperatures were more suited to work involving carbon-arc lamps which operate at temperatures above 3500 K and were particularly popular at the turn of the twentieth century, or work of an astronomical nature. Moon's completed set of tables was favorably reviewed in the periodical *Mathematical Tables and Other Aids to Computation* [55] and appear to have been extensively used.

The first set of tables calculated under Moon's direction in 1936 had been with the aid of two of his students. Each had produced a similar set of tables as part of their bachelor and master theses [266, 140]. The computations for the first and second parts of his 1937 and 1947 tables, on the other hand, had been made by students who were paid as part of the National Youth Administration, an economic program enacted in the United States as part of the so-called New Deal public works program. The National Youth Administration focused on providing education and work to Americans aged between 16 and 25 and ran between the years 1935 and 1943. In his tables of 1937, Moon thanked seven people for their help in their preparation. By 1947, his list of human computers had grown to twelve. While Moon's tables were the first to be produced with the aid of manpower mobilized and paid for by the New Deal public works program operating in the United States at the time, they would not be the last.

7.3   TABLES FROM 1940 TO 1954

Starting in the 1940s, the tables assembled were, on the whole, far more extensive in nature and produced a far greater degree of accuracy compared to those found in the period preceding it. In 1940 Arnold N. Lowan and Gertrude Blanch published their set of tables for the relative spectral radiant exitance and integrated fractional amount in both energetic and photonic units [399]. Like the 1937 tables of Moon before, the tables were the product of the Mathematical Tables Project created by the Work Projects Administration (WPA), the third of the New Deal public works program to be implemented in the United States in the late 1930s [269]. As a nation wide program, the WPA attempted to create jobs for "needy employable workers."

The Mathematical Tables Project started operations in January 1938 and at its height in 1941 had as many as 100 human computers working under its direction. The project sought to tabulate mathematical functions of fundamental importance in both pure and applied mathematics. The decision to tabulate four functions related to blackbody radiation, while not purely mathematical functions per se as each depends on the value chosen for the second radiation constant, had been prompted by a series of recent papers published on blackbody radiation. In the preceding two years, Archie G. Worthing of the University of Pittsburgh and Frank Benford of the General Electric Company had each published a number of papers where the radiation from thermal sources had been considered. In doing so, both men had identified a need for a well organized and accurate set of tables in this area. At the time, both Worthing and Benford were well-known individuals working in what was then the important field of illumination. Indeed, the scope and arrangement of the tables Lowan and Blanch finally produced closely followed the recommendations of these two men.

The first table comprised four functions. These were the fractional function of the first kind for the radiometric and actinometric cases and the ratio of the spectral exitance to its maximum spectral exitance at its peak wavelength in both energetic and photonic units. All were given as a function of the wavelength–temperature product from 0.100 to 2.00 cm·K for various interval sizes. The tabulations depended on the value chosen for the second radiation constant. Lowan and Blanch chose $c_2 = 1.436 \pm 0.001$ cm·K, a value based on those calculated from the fundamental physical constants proposed by H. T. Wensel in 1939 [679]. The second table gave the same four functions, except this time as a function of wavelength ranging from 0.50 to 20 μm at a temperature of $T = 1000$ K. These values depended on both radiation constants, with $c_1 = 3.732 \pm 0.006 \times 10^{-5}$ erg·cm^2·s^{-1} being used in this case. The third table tabulated values for the spectral photon exitance only at the six temperatures of 1000, 1500, 2000, 2500, 3000, 3500, and 6000 K for the more limited wavelength range of 0.25 to 10.00 μm. All values appearing in the first three tables were given to five significant figures. The fourth and final table provided a convenient means for correcting any of the tabulated values in the preceding three tables resulting from small changes in either of the radiation constants. In the first and second tables, first differences and second central differences were given for the two fractional functions and their spectral ratios. The authors suggested that interpolated values should be found using the Laplace–Everett formula[7] and the differences provided, as these were claimed to produce interpolated values of reasonable accuracy. An example of the first page of Table 1 from Lowan and Blanch's 1940 publication is shown in Fig. 7.1.

As all tables produced by the Mathematical Tables Project were prepared under the sponsorship of the National Bureau of Standards, a year later they were reissued by the Bureau as *Miscellaneous Physical Tables* [398]. Copies were available from the Bureau for a cost of $1.50. Though published in 1941, the work was not distributed until 1943 [69]. The first part of these tables were identical to the tables which had appeared in their 1940 publication.

PLANCK'S RADIATION FUNCTIONS   71

TABLE I. $\mathfrak{R}_\lambda = c_1\lambda^{-5}(e^{c_2/\lambda T}-1)^{-1}$; $\mathfrak{R}_{0-\lambda} = \int_0^\lambda \mathfrak{R}_\lambda d\lambda$; $N_\lambda = 2\pi c\lambda^{-4}(e^{c_2/\lambda T}-1)^{-1}$; $N_{0-\lambda} = \int_0^\lambda N_\lambda d\lambda$.*

λT IN CM K°	$\frac{\mathfrak{R}_{0-\lambda}}{\mathfrak{R}_{0-\infty}} = F \times 10^{-p}$		$\frac{\mathfrak{R}_\lambda}{\mathfrak{R}_{\lambda max}} = f \times 10^{-q}$				$\frac{N_{0-\lambda}}{N_{0-\infty}} = G \times 10^{-p}$		$\frac{N_\lambda}{N_{\lambda max}} = g \times 10^{-q}$			
	F	p	f	q	Δf	δ²f	G	p	g	q	Δg	δ²g
0.050	1.3652	9	3.1018	7	1.8320	0.6398	1.2379	10	4.7906	8	2.9818	1.0815
.051	2.2642	9	4.9339	7	2.7606	0.9286	2.0924	10	7.7724	8	4.5866	1.6048
.052	3.6788	9	7.6945	7	4.0844	1.3238	3.4638	10	1.2359	7	0.6924	0.2338
.053	5.8629	9	1.1779	6	0.5939	0.1855	5.6220	10	1.9283	7	1.0271	0.3346
.054	9.1749	9	1.7718	6	0.8497	0.2558	8.9571	10	2.9554	7	1.4983	0.4712
.055	1.4113	8	2.6216	6	1.1971	0.3474	1.4022	9	4.4537	7	2.1517	0.6534
.056	2.1358	8	3.8187	6	1.6623	0.4652	2.1590	9	6.6054	7	3.0447	0.8929
.057	3.1829	8	5.4810	6	2.2767	0.6144	3.2723	9	9.6501	7	4.2482	1.2035
.058	4.6745	8	7.7577	6	3.0781	0.8013	4.8865	9	1.3898	6	0.5849	0.1601
.059	6.7710	8	1.0836	5	0.4111	0.1033	7.1944	9	1.9747	6	0.7953	0.2104
.060	9.6798	8	1.4946	5	0.5426	0.1315	1.0451	8	2.7700	6	1.0685	0.2732
.061	1.3667	7	2.0372	5	0.7083	0.1657	1.4990	8	3.8385	6	1.4194	0.3509
.062	1.9069	7	2.7455	5	0.9150	0.2067	2.1242	8	5.2580	6	1.8654	0.4460
.063	2.6307	7	3.6606	5	1.1703	0.2553	2.9755	8	7.1234	6	2.4266	0.5612
.064	3.5907	7	4.8309	5	1.4827	0.3124	4.1225	8	9.5500	6	3.1261	0.6995
.065	4.8510	7	6.3135	5	1.8615	0.3789	5.6521	8	1.2676	5	0.3990	0.0864
.066	6.4902	7	8.1751	5	2.3172	0.4557	7.6722	8	1.6666	5	0.5048	0.1058
.067	8.6028	7	1.0492	4	0.2861	0.0544	1.0316	7	2.1714	5	0.6333	0.1285
.068	1.1302	6	1.3353	4	0.3505	0.0644	1.3744	7	2.8047	5	0.7882	0.1549
.069	1.4723	6	1.6858	4	0.4261	0.0757	1.8153	7	3.5929	5	0.9734	0.1853
.070	1.9025	6	2.1119	4	0.5145	0.0883	2.3778	7	4.5664	5	1.1935	0.2200
.071	2.4393	6	2.6264	4	0.6169	0.1024	3.0899	7	5.7599	5	1.4531	0.2596
.072	3.1045	6	3.2433	4	0.7349	0.1180	3.9847	7	7.2129	5	1.7573	0.3042
.073	3.9230	6	3.9782	4	0.8701	0.1352	5.1011	7	8.9703	5	2.1117	0.3544
.074	4.9236	6	4.8483	4	1.0240	0.1539	6.4848	7	1.1082	4	0.2522	0.0410
.075	6.1392	6	5.8723	4	1.1983	0.1743	8.1886	7	1.3604	4	0.2995	0.0472
.076	7.6070	6	7.0706	4	1.3947	0.1963	1.0274	6	1.6599	4	0.3536	0.0541
.077	9.3692	6	8.4653	4	1.6147	0.2200	1.2810	6	2.0134	4	0.4152	0.0616
.078	1.1473	5	1.0080	4	0.1860	0.0245	1.5877	6	2.4286	4	0.4850	0.0698
.079	1.3971	5	1.1940	3	0.2132	0.0272	1.9567	6	2.9136	4	0.5638	0.0788
.080	1.6923	5	1.4072	3	0.2433	0.0301	2.3981	6	3.4774	4	0.6523	0.0885
.081	2.0393	5	1.6506	3	0.2764	0.0331	2.9236	6	4.1297	4	0.7512	0.0989
.082	2.4453	5	1.9270	3	0.3127	0.0363	3.5462	6	4.8809	4	0.8613	0.1101
.083	2.9183	5	2.2397	3	0.3523	0.0396	4.2802	6	5.7421	4	0.9833	0.1220
.084	3.4668	5	2.5920	3	0.3953	0.0430	5.1418	6	6.7254	4	1.1180	0.1347
.085	4.1002	5	2.9874	3	0.4419	0.0466	6.1487	6	7.8434	4	1.2662	0.1482
.086	4.8287	5	3.4293	3	0.4922	0.0503	7.3204	6	9.1097	4	1.4287	0.1624
.087	5.6633	5	3.9215	3	0.5463	0.0541	8.6786	6	1.0538	3	0.1606	0.0177
.088	6.6159	5	4.4678	3	0.6043	0.0580	1.0247	5	1.2144	3	0.1799	0.0193
.089	7.6993	5	5.0721	3	0.6662	0.0619	1.2050	5	1.3944	3	0.2009	0.0210
.090	8.9269	5	5.7383	3	0.7322	0.0660	1.4117	5	1.5952	3	0.2235	0.0227
.091	1.0314	4	6.4705	3	0.8022	0.0700	1.6478	5	1.8188	3	0.2479	0.0244
.092	1.1874	4	7.2727	3	0.8764	0.0742	1.9165	5	2.0667	3	0.2742	0.0263
.093	1.3626	4	8.1491	3	0.9547	0.0783	2.2213	5	2.3410	3	0.3024	0.0282
.094	1.5586	4	9.1039	3	1.0372	0.0825	2.5660	5	2.6433	3	0.3325	0.0301
.095	1.7772	4	1.0141	2	0.1124	0.0087	2.9546	5	2.9758	3	0.3646	0.0321
.096	2.0204	4	1.1265	2	0.1215	0.0091	3.3914	5	3.3404	3	0.3987	0.0341
.097	2.2901	4	1.2480	2	0.1309	0.0095	3.8811	5	3.7391	3	0.4349	0.0362
.098	2.5885	4	1.3789	2	0.1408	0.0099	4.4284	5	4.1740	3	0.4733	0.0383
.099	2.9179	4	1.5197	2	0.1511	0.0103	5.0386	5	4.6473	3	0.5137	0.0404
.100	3.2804	4	1.6709	2	0.1618	0.0107	5.7171	5	5.1610	3	0.5564	0.0426

* NOTE: $\Delta f \times 10^{-q}$ and $\delta^2 f \times 10^{-q}$ are the first difference and the second central difference, respectively, of $f \times 10^{-q}$. Similar relations hold between $\Delta g \times 10^{-q}$, $\delta^2 g \times 10^{-q}$ and $g \times 10^{-q}$.

Figure 7.1 A typical example of a set of tables produced during the early table making era. Shown is the first page of Table 1 from Lowan and Blanch's publication of 1940. Reproduced with permission from Lowan, A. N. and Blanch, G., 1940 "Tables of Planck's radiation and photon functions," *Journal of the Optical Society of America* **30**(2), p. 71. Copyright 1940, OSA Publishing.

The introductory section was however, expanded to include details on how each tabulated quantity had been calculated. Each value for the spectral exitance for the radiometric and actinometric cases had been computed with the aid of the tables the Mathematical Tables Project had produced a year earlier for the exponential function [397] while the integrated fractional amounts were evaluated by means of an infinite series expansions identical with those developed in Chapter 3 (see Eqs (3.49) and (3.51)). For the latter, the exponential terms appearing in either of these infinite series were computed to ten significant figures with the aid of their tables for the exponential function. Enough terms in either series were then summed to ensure seven significant figure accuracy of the final result before being rounded off to five significant figures. For the greater part of the ranges considered, this meant between one to five terms in either series was sufficient. Values for large wavelengths did however, require more terms to be summed in order to retain sufficient accuracy. Lowan and Blanch seemed to have been aware that a more rapidly converging series, in a form employing Bernoulli numbers, could be used in such cases though it is not something they explicitly state. Rather, in the interest of uniformity in the computations, the exponential infinite series forms were used throughout. For the next twenty years their tables were widely used, extensively referred to in the literature, and had a lasting influence on future table making enterprises of this type. The tables were reviewed favorably a few years later [692] and, in a slightly more compact form, can be found in the first and second editions of Jean D'Ans and Ellen Lax's *Taschenbuch für Chemiker und Physiker* of 1943 and 1949 [167, 168].

Thirteen years later, the tables of Lowan and Blanch were updated and put into a form more suited to engineers by Robert V. Dunkle who was working at the University of California, Berkeley [198]. Values for the spectral radiant exitance and the in-band integrated amount as a function of the wavelength–temperature product from Lowan and Blanch's 1941 tables were corrected using the latest values for the two radiation constants recently published by DuMond and Cohen in 1953 [197] and converted to a set of units deemed more convenient for those working in the field of radiative heat transfer. In this system of so-called "engineering units," Dunkle used Rankine [R] for temperature, British thermal units per hour per square foot per micron [$Btu \cdot hr^{-1} \cdot ft^{-2} \cdot \mu m^{-1}$] for spectral radiant exitance, and British thermal units per hour per square foot [$Btu \cdot hr^{-1} \cdot ft^{-2}$] for the total and in-band integrated fractional amount.

The ninety-three values he gave for the spectral radiant exitance normalized relative to the quantity σT^5, and the integrated fractional amount as a function of the wavelength–temperature from 1000 to 100 000 cm·R would prove to be incredibly successful. They in turn were widely used and referred to in the literature and were often reproduced by authors of texts on radiative heat transfer [129, 614]. Some years later they were re-evaluated using a digital computer [683], demonstrating the need engineers had for using

quantities in a system of units many were more familiar with. Of the
two reasons Dunkle gave for computing his tables, the second, to put the
"...information in a readily usable form for engineers," he certainly achieved.
The first was to make use of the latest accepted values for the two radiation
constants.

By the late 1930s the American Meteorological Society had been urging the
American Physical Society for a number of years to encourage its members to
undertake research into infrared radiation, particularly as it related to the atmosphere, but to no avail [577]. Help would finally come to the Meteorological
Society's clarion, though what was by now becoming increasingly desperate,
call from an atomic physics émigré escaping from Hitler's fascist Germany.
During the years 1937 until 1941, shortly after joining the recently established Meteorology Department at CalTech in California, Walter M. Elsasser,
despite having no prior experience in meteorology, undertook a systematic
analysis of many of the properties of far-infrared atmospheric radiation.

During the course of these investigations, a need to better understand the
transfer of heat by infrared radiation in the atmosphere led to a need for values
of the integrated in-band amount and its temperature derivative[8] within the
linear frequency representation. These he wrote as

$$\int_0^\nu M(\nu, T)d\nu = \frac{c_1}{c_2^4}T^4\phi(x),\tag{7.15}$$

and

$$\int_0^\nu \frac{\partial M(\nu, T)}{\partial T}d\nu = \frac{c_1}{c_2^4}T^3\phi'(x).\tag{7.16}$$

Tabulated were the dimensionless quantities

$$\phi(x) = \int_0^x x^3(e^x - 1)^{-1}dx,\tag{7.17}$$

and

$$\phi'(x) = \int_0^x x^4e^x(e^x - 1)^{-2}dx,\tag{7.18}$$

as a function of the dimensionless parameter $x = h\nu/(k_B T)$.

Having left CalTech shortly after their completion and headed east to the
Blue Hill Meteorological Observatory at Milton, Massachusetts, the tables
were published in 1942 as part of the monograph Elsasser wrote on atmospheric radiation which appeared under the *Harvard Meteorological Studies*
series [211]. The monograph contained a total of 102 values for x ranging between 0.1 and 15.0. The values for $\phi(x)$ are given to four decimal places while
those for $\phi'(x)$ are given to three. The integrated fractional amount expressed
as a percentage to three decimal places is also given.

An updated version of Elsasser's 1942 monograph appeared in 1960 [212].
Working now with Margaret F. Culbertson from the University of California in La Jolla, a number of additional quantities were given that had been

found to be useful for calculations relating to radiative transfer in the far infrared spectrum of the atmosphere since the appearance of Elsasser's initial monograph. Previous tabulations for the integrated quantities $\phi(x)$ and $\phi'(x)$ no longer appeared. Instead, the only quantity related directly to blackbody radiation that was given this time was for the temperature derivative of the spectral radiant exitance in the linear wavenumber $(\bar{\nu})$ representation. Recognizing that

$$\frac{dM_{e,\bar{\nu}}^b}{dT} = \frac{c_1 T^2}{c_2^3} \frac{x^4 e^x}{(e^x - 1)^2}, \tag{7.19}$$

where $x = c_2\bar{\nu}/T$, values for the temperature derivative of the spectral radiant exitance were tabulated for wavenumbers ranging from 40 to 2400 cm^{-1} at 40 cm^{-1} intervals and temperatures from -80 to $+40\,°C$ at $10\,°C$ intervals. All tabulated values were given to between one and four significant figures and were quoted in meteorological units of gram calories per square centimeter per day [gcal·cm^{-2}·day^{-1} ($\times 10^{-6}$)]. The 1955 values of 3.7413×10^{-5} erg·cm^2·s^{-1} and 1.4389 cm·K for the first and second radiation constants as given by E. R. Cohen, J. W. M. DuMond, T. W. Layton and J. S. Rollett [142] were used.

Motivated by interest in a range of particular problems, the period under consideration starts to see the extension and the filling in of gaps found in previously produced tables. For example, R. Stair and W. O. Smith in their 1943 work on tungsten filament in a quartz lamp required the spectral radiant exitance of a blackbody in the ultraviolet region [618]. From work with their lamps, they produced a table for the relative spectral radiant exitance at wavelengths from 2300 to 3500 at 50 intervals for temperatures in the range from 2500 to 2900 K at 50 K increments. Relative values were given relative to the value of the spectral radiant exitance at a wavelength equal to 3500 to four significant figures using Wensel's most probable value for the second radiation constant of 1.4360 cm·K from 1939 [679]. In regards to how the computations were made, little is revealed. The authors do, however, mention the value for the exponential term appearing in the spectral radiant exitance had been calculated using values taken from tables recently published by the Mathematical Tables Project [397].

Elliot Q. Adams and William E. Forsythe were also led in 1943 to filling in a gap in values for the spectral radiant exitance. As a result of their work on tungsten filament lamps [41], values for the spectral radiant exitance in the visible portion of the spectrum for temperatures between 2800 to 3800 K were needed. Between the highest temperature of 3120 K for the spectral radiant exitance as tabulated by Skogland (1929) and the lowest temperature of 3500 K as tabulated by Moon (1937) was a rather inconvenient temperature gap. Using a value for the second radiation constant of 1.4350 cm·K, the spectral radiant exitance (actually $10^6 M_\lambda/c_1$ on account of the uncertainty in the first radiation constant and so as to avoid non-significant zeros) as a function of wavelength between 380 to 760 nm at 10 nm intervals for temperatures between 2800 to 3800 K at 100 K intervals were tabulated.

Most of the values Adams and Forsythe calculated had been made using Wien's equation for the spectral radiant exitance. For values which fell below a wavelength–temperature product of $2500\,\mu\text{m}\cdot\text{K}$, an additional logarithmic correction term was used. At the time, these logarithmic correction factors had been tabulated but remained unpublished. As the correction values did not depend on the values for c_2, λ, and T, their possible general usefulness to others was recognized and they were duly published several years later [40].[9] Finally, the year 1943 saw the astronomers Frederick H. Seares and Mary C. Joyner produce a brief set of tables for the spectral radiant exitance relative to its maximum value at the spectral peak for wavelengths ranging from 0.30 to $0.60\,\mu\text{m}$ at $0.01\,\mu\text{m}$ steps for eleven temperatures ranging from 3000 to $22\,000\,\text{K}$ [593]. Their tabulations were made using an interpolation from Fabry's earlier dimensionless single argument tables. A value of $1.433\,\text{cm}\cdot\text{K}$ for the second radiation constant was used while the tabulated values were given to four significant figures.

In 1941 a table of values for thermal quantities mainly of astrophysical interest were given by Hans G. Kienle [350]. Accompanying the graph he produced as an aid to estimating the absolute gradient $\Phi(\lambda, T)$ used in order to determine the color temperature of a star (see page 174) was a modest tabulation for this quantity as a function of c_2/T and wavelength. Wavelengths from 250 to 10 000 at 25 intervals were considered while the quantity c_2/T ranged from zero to $4.00 \times 10^{-4}\,\text{cm}$ at $0.20 \times 10^{-4}\,\text{cm}$ intervals. All tabulated values were given to four significant figures while the value for the second radiation constant used was $1.4320\,\text{cm}\cdot\text{K}$. Kienle noted that many of the tables that existed for the spectral radiant exitance of a blackbody up until the time of 1941 did not meet the needs of most astronomers in terms of the wavelength and temperature ranges considered. This was certainly true. Kienle writes he had also found values for the logarithm of the spectral radiant exitance of a blackbody with c_2/T set equal to unity and normalized relative to the value of the spectral radiant exitance when the wavelength was set equal to 4000 particularly useful, though no tabulation for this quantity is given.

At the same time as Moon was completing his tables for the spectral radiant exitance for a blackbody, his work in the areas of photometry and colorimetry with his collaborator Domina E. Spencer saw a detailed table for the spectral radiant exitance as a function of wavelength at two particular temperatures produced [460]. Temperatures of 2842 and $7000\,\text{K}$ corresponding to Standard Illuminates A and C were used. Standard Illuminate A is defined to represent typical tungsten-filament lighting while C corresponds to average daylight lighting. The spectral distribution profile for each standard is taken to be that of a blackbody at some agreed temperature (presently these are the temperatures 2856 and $6774\,\text{K}$, respectively). Adopting a value of $c_2 = 1.4320\,\text{cm}\cdot\text{K}$, at these two temperatures the spectral radiant exitance is given for wavelengths ranging from 0.360 to $0.800\,\mu\text{m}$ at intervals of $0.001\,\mu\text{m}$. The computations were made to ten significant figures using three human computers. Values were checked using third differences before being rounded off to eight significant figures, though given the accuracy in the two radiation constants known at the time, how meaningful those significant figures past the first five were is debatable. Moon had already given values for the spectral radiant exitance at a temperature of 7000 K in his tabulation of 1937 [456]. There it had been given for larger wavelength range and interval sizes with values being tabulated to five significant figures only.

Until now, the output of tables by French authors had been limited to those produced by Fabry which he gave in his text *Photométrie* of 1927, and it would be some twenty years before a volume containing tabulations of any description for a number of quantities relating to blackbody radiation next appeared [91]. It was authored by Marcel Boll, who at the time was a professor of chemistry and electricity at the École des Hautes Études Commerciales in Paris. His volume of tables was a compilation of previously published tables together with many new tables. It included not only tables of purely mathematical functions, but also contained many tables that would be of use to the physicist and chemist. Of the latter, a number of tables relating to blackbody radiation were given. The first set of tables gave the spectral radiant exitance in the linear wavelength representation normalized relative to the maximum amount at the spectral peak, and the integrated fractional amount as a function of the dimensionless variable $x = \lambda T / c_2$ from 0.10 to 50, for various step sizes with values given to three or four significant figures. The second table was the same as the first, except this time for the spectral radiant exitance in the linear frequency representation, while the third table was for the spectral photon exitance in the linear frequency representation. In all cases, graphs illustrating each quantity were also given. A table for the corresponding peak values (wavelength, wavenumber, and frequency) for each of the three different representations considered as a function of temperature were produced. Temperatures were given in Celsius from -200 to $20\,000\,°C$ at various step sizes and ran to an incredible twenty-five pages, compared to the two pages each that Boll devoted to the first three of his tables. All are given to five significant figures. Running alongside these peak values the total radiant exitance at each given temperature is given to four or five significant figures. Exactly why Boll devoted so much space to these latter peak values, tabulations which in themselves were not that common, is not known. Maybe it was a demonstration of the usefulness he wished to place on his tables in the eye of the working scientist. In the end, his volume of tables managed to pass through three editions, suggesting that they were on the whole generally well received, but were rarely if ever referred to by those working in the infrared.

The first set of tables produced specifically with the infrared in mind were probably those due to the German physicist Werner Brügel. He produced these as an appendix to his 1951 text *Physik und Technik der Ultrarotstrahlung* [108]. Brügel was an industrial physicist working at BASF in Ludwigshafen am Rhein and his text on the infrared was as much about the technology of the radiation as it was of the physics. As such, Brügel had applications very much in mind and that probably explains the inclusion of three tables relating to blackbody radiation in his text. In the first of his tables, moderately low temperatures from 800 to 3000 K at 200 K intervals for twenty-eight wavelengths from 0.4 to 50.0 μm, a wavelength range extending from visible light to deep into the infrared, for the spectral radiant exitance are given. Values are tabulated to three signifiance figures.

Calculations we have made using the values Brügel used for the first and second radiation constants of 3.7413×10^{-12} W·cm^2 and 1.4388 cm·K suggest the values are only accurate to the second significant figure at best. In the second table of Brügel's is given the spectral radiant exitance normalized relative to the spectral radiant exitance at its peak as a function of the wavelength–temperature product. Its range extended from 0.0800 to 3.6 cm·K. Values are given to two or three significant figures, but again calculations we have made suggest that only the first figure can be trusted. The third table is a tabulation for the integrated fractional amount as a function of the wavelength–temperature product. It contains a selection of roughly half the number of values for the wavelength–temperature product between 0.050 to 2.00 cm·K from Lowan and Blanch's tables of 1940. The errors in the first two tables seemed to have gone unnoticed, as they remained uncorrected in the second edition of Brügel's text published ten years later [109].

Shortly after Brügel's tables appeared, the Belgian astronomer Edgard Vandekerkhove, who was attached to the Royal Observatory of Belgium for most of his professional life, produced in 1953 a number of tables and graphs for various quantities related to blackbody radiation [654]. These were produced using values for temperature typically encountered in astrophysics. The spectral radiance and a quantity related to the gradient of the spectral radiance given by

$$\phi_\lambda(T) = \frac{c_2}{T} \left[1 - \exp\left(-\frac{c_2}{\lambda T} \right) \right]^{-1} = 5\lambda - \frac{d(\ln[L_\lambda(T)])}{d(1/\lambda)},$$

as a function of wavelength and temperature are given. Wavelengths predominantly in the visible portion of the spectrum between 3000 to 8000 at 500 intervals are given for temperatures from 1000 to 150 000 K at progressively increasing interval sizes. Vandekerkhove uses DuMond and Cohen's 1948 values for the two radiation constants of $c_1' = 8\pi hc = 4c_1/c = 4.990\,45 \times 10^{-15}$ erg·cm and $c_2 = 1.438\,53$ cm·K [196]. Each was tabulated to three or four significant figures. A table for the total radiant exitance at the same temperatures as those used for the previous two tables was also given, along with values for the peak wavelength for twenty different temperatures from 50 to 1000 K.

The ninth and last of the revised editions for the *Smithsonian Physical Tables*, appeared in 1954 [232]. Prepared now by William E. Forsythe, the tables essentially followed those found in the Eighth Revised Edition of 1933. Values used for the two radiation constants and the Stefan–Boltzmann constant were updated and all tabulations were given to either four or five significant figures. The temperature range for the spectral radiant exitance was extended to an upper limit of 25 000 K compared to the 20 000 K limit used in the Eighth Revised Edition while the final table found in the 1933 edition for the spectral radiant exitance relative to the spectral radiant exitance at a wavelength equal to 0.59 μm, which at the time had been an abridgment of Skogland's 1929 paper, was removed.

As part of his thesis work on the emissivity of tungsten ribbon in 1953, Jan Cornelis de Vos produced a sixteen page table for the spectral radiance as a function of wavelength and temperature [181]. In photometry, colorimetry, and pyrometry, tabulated values for the spectral radiance of a blackbody were often needed. Since the revision in 1948 of the international temperature scale when a new value for the second radiation constant of $c_2 = 1.438$ cm·K was adopted [278], as no new tables for Planck's equation using the revised value for c_2 had appeared, compelled de Vos to proceed with their tabulation. Despite the existence of tables for the spectral radiant exitance and radiance in dimensionless form, de Vos makes it clear he did not find such tables very useful, as additional computations were always required in order to find the spectral radiance from a given temperature and wavelength. In calculating the spectral radiance, as de Groot had done earlier, de Vos converted from base e to base 10 so as to take full advantage of the many existing tables for logarithms of base 10. In calculating, he first tabulated $1/\lambda$ and Mc_2/T for a great many wavelengths ranging from 0.18 to 15 μm for forty-two different temperatures between 500 to 3200 K. Here M was a mathematical constant equal to $\log_{10}$ e. The range in wavelengths and temperatures chosen by de Vos where typical of those of interest to the working pyrometrists. Values for the logarithmic term could then be easily read from a table of logarithms, and after having subtracted one from it, were tabulated. The term c_1/λ^5 was then tabulated using a value of $c_1 = 11.907 \times 10^{10}$ erg·cm^3·s^{-1}·sr^{-1}·μm^{-1}. All necessary multiplications and divisions required throughout the tabulation were performed using an electrically-driven analogue calculating machine. Each auxiliary tabulation was performed to a sufficient number of significant figures so the final result for the spectral radiance would be accurate to four significant figures. In most cases this was true. A small random sample we checked did however, show a few of the tabulated values were correct to the first three significant figures only. Interestingly, values for the spectral radiance which fell below 1.0×10^{-3} erg·cm^3·s^{-1}·sr^{-1}·μm^{-1} were not calculated since at the time these values represented the minimum detectable radiation yielding a measurement uncertainty greater than five per cent.

Work on the calculation of colors of materials by colorimetrists on the CIE (Commission Internationale de l'Éclairage) trichromatic system had always had a need for the tabulation of the relative spectral radiant exitance. As part of their colorimetric investigations at the National Physical Laboratory in Teddington, England, H. G. W. Harding, together with his collaborators, had by the late 1940s developed and been using an extensive set of tables for a number of years. They were however, reluctant to publish them due to the general lack of international agreement on a value for the second radiation constant, being dependent on this value as they were. Despite this, many requests from colorimetrists for such a set of tables had been received. After settling on a value of 1.4350 cm·K they thought "most satisfactory," in 1947 Harding along with R. B. Sisson duly published [280]. The spectral radiant exitance relative to its value at a wavelength equal to 0.56 µm for wavelengths ranging from 0.38 to 0.77 µm at 0.01 µm intervals as a function of temperature from 1500 to 3500 K at 250 K steps were given to an incredible seven significant figures. All values had been checked by forming fifth differences and were estimated to be correct, as far as the value for c_2 was correct, to about one part in a million.

Six years later, using the internationally agreed 1951 value for c_2 of 1.438 cm·K by the CIE, and now with access to one of Britain's earliest computers, the *Automatic Computing Engine*, or ACE for short, Harding together with T. Vickers calculated values for the same relative spectral radiant exitance for wavelengths 0.350(0.005)0.800 µm and for temperatures 1000(250)3500 K, and at eight other special temperature values [281]. All tabulated results were given to a staggering nine significant figures, though given the accuracy to which c_2 was known it was recognized that the last significant figure was of little value. The absolute value for the spectral radiant exitance at 0.560 µm for each temperature was also given to five significant figures. It had been calculated using a value for the first radiation constant equal to 3.7407×10^{-5} erg·cm^2·s^{-1}.

ACE was an early electronic stored-program computer designed and built at the National Physical Laboratory. It ran its first program in May of 1950 [380]. The machine program had been designed to be sufficiently flexible to allow for changes in the values of the second radiation constant, the wavelength, and the temperature to be readily made. For a given temperature run, values for the relative spectral radiant exitance at the ninety-one different wavelength values considered could be computed in about two minutes and were available on punched cards for subsequent calculations. The punch cards were, in turn, converted into printed form using a card-controlled electronic typewriter. Considering what had passed for table making in the past, the time savings to be had from using electronic computers compared to existing human computing power was simply incomparable. The age of the electronic computer had finally arrived. The computation of tables would never be the same again.

7.4   TABLES FROM 1955 ONWARDS

While a small number of the earlier tables to appear in this period were still produced using humans as computers, the vast majority were produced using digital computers. Development of the electronic computer in the years immediately following the Second World War was intense. By the early 1950s a number of computers in countries such as England and the US were in operation and had slowly begun to spread. The tables produced by Harding and Vickers in 1953 to an unprecedented number of significant figures in a relatively short period of time led the way in showing what was possible with the arrival of this new electronic marvel. The era of calculating tables using an electronic computer had begun.

No more so was this in evidence than in a 1957 paper by Karl-Heinz Böhm and Bodo Schlender where both authors made it clear exactly how they performed their calculations [90]. As astrophysicists, Böhm and Schlender were well aware of the two integrals relating to the integrated fractional amount in both energetic (which they denoted $\mathfrak{I}_2(x)$) and photonic (denoted by $\mathfrak{I}_1(x)$) units that often arose in the study of gaseous nebulae. As noted earlier, Zanstra had already given a very short table for each of these integrals as early as 1931 (see page 253), but for only nineteen values of the dimensionless argument $x = h\nu/(k_B T)$. A somewhat more detailed tabulation for $\mathfrak{I}_1(x)$ had been given by Karl Wurm in his text *Die Planetarischen Nebel* of 1951, but these were given to only two or three significant figures and for forty values of x between 0.56 to 11.0 [714].

As part of a training course organized by IBM Germany using one of their 650 computers, Böhm and Schlender evaluated both integrals for 106 arguments ranging from 0 to 12.0 with the resulting values given to five significant figures. The 650 was one of IBM's early electronic computers and the world's first mass-produced computer. Since

$$\mathfrak{I}_1(x) = \int_x^\infty \frac{x^2}{e^x - 1} dx = 2\zeta(3) - \int_0^x \frac{x^2}{e^x - 1} dx, \qquad (7.20)$$

and

$$\mathfrak{I}_2(x) = \int_x^\infty \frac{x^3}{e^x - 1} dx = \frac{\pi^4}{15} - \int_0^x \frac{x^3}{e^x - 1} dx, \qquad (7.21)$$

the complement of each integral from zero to an upper limit of $x = 0.2, 0.4, 0.6, \ldots$ were first evaluated numerically using Simpson's rule employing a step size of $\Delta x = 0.1$. Function values for $x = 0.3, 0.5, 0.7, \ldots$ were subsequently obtained by interpolation using Bessel's interpolation formula [286]. The way was now open for the production of ever more extensive tabulations at ever smaller interval sizes to an ever greater number of significant figures — advances that ultimately saw tables running to hundreds of pages being produced.

In 1955 Clabon W. Allen published the first edition of his *Astrophysical Quantities* [44]. The text consisted of a compilation of numerical data of astrophysical interest. It was presented in a form that could be readily used by those working in the field and led to it becoming one of the most widely quoted texts in astrophysics. In the first edition, under the section entitled "Black Body Radiation," Allen gives all relevant equations relating to a blackbody together with a short, single page table. In this table he managed to pack in values for the following five quantities as a function of the wavelength–temperature product: spectral radiant exitance in the linear wavelength representation normalized relative to the peak value and its corresponding integrated fractional amount, spectral radiant exitance in the linear frequency representation normalized relative to the peak value, spectral photon exitance in the linear wavelength representation normalized relative to the peak value and its corresponding integrated fractional amount. The wavelength–temperature products ranged from 0.01 to 100 cm·K and all values were given to five significant figures. Values used for the two radiation constants were 3.7403×10^{-5} erg·cm^2·s^{-1} and 1.438 68 cm·K respectively. By the time the third edition came round in 1973 the number of values for the wavelength–temperature product within the same range had grown from thirty-eight in the first edition to seventy by the third [45]. All tabulated values were once more given to five significant figures. Revised values of $3.741\,85 \times 10^{-5}$ erg·cm^2·s^{-1} and 1.438 83 cm·K were used for the two radiation constants while Allen acknowledges G. N. Cooke for assistance in programming, suggesting values in the third edition had been calculated using a digital computer. For the fourth edition, published thirteen years after Allen's death in 2000 and retitled *Allen's Astrophysical Quantities*, these tables had all but disappeared [148].

In North America, the astronomer Jean K. McDonald published in 1955 an extensive tabulation for the spectral radiance in the linear frequency representation over a range of frequencies and temperatures of typical astronomical interest [436]. A frequency range from 0.001315 to 0.300000×10^{17} s^{-1} was chosen. The choice in the range of frequencies was made so as to include the first five hydrogen spectral series (Lyman, Balmer, Paschen, Brackett, and Pfund) frequency limits. Temperatures from 15 000 to 26 000 K at 200 K intervals and from 26 000 to 50 000 K at 500 K intervals are given. All tabulations were given to four significant figures and had been calculated on Ferut, a digital computer located at the University of Toronto and made by the UK-based electrical engineering and equipment firm Ferranti. The tabulations themselves were the by-product of a numerical investigation into the transfer of radiation in the atmosphere of a star and were made available in the hope others may find them of some use.

Some years later Seán A. Twomey, who at the time was working with the U.S. Weather Bureau in Washington, DC, also made a tabulation for the spectral radiance in the linear frequency representation [650]. However, unlike McDonald's tabulations, Twomey's are given for a range of temperatures and wavenumbers useful for terrestrial radiation. Temperatures ranged between 180 to 315 K at intervals of 5 K for wavenumbers from 20 to 3400 cm^{-1} at 20 cm^{-1} intervals are used. The tabulations were made using an IBM 7090 computer and given to between six to eight significant figures, though users were advised to read no more than the first five significant figures, given the accuracy of the physical constants used. Another very brief set of tables of astronomical interest for the spectral radiant exitance in dimensionless form as a function of wavelength from 0.30 to 0.60 µm at 0.01 µm intervals for eleven different temperatures from 3000 to 22 000 K was given by A. A. Mikhailov in his astrophysical text of 1969 [448].

As we have already seen, the field of meteorology was an area where tabulations for various quantities relating to blackbody radiation were made. In 1959 Champ B. Tanner and Stephen M. Robinson produced an extensive set of tables for the total radiant exitance of a blackbody at various temperatures in a number of system of units commonly used by meteorologists [640]. At the time Tanner was a Professor in the Department of Soils at the University of Wisconsin while Robinson was a sixteen year-old first-year undergraduate student looking to pay his way through university by writing computer programs for others [565]. With so few programmers around in the late 1950s Robinson recalls wages were good [565]. Using the Wisconsin Integrally Synchronized Computer located in the Numerical Analysis Laboratory at the university, Robinson coded and ran the calculations prepared by Tanner. Tables for the total radiant exitance in units of langley per minute [ly/min],[10] langley per day [ly/day], evaporation rate of water in millimeters per hour [mm water/hr] and millimeters per day [mm water/day], and watts per square metre [W·m^{-2}] as a function of temperature in units of Celsius [°C] and Fahrenheit [°F] are given. All tabulated values were given to between four to six significant figures, but were only reliable to the fourth. Similar tables for the total radiant exitance for a range of typical terrestrial temperatures of interest to meteorologists were later given by others [106, 192].

Two other more modest tabulations at temperatures of terrestrial interest can be found in the book of Richard M. Goody [263] and in the Sixth Revised Edition of the *Smithsonian Meteorological Tables* prepared by Robert J. List [392]. In addition to giving the ratio of the spectral radiant exitance relative to its maximum value in both the linear wavelength and frequency representations, and the integrated fractional amount in the linear wavelength representation for 38 values of the wavelength–temperature product, Goody also gives values for the ratio of the product of the spectral radiant exitance with wavelength relative to the maximum value of this product. By the time the second edition appeared in 1989, a reduction in the number of values for each of these quantities is found while a second table for values of the derivative of the spectral radiant exitance in the linear frequency representation with respect to temperature at ten different frequencies from 1.0 to 70.0×10^{-12} Hz at the four temperatures of -80, -40, 0 and 40 °C appears [264].

The addition of tables relating to a blackbody in the 1949 Sixth Revised Edition of the *Smithsonian Meteorological Tables* was completely new. Three tables were given. In the first the integrated fractional amount as a function of wavelength–temperature product from 0.050 to 2.00 cm·K is given to five significant figures. In the second, in common with the tables of Tanner and Robinson and of others later, the total radiant exitance in the terrestrial range of 170 to 379 K at 1 K intervals are given to four significant figures in units of calories per square centimeter per fourth power of Kelvin per minute [cal·cm^{-2}·K^{-4}·min^{-1}]. The third table was somewhat unusual. It gave the in-band integrated amount for thirty-six different wavelength intervals at three typical terrestrial temperatures of 200, 250, and 300 K in units of calories per square centimeter per minute [cal·cm^{-2}·min^{-1}] and was calculated using data from the previous two tables.

    The first of many tables to come that were directly applicable to radiometric work in the infrared produced in North America using a digital computer were those published by Robert L. LaFara, Edward L. Miller, Walter E. Pearson, and J. Fred Peoples in late 1955 [367], and by Forrest R. Gilmore in July the following year [258]. Gilmore tabulates two dimensionless functions related to Planck's equation when it is considered in the linear frequency representation. In the first, on setting $x = h\nu/(k_B T)$ Planck's equation in this representation can be re-written as

$$M^{b}_{e,\nu}(\nu, T) = \frac{2\pi(k_B T)^3}{(ch)^2} \frac{x^3}{e^x - 1},$$ (7.22)

and it was the dimensionless function $x^3/(e^x - 1)$ as a function of x one finds tabulated. In the second, the integrated fractional amount radiated into the spectral band from zero up to some frequency ν, namely

$$\mathfrak{F}_{e,0 \to \nu} = \frac{15}{\pi^4} \int_0^x \frac{x^3}{e^x - 1} dx,$$ (7.23)

is given. Eq. (7.23) is nothing more than the complement of the corresponding integrated fractional amount found when the linear wavelength representation is used (see Eq. (2.118) on page 73). The dimensionless parameter x ranged from 0 to 10.00 at intervals of 0.02 and all values were given to five significant figures. For x from 10.0 to 20.0, the dimensionless form for the spectral radiant exitance is again given together with the complement of the fractional amount at 0.1 intervals. In both cases the integrals were calculated numerically using Simpson's rule, and rather surprisingly, were checked using both exponential and Bernoulli infinite series forms.

    Gilmore seems to have been motivated to produce his table of values as many of the earlier tables, particularly those he was aware of from the United States, were given in dimensional form using values for the physical constants which by the mid-1950s were very much out of date. Also, most of the previous tabulations had been made using evenly spaced values for the wavelength, which he thought rather inconvenient compared to evenly spaced values for the frequency, as these completely omitted tabulations at particularly low frequencies.

    After Gilmore had computed his tables, they were used internally at the RAND Corporation in Santa Monica, California, for a period of time. RAND Corporation was the research and development think tank created in 1948 by

Douglas Aircraft Company to advise the U.S. Armed Forces. The issuance of a Naval Ordnance report in November 1955 by LaFara and co-workers prompted Gilmore to publish his own tables as a Research Memorandum for the RAND Corporation. Gilmore also felt that as the Naval Ordnance (NAVORD for short) report by LaFara was not widely available, a statement we tend to agree with as we ourselves found locating a copy of the report particularly difficult, his own tables would be found to be more convenient for most practical applications. The tables given by LaFara and co-workers in the NAVORD report were for the integrated fractional amount and its complement as a function of the dimensionless parameter $x = c_2/(\lambda T)$. As a further aid to users, using a value of 1.436 cm·K for the second radiation constant, the corresponding wavelength–temperature product for each x was also given. The range in the dimensionless parameter considered was from 0.10 to 65.0 and corresponded to a wavelength–temperature product range of 0.072 to 14.36 cm·K, which for low temperatures, covered blackbodies radiating in the infrared at mid to long wavelengths. All tabulations were given to eight significant figures. The integrals for the integrated fractional amount and its complement were calculated using both the exponential and Bernoulli infinite series forms. Computationally, no details about the computer they must have used to gain eight significant figure accuracy is given.

In 1957, two modest tables related to the radiation from a blackbody appeared in the section on radiometry in the *American Institute of Physics Handbook* [556]. Tabulated by M. M. Reynolds, R. J. Corruccini, M. M. Fulk, and R. M. Burley, the spectral radiant exitance relative to its maximum amount at the spectral peak and the integrated fractional amount as a function of the wavelength–temperature product are given to four significant figures. Values for the product ranged from 0.050 to 2.0 cm·K. In the second table, the total radiant exitance and the maximum spectral radiant exitance at its peak are given to four significant figures as a function of temperature from 1 to 10 000 K. Reynolds used 3.741×10^{-16} W·m^2 and 1.438 cm·K for the two radiation constants, the latter being based on the value established as part of the International Temperature Scale revision of 1948 [278].

Not satisfied with the values Reynolds had chosen for the radiation constants, Earle B. Mayfield, as part of his PhD work in 1959, selected values of 3.7413×10^{-16} W·m^2 and 1.43884 cm·K for the two radiation constants. These he used to calculate the spectral radiant exitance using an IBM 704 computer [433]. Introduced in early 1954 the IBM 704 was the first mass-produced computer that could handle floating-point arithmetic. Mayfield's principal reason behind his tabulation seems to have been the concern he had with the accuracy of the values found in the *American Institute of Physics Handbook* tables. Not being accurate enough for the type of work he was interested in doing, using the IBM 704 located at the Naval Ordnance Test Station where the experimental work for his thesis was undertaken, Mayfield tabulated to six significant figures the spectral radiant exitance as a function of wavelength

0.500(0.020)0.800 µm at temperatures of 2000(200)4000 K [435]. A more extensive set of tabulations separate from his thesis was given in a technical report written for the Naval Ordnance Test Station [434], a copy of which we have not managed to locate. Somewhat surprisingly, for a standard work of reference, the tables that appeared in later editions of the *American Institute of Physics Handbook* held fast to those given in the first edition, so by the time the third and final edition appeared in 1972 [339], the tabulated values had become woefully inaccurate.

In the 1960s, as digital computers started to become more widespread in universities, national laboratories, and other research organizations, the number of tables produced relating to blackbody radiation rose sharply. For a time it seemed the smallest change in value for either radiation constant, or a specific problem requiring a particular wavelength and temperature range, was enough to spark off a new wave of table making. These tables appeared as either technical reports, appendices to specific texts, or for the first time, as entire books devoted to such tabulations.

In Thomas R. Harrison's 1960 text *Radiation Pyrometry and its Underlying Principles of Radiant Heat Transfer* the first twenty-seven pages of chapter six are devoted to a tabulation of the spectral radiant exitance relative to its spectral peak, and the corresponding integrated fractional amount, as functions of the wavelength–temperature product [282]. Harrison uses $c_2 = 1.4388$ cm·K and all values are tabulated to five significant figures. The range in wavelength–temperature product selected is extensive, running from 500 to 250 000 µm·K. Harrison himself did not calculate the values he presents. Instead, these were calculated by William F. Roeser of the National Bureau of Standards and were meant to serve as an updated version to those produced by the Bureau back in the time of Frehafer and Skogland. Unfortunately details concerning how the tabulations were made are not given.

An extension to the two tables found in the 1957 edition of the *American Institute of Physics Handbook* was given by Gary T. Stevenson in September of 1961 [625]. Appearing as a technical report for the U.S. Naval Ordnance Test Station, China Lake, California, it gave the spectral radiant exitance relative to its maximum at the spectral peak, and the corresponding integrated fractional amount, for wavelength–temperature products of 0.0500(0.0001)1.0003 cm·K and 1.0000(0.0100)6.2700 cm·K. Stevenson hoped the very small interval sizes chosen would reduce the need for interpolation between adjacent values. A value of 1.43886 cm·K was used for the second radiation constant and was tabulated to either six or seven significant figures. Stevenson's values differed with those given in the *Handbook* in the third significant figure due to a value of 1.438 cm·K being used in the latter case. Stevenson notes the tabulated values could only be considered accurate to the sixth place if the value for the second radiation constant was taken as exactly 1.438 8600 cm·K. All computations were performed on an IBM 709 computer; an improved version of the 704 computer Mayfield had used just a few years earlier at the same location, using the relatively new programming language of FORTRAN. In calculating both tables, infinite series expansions in terms of exponentials were used in their evaluation. A second printing of the report correcting a number of errors found in the first was issued in May 1963 [626].

A year later, an extension to Stevenson's tables (second printing) was given by Edward J. Pisa [500]. Pisa, like Stevenson, was working at the U.S. Naval Ordnance Test Station at China Lake in California. With programming assistance from Stevenson himself and William Clelland III, using an IBM 7094 computer and FORTRAN IV, Pisa tabulated three quantities related to blackbody radiation in two tables. In the first table the spectral radiant exitance relative to its maximum at the spectral peak $M(\lambda, T)/M(\lambda_{\max}, T)$ and

$$\frac{\partial \left(\dfrac{M(\lambda, T)}{M(\lambda_{\max}, T)} \right)}{\partial (\lambda T)} \bigg/ \left(\frac{M(\lambda, T)}{M(\lambda_{\max}, T)} \right),$$

as a function of the wavelength–temperature product ranging from 200 to $120\,000\,\mu\text{m·K}$ for four different interval sizes were given. The second table gave the value for the maximum spectral radiant exitance at the peak wavelength as a function of temperature from 50 to $10\,000\,\text{K}$ at interval steps of 1, 2, and $5\,\text{K}$. All values in each table were given to six significant figures. The values used for the radiation constants were $c_1 = 3.739\,80 \times 10^{-12}\,\text{W·cm}^2$ and $c_2 = 1.43847\,\text{cm·K}$. The value for the second radiation constant therefore differed to that used by Stevenson in the fourth decimal place. The computation for each quantity was made using infinite series expansions in terms of exponentials.

Using Pisa's two tables the following three functions related to blackbody radiation could then be found

$$M(\lambda, T), \quad \frac{1}{M(\lambda, T)} \frac{\partial M(\lambda, T)}{\partial T}, \quad \text{and} \quad \frac{1}{M(\lambda, T)} \frac{\partial M(\lambda, T)}{\partial \lambda}.$$

The first of these quantities was readily found by multiplying the value for the ratio found in Table 1 by the value given in Table 2 at the temperature of interest. The second and third quantities, on the other hand, were found using the following relations that can be readily shown to be valid

$$\frac{1}{M(\lambda, T)} \frac{\partial M(\lambda, T)}{\partial T} = \frac{5}{T} + \lambda \left[\frac{\partial \left(\dfrac{M(\lambda, T)}{M(\lambda_{\max}, T)} \right)}{\partial (\lambda T)} \bigg/ \left(\frac{M(\lambda, T)}{M(\lambda_{\max}, T)} \right) \right],$$

and

$$\frac{1}{M(\lambda, T)} \frac{\partial M(\lambda, T)}{\partial \lambda} = T \left[\frac{\partial \left(\dfrac{M(\lambda, T)}{M(\lambda_{\max}, T)} \right)}{\partial (\lambda T)} \bigg/ \left(\frac{M(\lambda, T)}{M(\lambda_{\max}, T)} \right) \right],$$

respectively.

Throughout the 1960s many other tabulations for various functions related to blackbody radiation appeared as technical reports. In September 1962, Russell G. Walker published a table for the spectral radiance in the linear wavenumber representation as a function of wavenumber and temperature [665]. Temperatures ranged from 77 to 30 000 K for five different interval sizes while the wavenumber ranged from 50 to 37 500 cm^{-1} for three different interval sizes. Values used for each radiation constant were $c_1' = c_1/\pi = 1.1909 \times 10^{-12}$ W·cm^2·sr^{-1} and $c_2 = 1.4380$ cm·K respectively. The tabulations were performed on the Philco 2000 Model 212 computer that at the time was in use at the Air Force Cambridge Research Laboratories located in Bedford, Massachusetts. Given to five significant figures, the tabulations ran to 175 pages.

The same month of September also saw an identical publication for the spectral radiance in the linear wavenumber representation by Carmine C. Ferriso [225]. Appearing as it did in the same month as Walker's tabulations for an identical quantity suggests that neither knew of the other's work. Compared to Walker's publication, Ferriso's is far more extensive in the range considered. It ran to an incredible 650 pages and would stand as the most voluminous of all the tables ever produced. At the time, Ferriso was working on the temperature of flames. Temperatures ranging from 5 to 1×10^8 K at various interval sizes were considered while wavenumbers ranged from 5 to 100 000 cm^{-1} for four different intervals sizes. Values for the spectral radiance that were less than 1×10^{-35} W·cm^{-2}·sr^{-1}·cm^{-1} were not included in the tables. Ferriso used the 1955 adjusted values for the radiation constants of $c_1 = 3.74126 \times 10^{-12}$ W·cm^2 and $c_2 = 1.4388$ cm·K as give by E. Richard Cohen, Jesse W. DuMond, Thomas W. Layton, and John S. Rollett [142]. All tabulations were made on an IBM 7090 computer located at General Dynamics/Astronautics, a division of the General Dynamics Corporation but surprisingly were only given to four significant figures, one less than the tabulations of Walker.

A third set of tables for the spectral radiance as a function of wavenumber and temperature was published by L. H. Byrne in January 1965 [114]. As these were prepared at the Goddard Space Flight Center in Greenbelt, Maryland, it suggests their intended use was for meteorological applications and is reflected in the range of wavenumbers and temperatures considered. Temperatures from 150 to 350 K at intervals of 2 K and from 250 to 2745 cm^{-1} at 5 cm^{-1} intervals are given. The very small interval sizes considered meant the tabulations ran to 202 pages. Values used for the two radiation constants were those recently recommended in 1963 by the National Academy of Sciences, National Research Council Committee on Fundamental Constants [14]. These were $c_1' = c_1/\pi = 1.906 \times 10^{-12}$ W·cm^2·sr^{-1} and $c_2 = 1.43879$ cm·K. The calculations were performed on an IBM 7094 Model II computer using single-precision FORTRAN with all values quoted to five significant figures. The report is labelled *Volume 1 - For Wavenumbers from 250 to 2745 cm^{-1}*

and Temperatures from 150 to 350 K suggesting it was the first of others to come. As far as we are aware, no further volume of tables from Byrne ever appeared.

As part of his spectroscopic studies of the radiation emitted in the visible portion of the spectrum by pyrotechnic compositions containing magnesium and calcium, Robert M. Blunt in March 1967 published separately as a technical report an extensive set of tables for a number of quantities related to blackbody radiation [87]. At each temperature from 1500 to 4050 K, at 50 K intervals Blunt gave values for the total radiant exitance, wavelength at the spectral peak, the spectral radiant exitance at the peak wavelength, and the in-band fractional amount radiated into the visible portion of the spectrum,[11] and as a ratio when expressed relative to the total radiant exitance. Furthermore, for each temperature as a function of wavelength ranging from 0.350 to 12.0 μm at various intervals, the spectral radiant exitance, the spectral radiant exitance relative to the maximum spectral radiant exitance at its spectral peak, and the fractional amount as a ratio relative to the total radiant exitance were also tabulated and given to between four and eight significant figures. Computations were performed at the University of Denver's Computer Center using B-5500 Extended ALGOL as the programming language. Blunt acknowledges programming assistance he received from Mrs Anita West from the Denver Research Institute at the University of Denver, but gives no mention of the type of computer the computations were performed on. The values Blunt used for the two radiation constants and the Stefan–Boltzmann constant were 3.7405×10^{-12} W·cm^2, 1.4385 cm·K, and 5.6697×10^{-12} W·cm^{-2}·K^{-4}, respectively.

In a NASA Technical Note published in March 1968, Thomas E. Michels gives a detailed account showing how the integral appearing in the integrated fractional amount could be calculated and the size of its error estimated [443]. He evaluates the integral using the exponential infinite series form (see Eq. (3.49)). Tabulations for the spectral radiance in the linear wavenumber representation normalized relative to the maximum value at its peak and the corresponding fractional amount as functions of the ratio of wavenumber to temperature are given in the first table he presents. The wavenumber to temperature ratio ranged between 0.001 to 30.000 cm^{-1}·K^{-1}. In his second table the total radiance and the spectral radiance at its spectral peak as functions of temperature from 1.0 to 25 000 K are tabulated. Values Michels uses for the first radiation constant and the Stefan–Boltzmann constant are $c_1' = c_1/\pi = 1.1909 \times 10^{-12}$ W·cm^2·sr^{-1} and $\sigma' = \sigma/\pi = 1.80466 \times 10^{-12}$ W·cm^{-2}·K^{-4}·sr^{-1}. In presenting the first of his tables as a function of the ratio of wavenumber to temperature, the type of user Michels was most directly appealing to were those working in the areas of pyrometry, spectroscopy, and radiative transfer.

Starting in 1968 and continuing into 1969, Radames K. H. Gebel, who was working at the Aerospace Research Laboratories in Ohio, made tabulations for a large number of quantities relating to blackbody radiation [246, 247].

Gebel was concerned with the fundamental infrared sensitivity limit for thermal image detection in a detector resulting from a source that is perfectly black [248]. In four sets of tables Gebel tabulated the following quantities in both energetic and photonic units: (i) the spectral radiance normalized relative to its spectral peak value, (ii) the integrated radiance from zero up to some finite wavelength, (iii) the integrated fractional radiance from zero up to some wavelength normalized relative to the total radiance, and (iv) the integrated fractional radiance from zero up to some finite wavelength normalized relative to the integrated fractional radiance from zero up to the value for the peak wavelength in energetic units. All tables were given as functions of the dimensionless ratio between the wavelength and the peak wavelength value corresponding to the energetic case, namely $\lambda_n = \lambda/\lambda_{\max}$. The dimensionless parameter λ_n ranged from 0.20 to 140 for various interval sizes. All tabulated values were given correct to six decimal places. Appearing in an appendix, an extension to the third table was given. This time the dimensionless parameter λ_n ranged from 0.100 to 140.0 with a far greater number of values being given for far smaller interval sizes. The tabulation ran for thirty pages with values given to seven significant figures. Values of $c_1 = 3.7405 \times 10^{-16}\,\text{W·m}^2$ and $c_2 = 1.43879 \times 10^{-2}\,\text{m·K}$ for the radiation constants were used while all computations were performed on IBM 1620 and 7094 computers.

As a by-product of very accurate work on a new spectral radiometric standard, Japanese authors Mamoru Suzuki, Mitsuhiro Habu and Takehiko Nagasaka published in December 1970 a number of tables for several radiometric, photometric, and colorimetric quantities [638]. The bulk of the tabulations given were for the spectral radiance as a function of wavelength and temperature. The region spanning the ultraviolet to the infrared corresponding to wavelengths from $0.1\,\mu\text{m}$ to $15\,\mu\text{m}$ at temperatures ranging from 100 to $100\,000\,\text{K}$ are considered. It appeared as the first table with values for the spectral radiance given to seven significant figures. The new value for the second radiation constant of $1.4388 \times 10^{-2}\,\text{m·K}$ adopted in 1968 by the International Practical Temperature Scale [63] is used while an earlier 1963 value of $3.7415 \times 10^{-16}\,\text{W·m}^2$ for the first radiation constant as recommended by the National Academy of Sciences, National Research Council Committee on Fundamental Constants is used [14].

The second of the tables was a little unusual. It gave the number of photons per second for monochromatic radiant flux of one watt as a function of wavelength in units of photons per second per watt [photon·s^{-1}·W^{-1}]. It corresponds to $L_{\text{q},\lambda}/L_{\text{e},\lambda} = M_{\text{q},\lambda}/M_{\text{e},\lambda} = \lambda/(hc)$. The same wavelength range as that considered in the first table is considered while all values are quoted to six significant figures. As the quantity tabulated was independent of temperature, compared to the first table which ran to 104 pages, the second ran to only two pages. The third table gave four quantities related to photometry and colorimetry while the fifth gave the in-band amount within the visible portion of the spectrum which was considered to lie between wavelengths of 380 to

780 nm. It was given to six significant figures for a range of temperatures that spanned from 500 to 100 000 K. Table four gave chromaticity coordinates for a blackbody as a function of temperature.

Table five contained a tabulation for the base 10 logarithm for the spectral radiance relative to the spectral radiance at the gold-point temperature as a function of wavelength and temperature. The gold-point temperature corresponds to the freezing point of gold at standard pressure and was assigned the value of 1337.58 K by the International Practical Temperature Scale in 1968 [63]. Values within the very narrow wavelength range of 0.648 to 0.666 μm at 0.001 μm intervals and for temperatures from 1300 to 4000 K at 10 and 20 K intervals to eight significant figures are given. Table six gave values for the derivative of the spectral radiance with respect to temperature relative to the spectral radiance as a function of temperature from 550 to 50 000 K at the two wavelengths of 0.650 and 0.660 μm to six significant figures. Table seven was a simple table that could be used to convert values of wavelength in units of micrometers into wavenumber in units of per centimeter. For interpolation between adjacent values the authors recommended a method due to Y. Ishida using interpolation coefficients, a summary of which appeared in the first appendix of their report. The authors also mentioned that if a large number of calculations were needed, these could be more efficiently calculated directly provided ready access to a digital computer could be had, something which by 1970 was on the cusp of becoming a reality.

Not satisfied with some of the earlier tables that gave tabulations for the integrated fractional amount as a function of the wavelength–temperature product, Stephen L. Sargent in 1972 produced a compact set of tables for this function [587]. Sargent noted that on the one hand many of the tables which existed in the literature for this function, such as those we have seen by Dunke (1954), Allen (1955), and Goody (1964), used interval sizes that tended to be fairly large, necessitating frequent interpolation. On the other hand, the tables by Pivovonsky and Nagel (1961), to be discussed shortly, were made with very small interval sizes that meant their tabulation ran to almost a hundred pages and made it, in the opinion of Sargent, somewhat unwieldy to use. In his estimation what was needed was a table with relatively small interval sizes in the wavelength–temperature product so that interpolation was not necessary in most cases, yet was sufficiently compact in form. These he produced for wavelength–temperature products in units of micron–Kelvin (500–100 000 μm·K) and micron–Rankine (1100–150 000 μm·R). The fractional amount was given to six decimal places, spanned a modest two pages for each different product, and was available upon request in the form of a technical note from the Center for Climatic Research, University of Wisconsin [586].

Other tabulations which appeared as technical reports exist. These include sets made by Arthur G. DeBell in 1959 [183] and S. A. Golden [261] in 1960. Both men at the time were working at the Rocketdyne division of North

American Aviation in California. DeBell, like Ferriso at around the same time, was working on flame temperatures [301]. Requiring an improved set of tables compared to those in current existence, DeBell undertook his calculations on an IBM 704 computer using the most recent though not currently accepted value for the second radiation constant of 1.4388 cm·K. What quantities he tabulated has not been recorded in the secondary literature, but judging from the title of the work must have included tabulations for the spectral radiant exitance of a blackbody. Golden's tabulations consisted of two separate tables. The first table gave the ratio of the spectral radiant exitance relative to the maximum spectral radiant exitance at the peak wavelength, the integrated fractional amount, and their first derivatives as functions of the dimensionless variable $c_2/(\lambda T)$ from zero up to 50 for a range of interval sizes. The second table gave values for the maximum spectral radiant exitance at its peak and the total radiant exitance as functions of temperature from zero to 10 000 K at 10 K intervals [696]. Values of $3.7413 \times 10^{-5}\,\mathrm{erg \cdot cm^2 \cdot s^{-1}}$ and 1.4388 cm·K were used for the two radiation constants. We have however, not been successful in locating copies of either of these reports.

A detailed tabulation for the spectral radiance as a function of the temperature and wavelength was made by T. Roy Bowen in May of 1963 for the U.S. Army Missile Command at Huntsville in Alabama [94]. The tables were prepared in response to a need within the group Bowen found himself working in at the time, the Re-Entry Physics Branch, for a convenient set of tabulations for the spectral radiance of a blackbody after having found the large number of repetitive calculations they faced was simply not feasible by hand. Appearing as a Technical Note, the tables were published in the hope that others might find them of some use. The great difficultly we had in acquiring a copy suggests that the number produced at the time was probably no more than a few hundred, and most of these did not find their way into conventional academic libraries.

In Bowen's tables the spectral radiance in units of watts per square centimeter per steradian per micrometer $[\mathrm{W \cdot cm^{-2} \cdot sr^{-1} \cdot \mu m^{-1}}]$ was tabulated for the following wavelengths: $0.10(0.02)1.0(0.2)10(2)30\,\mu\mathrm{m}$ at sixty-four different values of the temperature given by $300(50)1000(100)5000(500)10\,000\,\mathrm{K}$. The units selected reflect those that were in general use at the time within Bowen's group. Values for the spectral radiance at each temperature considered were given on a page of its own meaning the tabulations ran to a total of sixty-four pages. They were calculated using a computer and expressed to five significant figures, though it should be noted that the values were truncated rather than rounded at the fifth significant digit. In order to avoid the possibility of errors being introduced during transcription, the tables were reproduced directly from printouts received from the computer. Values that Bowen uses for the two radiation constants were $1.1906 \times 10^4\,\mathrm{W \cdot \mu m^4 \cdot cm^{-2} \cdot sr^{-1}}$ and $1.4388 \times 10^4\,\mu\mathrm{m \cdot K}$, respectively. Following the tables, four graphs for the spectral radiance as a function of wavelength at forty-seven different temperatures ranging from 300 to 10 000 K were presented, having been produced from data from the foregoing tabulations.

A final technical report consisting of tabulations of a number of quantities related to blackbody radiation was given in 1971 by John Roberts [564]. Roberts was a lecturer in the Department of Chemical Engineering at the University of Newcastle in Australia. At the time, the department was involved in a number of projects that were concerned with the analysis of radiation emitted from industrial flames. The nature of this work frequently required reference to a number of quantities related to blackbody radiation. It quickly become clear having access to a permanent set of tabulations of the most commonly needed functions would be useful. Roberts obliged by organising and arranging for their preparation and publication. Having been found useful internally, additional copies of the tables were arranged and issued as a *Departmental Bulletin* in the hope they might prove of equal value to others.

Four quantities were tabulated. These were $M_{e,\lambda}^{b}/(\sigma T^5)$ in units of per micrometer per Kelvin $[(\mu m \cdot K)^{-1} \times 10^5]$ and the three dimensionless quantities of: $M_{e,\lambda}^{b}/M_{e,\lambda_{max}}^{b}$ the ratio of the spectral radiant exitance to its value at its peak, the fractional function of the first kind for the radiometric case, and what Roberts termed an auxiliary function $\psi(x)$, viz.,

$$\psi(x) = \frac{x}{1 - e^{-x}} \qquad (7.24)$$

where $x = c_2/(\lambda T)$ and was a function that could be related to the derivative of the three previous quantities with respect to the temperature, wavelength, and the second radiation constant. The last of these derivatives with respect to the second radiation constant could be used to correct any of the first three quantities for small changes in its value. Each quantity was tabulated as a function of the wavelength–temperature product extending over a range from 250 K to 30 000 $\mu m \cdot K$ for six different interval width sizes ranging from 5 to 20 $\mu m \cdot K$.

Working in extended precision, Roberts performed all calculations on an IBM 1130 Series II computer located at the university. The first of the tabulated quantities, $M_{e,\lambda}^{b}/(\sigma T^5)$, was given to between five to eight significant figures, the ratio of the spectral radiant exitance to its spectral peak to six to eight significant figures, the fractional function to four to seven while the auxiliary error function to either four or five. Rather than using the familiar power series expansion for the fractional function, Roberts evaluated it numerically using Simpson's rule where an automated interval halving technique was used until a predetermined error tolerance was reached. Values used for the two radiation constants were $3.741\,473\,5 \times 10^4$ W·cm^{-2}·μm^4 and $14\,387.89\,\mu m \cdot K$, while a value of $5.667\,31 \times 10^{-12}$ W·cm^{-2}·K^{-4} was used for the Stefan–Boltzmann constant, and 2897.796 $\mu m \cdot K$ for Wien's displacement constant. These values Roberts derived from values published for the fundamental constants by the National Bureau of Standards in the United States which had been adopted in October of 1963 based on the recommendations of the National Academy of Sciences, National Research Council Committee on Fundamental Constants [14].

Unlike other compilations, Roberts gives a very brief comparison between his values for $M_{e,\lambda}^b/(\sigma T^5)$ and the fractional function at nine different wavelength-temperature products between 1000 and 15 000 µm·K with those of Harrison (1960), Pivovonsky and Nagel (1961), and Wiebelt (1966). Roberts thought the rounding error in his tabulated values would not exceed 0.6 units in the seventh significant figure. In addition to their improved accuracy, he also felt his tabulations were more suited to engineering purposes. As an application of his tables, four examples making use of them were given.

How widely Roberts' *Departmental Bulletin* was circulated and used by others is difficult to say. Being an extremely rare document, coupled to the fact it was never referred to or quoted in the literature, suggests its existence was most likely only known to a very small circle of people connected with the Chemical Engineering department where Roberts worked. So while their intention may have been to be of equal value to others involved in radiative transfer computations, we suspect this goal was never achieved.

The problem many of the tabulations produced as technical reports faced was one of accessibility. Limited distribution often meant many of these tables simply went unnoticed and that probably explains why the total number of citations to this body of work is small. While attention was occasionally drawn to a few of these tables in the literature [301, 15, 696, 93], all this was about to change. Starting in 1961, in the decade that followed no less than seven books devoted entirely to the tabulation of functions related to blackbody radiation appeared. With their wide availability, two of these texts in particular would prove to be immensely popular. Widely cited, and no doubt even more widely used than the number of actual citations suggest, each would go on to become classics within the field. Both appeared in 1961 and both had their origins in Germany.

The first of the two was by Marianus Czerny and Alwin Walther [164]. Czerny, of "System Czerny" slide rule fame, had already developed the first radiation slide rule prototype some 17 years earlier and was by now well known and established for his work in the infrared. Walther, on the other hand, was an applied mathematician. He headed the Institut für Praktische Mathematik at the Technische Hochschule in Darmstadt, Germany, from the time of its establishment in 1928 until his death in 1967 [175]. The institute was a very important early German center for computational mathematics and in the application of mathematics to applied problems and it was to Walther that Czerny turned to provide the all important computational muscle that was going to be needed for such an ambitious project. Their text was bilingual, appearing in both German and English, and in English went by the name of *Tables of the Fractional Functions for the Planck Radiation Law* [164].[12]

As the values for both radiation constants in the late 1950s had yet to be firmly established, the tabulations Czerny and Walther gave were in dimensionless form. Five dimensionless functions were tabulated. Showing how each was related to the radiation emitted by a blackbody, these were

$$\int_0^\lambda L(\lambda, T)\, d\lambda \;=\; \frac{\sigma}{\pi} T^4 B(v), \tag{7.25}$$

$$L(\lambda, T) \;=\; \frac{\sigma}{\pi c_2} T^5 B'(v), \tag{7.26}$$

$$\frac{\partial L(\lambda, T)}{\partial \lambda} \;=\; \frac{\sigma}{\pi c_2^2} B''(v), \tag{7.27}$$

$$\int_0^\lambda \frac{\partial L(\lambda, T)}{\partial T}\, d\lambda \;=\; \frac{4\sigma}{\pi} T^3 B^*(v), \tag{7.28}$$

$$\frac{\partial L(\lambda, T)}{\partial T} \;=\; \frac{4\sigma}{\pi c_2} T^4 B^{*\prime}(v). \tag{7.29}$$

Here $v = \lambda T / c_2$. The function $B(v)$ was equal to

$$B(v) = \frac{15}{\pi^4} \int_0^v \frac{u^{-5}}{e^{1/u} - 1}\, du, \tag{7.30}$$

and was referred to, for the first time, as the fractional function of the first kind while $B'(v)$ and $B''(v)$ were the first and second derivatives of the fractional function of the first kind with respect to v. Czerny also showed the function $B^*(v)$ could be expressed in terms of $B(v)$ and its derivative (see page 75). The result is

$$B^*(v) = B(v) + \frac{v}{4} B'(v), \tag{7.31}$$

and was referred to by them as the fractional function of the second kind, while $B^{*\prime}(v)$ was its derivative with respect to v. Compared to the fractional function of the first kind, very few later tabulations for the fractional function of the second kind appeared, and those which did would be very brief in form [203, 471].

As early as 1954, in a paper on the problem of integrating Planck's equation, Czerny defines each fractional function and presents a modest table consisting of thirty-one entries for the functions $B(v)$, $B'(v)$, and $B^*(v)$ [160]. Work done by Czerny and co-workers for the *Deutsche Glastechnische Gesellschaft* (German Technical Glass Society) where problems related to the transfer of radiative heat within the interior of molten glass [161, 162] had suggested the possible usefulness for such a set of tables. Preliminary work on the tables was begun late in 1955 by Inge Möll-Franz using a mechanical desk calculator. Subsequent calculations were performed on an IBM 650 digital computer in the summer of 1956, first at the Computational Center of IBM Germany in Sindelfingen by H. Schappert, and later at Walther's home institution in Darmstadt by H. Schappert and G. Hund.

Initially the tables appeared in 1957 as a technical report for the *Deutsche Glastechnische Gesellschaft* [163], but it was not long before their general utility to a wider audience was recognized. They were picked up by the publisher Springer-Verlag and in a slightly expanded form were duly published in 1961. After an introductory section the text contains four tables. In Table 1 the more common fractional functions $B(v)$, $B'(v)$, and $B^*(v)$ as a function of v ranging from 0.040 to 3.00 for various step sizes were tabulated. For values of v between 3.0 and 100 the functions $[1 - B(v)]$, $[1 - B^*(v)]$ were given instead of $B(v)$ and $B^*(v)$. In general, all values were given to five significant figures though some, particularly for large v, were given to six and even seven significant figures. Table 2 contains tabulations for the two less frequently used fractional functions $B''(v)$ and $B^{*\prime}(v)$ as a function of v. The range in v ran from 0.040 to 100 but this time the number of entries considered was a tiny fraction of those considered in Table 1. For comparison, Table 1 consisted of thirty-four pages while Table 2 runs to a mere two pages.

As an aid to calculating the physical quantities of interest from the tabulated dimensionless form, two auxiliaries tables were provided by Czerny and Walther. The first of these, Table 3, gave the third, fourth, fifth, and sixth powers of the temperature for 0 to 1000°C and 1000 to 10 000 K. For the temperatures given in Table 3, Table 4 tabulated values of v for various wavelengths ranging from 0.3 μm to 30 μm. Their calculation required a specific choice in the value for the second radiation constant, and was the only table of theirs that explicitly depended on the value for this constant. They chose a value of 1.438 cm·K.

The second book which appeared in 1961 was by Mark Pivovonsky and Max R. Nagel. Entitled *Tables of Blackbody Radiation Functions*, it was as long and voluminous as the book by Czerny and Walther was short and succinct [501]. The compilation was initially begun by Nagel in Germany as part of a wartime project designed to aid work in the development of artificial light sources for photographic, photometric, and colorimetric research. In the beginning, Nagel saw to all the computations himself, using nothing more than a mechanical hand calculator as his only aid. His one-man, privately pursued program would in time go through a series of agonizing setbacks. Fire, brought about by an Allied Forces bombing raid, saw the first hundred of his completed calculation sheets destroyed. Continued interruptions by war service, and later post-war writing paper rationing, further delayed Nagel's project. The decisive blow however, came in 1948 when the International Temperature Scale adopted a value for the second radiation constant of 1.438 cm·K. As all the previous work on Nagel's tables was based on a value of 1.432 cm·K it meant his tables were already obsolete by the time they finally appeared in 1952 [462]. Unperturbed, even before they finally appeared in print, detailed preparations had already begun for a new set of tables. With funding now coming from the United States, the project was now assured and Nagel teamed up with Pivovonsky and other staff members from the Computation Laboratories

of Harvard University, Cambridge, Massachusetts, to re-complete the task he
had started a decade earlier.

To enhance their usefulness, proposed plans for the tables were initially
sent to a number of experts, observatories, and research organizations around
the world for comment. With the response and interest in the proposal being encouraging and mostly favorable, Nagel decided to broaden the scope
of the tables so that they would be useful outside of the conventional photographic spectral region and temperature ranges. Moreover, as the tables of
Pivovonsky and Nagel tabulated values for a number of quantities related to
blackbody radiation directly, in order to diminish the danger of obsolescence
it was decided to include a number of auxiliary correction tables which could
be used in the event of any future changes in the value for the two radiation
constants. All numerical work for the tables was performed on the Harvard
Mark I automatic digital computer located at the university's computational
laboratory. Completed in 1957, a year later the tables appeared in loose-leaf
form as part of a progress report to the contract which had helped fund the
project. As only a very limited number of these reports were printed, and as
demand for them quickly outstripped available supply, the decision was made
to make the tables more widely available in permanent book form. These duly
appeared in 1961.

Pivovonsky and Nagel's text contains eight sets of tables. The first tabulates the spectral radiance, the ratio of the spectral radiance compared to
its value at a wavelength of 560 µm, and the corresponding integrated fractional amount as functions of wavelength and temperature to five significant
figures. Wavelength ranges between 200 to 590 µm at 5 µm intervals and from
590 to 1200 µm at 10 µm intervals for temperatures between 800 to 40 000 K
at various interval sizes are used. Table 2 is a continuation of the first table for the first and third quantities tabulated for wavelengths between 1.1
and 1000 µm at temperatures between 20 and 13 000 K. Values were given to
four significant figures. The third table tabulated the ratio of the spectral
radiance compared to its maximum value at its spectral peak, the integrated
fractional amount, and an auxiliary function related to both the temperature
or wavelength derivative of Planck's equation, as functions of the wavelength–
temperature product in the range 0.01 to 0.99 cm·K. The fourth table consisted of tabulations for the total radiance, the value of the maximum spectral
radiance at its spectral peak, and the corresponding value for the wavelength
at the spectral peak as functions of temperature. All quantities were given
to five significant figures and for the following temperatures: 1000(2)2500 K,
2500(5)5500 K, and 5500(1)10 000 K. Table 5 was identical to Table 3 except
the values were tabulated as a function of the ratio of wavenumber to temperature instead of the wavelength–temperature product. Table 6 was similar
to Table 4 except this time the value of the maximum spectral radiance at
its spectral peak and the corresponding value for the maximum wavenumber
where this occurs in the linear wavenumber spectral representation were given.

Lastly, Table 8 was an auxiliary temperature correction table to be used for any future change in the value of the second radiation constant.

In most cases using linear interpolation to interpolate between values found in Tables 1 and 2 was not possible. If interpolation was required, the authors suggested using Tables 3 through 6. As the interval size between consecutive values in these tables was small, it was thought that linear interpolation would not lead to errors in the calculated values greater than one unit in the last digit. One reviewer who used linear interpolation between adjacent temperature values in Tables 1 and 2 discovered that the values calculated were rarely accurate to more than three significant figures and suggested the F5100 radiation slide rule made for the Admiralty would be a useful adjunct to Pivovonsky and Nagel's tables if considerable interpolation was required [340]. Compared to the tables by Czerny and Walther, Pivovonsky and Nagel's tables were far easier to use as they gave the quantity one was interested in directly rather than in terms of a dimensionless parameter related to the quantity of interest. The former convenience would only prevail as long as the values adopted for the two radiation constants remained unchanged; a situation which was unlikely to last. Despite this, the tables were widely used for many years.

At the time each appeared, both books received comprehensive reviews in the literature [528, 340, 301, 251, 107, 694, 422]. In time, both texts proved to be highly successful. They were widely used and extensively referred to by authors in the literature. They were by far the two most popular set of tabulations ever made for quantities related to blackbody radiation and those produced by Czerny and Walther remain as valid today as when they first appeared, due to their dimensionless tabulated form. As a further measure of their success, Pivovonsky and Nagel's text was re-issued in 2013 [502].

Several other books devoted entirely to the tabulation of a number of quantities relating to blackbody radiation appeared throughout the 1960s and into the 1970s. From Russia two such books were published. In 1961, a text by P. A. Apanasevich and V. S. Aizenshtadt, consisting of tabulations for various blackbody radiation quantities in both energetic and photonic units in the linear frequency representation, appeared first in Russian [53], then in English translation a few years later [54]. At the time, very few tables relating to a blackbody existed in the Russian literature and those which did were small and by the early 1960s hopelessly out of date [318]. And unlike the tables which did exist outside of Russia, Apanasevich and Aizenshtadt were interested in producing a set of tabulations in the linear frequency rather than the linear wavelength representation. While McDonald's tables of 1955 were more up to date, the temperatures he considered ranged from $15\,000$ to $50\,000\,\mathrm{K}$ making them suitable for astrophysical work only. Apanasevich and Aizenshtadt's tables, on the other hand, covered temperatures from 25 to $1 \times 10^7\,\mathrm{K}$. The tables of Apanasevich and Aizenshtadt were also unique in one other important regard. All tabulations were given in terms of the spectral energy density $u(\nu, T)$ rather than the more common radiometric quantities of spectral radiant exitance or spectral radiance. They considered this to be the more fundamental of the three and it is related to the other two quantities by

$$u(\nu, T) = \frac{4M(\nu, T)}{c} = \frac{4\pi L(\nu, T)}{c}. \tag{7.32}$$

The first of their tables gave the ratio of the relative spectral energy density in the linear frequency representation to the value at its peak in energetic units as a function of the dimensionless parameter x which was equal to the ratio of the frequency to its peak frequency value for both the energetic and photonic cases. Here $x = 0.005(0.005)0.10(0.02)1.50(0.05)7.0$. Table 2 was the same as Table 1 except for the linear wavelength representation and in energetic units only. The dimensionless parameter was now z, the ratio of the wavelength to its peak wavelength value, and was equal to $0.200(0.005)0.700(0.01)1.50(0.05)4.00(0.2)16(1.0)25$. Table 3 was the same as Table 1 except for the logarithmic frequency (wavelength) representation and in energetic units only. The dimensionless parameter was now y, the ratio of the frequency (wavelength) to its peak value in the logarithmic frequency representation, and was equal to $0.005(0.005)0.10(0.02)2.50(0.05)5.50$. Table 4 was the same as Table 1 in the linear frequency representation for photonic units only, except this time the dimensionless parameter x' was for the ratio of the frequency to its peak in photonic units. Here $x' = 0.005(0.005)0.10(0.02)1.50(0.05)7.0$. Table 5 gave the integrated fractional amounts for both the radiometric and actinometric cases as a function of the dimensionless variable x which corresponded to the ratio of the frequency to its peak value in energetic units. Here, values for x considered were $0.01(0.01)3.00(0.02)4.00(0.1)6.0(0.2)7.0$. Table 6 gave the spectral energy density as a function of frequency and temperature. Temperatures in the range from 25 to 1×10^7 K were covered at seven different interval sizes, while the values for the frequencies were chosen so as to cover the most important values in Planck's equation within the temperature range being considered.

Tables 1 to 5 in Apanasevich and Aizenshtadt's text cover between one to three pages each. Table 6, on the other hand, runs for a staggering 186 pages. Tables 1 to 4 and Table 6 were all given to four significant figures while table five was given to five significant figures. For the two radiation constants, Apanasevich and Aizenshtadt chose the values that had been adopted by the International Congress on Universal Constants held in Turin in 1956 [71]. These values were $1.24795 \times 10^{-15}\,\mathrm{erg \cdot cm^2 \cdot sr^{-1}}$ and $1.43880\,\mathrm{cm \cdot K}$, respectively. The computations were performed on one of the early *Ural* digital computers located at the Institute of Mathematics and Computer Technology of the Belarussian Academy of Sciences in Minsk. The Ural computers were a series of mainframe computers first developed and built in 1959 in the former Soviet Union and were extensively used within the former Soviet and Eastern Block states.

A second text to come out of Russia appeared in 1964 by Mikaél' A. Bramson [98]. It appeared as a companion volume to his text on infrared radiation [97]. In 1968, it appeared in English translation as *Infrared Radiation. A Handbook for Applications, with a Collection of Reference Tables* with the two Russian texts being combined into a single volume [99]. Bramson's text was written principally for those involved in the design and development of a variety of infrared technologies. A unified approach throughout the text was taken. Thorough descriptions of the computational methods employed are presented. Tabulations for many of the basic quantities difficult to calculate otherwise are given. The tables Bramson produced are one of the most comprehensive sets ever assembled for the infrared portion of the spectrum at temperatures of interest to those working in infrared technology. Bramson believed the tables produced for various dimensionless quantities relating to blackbody radiation by Fabry (1927) and by Jahnke and Emde (1933) were far too brief, thereby compromising accuracy. Other tables for quantities in dimensional form, such as those by Pivovonsky and Nagel (1961), he considered too broad a spectral range had been used, while those by Apanasevich and Aizenshtadt (1964) covered too broad a temperature range, making neither particularly suited to detailed work in the infrared.

The tables Bramson gives are extensive. He presents 204 tables divided into nine parts. Parts 2 to 6 contain tabulations relating to blackbody radiation. In Part 2 the spectral radiant exitance, total radiant exitance, and value for the peak wavelength, for both the radiometric and actinometric cases are given for temperatures between 90 to 6000 K and at various intervals. Part 3 tabulations for the spectral radiant exitance relative to the maximum amount at the spectral peak and the integrated fractional amount for both the radiometric and actinometric cases as a function of the wavelength–temperature product ranging from 0.05 to 4.50 cm·K are given. Part 4 gives the following quantities for both the radiometric and actinometric cases as functions of the wavelength and temperature: spectral radiant exitance, ratio of the spectral radiant exitance relative to its peak value, and the integrated fractional amount up to some arbitrary wavelength. Temperatures between 100 and 6000 K and wavelengths between 0.75 to 15.0 μm for a range of interval sizes are considered in a total of 108 separate tables. Part 5 contains tabulations for quantities not usually found elsewhere. These are for quantities related to spectral and integrated radiance contrasts for small temperature differences. These are given as functions of temperature and wavelength. Wavelengths between 2.0 to 14.0 μm and temperatures over the very small range of 250 to 300 K are considered. Lastly, Part 6 tabulates a number of quantities related to blackbody radiation in completely dimensionless form. Both the radiometric and actinometric cases are considered in both the linear wavelength, wavenumber, and logarithmic scales. In almost all instances values are given to five significant figures. When values for the various physical constants are needed, those adopted in 1956 by the International Congress on Universal Constants in Turin are used. Like Apanasevich and Aizenshtadt before him, the computations were performed on an Ural computer located at the Academy of Sciences of the USSR in Moscow. Interestingly, the values tabulated in Part 3 were not calculated directly. Instead they were found by interpolation using values given in the 1941 tables of Lowan and Blanch.

A slim bilingual text entitled *Seven-place Tables of the Planck Function for the Visible Spectrum* came out of Germany in 1964 [277].[13] Authored by Dietrich Hahn, Joachim Metzdorf, Ulrich Schley and Joachim Verch, who were all working at Germany's national metrology institute at the Physikalisch-Technische Bundesanstalt in Braunschweig, their text was intended to make tabulations for Planck's equation more readily accessible to a wider German audience as these, according to the four authors, were not so easily found in the German literature. Apart from several older tables by Moon (1936, 1937, 1938, 1946, 1947, 1948) and by Lowan and Blanch (1940, 1941), and the more recent tables by de Vos (1954), Czerny and Walther (1961), and Pivovonsky and Nagel (1961), the authors felt that in fields such as photometry, pyrometry, and colorimetry, where an additional need to know the energy distribution of a light source evaluated with respect to the luminosity function of the human eye within the visible spectrum and to high accuracy, was largely lacking. It was this gap that their tables were intended to fill.

A total of five tables are given, and as the title of their text suggests, all values ware given to seven significant figures. In 1964, as it seemed more than likely a change in the value for the second radiation constant was imminent, the tables were prepared using the two values of 1.4380 cm·K and 1.4420 cm·K for c_2. The first was the value currently accepted by the International Practical Temperature Scale, while it was hoped that a new value for this constant would be fixed somewhere between the currently accepted value and the later selected value. The value adopted by the International Practical Temperature Scale of 1968 was 1.4388 cm·K [63]. Thus their selection proved to be a little on the generous side. Of the tables relating to the radiation emitted by a blackbody, Table 2 gives the ratio of the spectral radiant exitance relative to the first radiation constant as a function of temperature and wavelength for each of the two values for the second radiation constant. Temperatures from 1000 to 15 000 K, at interval sizes ranging from 100 to 1000 K, are given at wavelengths ranging from 350 to 845 nm at 5 nm intervals. Table 3 contained values for the integrated amount radiated into the spectral band between a wavelength of 350 to 845 nm at 60 different temperatures, ranging from 973.15 to 15 000 K. Table 4 was identical to Table 3, where the values tabulated were the base 10 logarithm of the values tabulated in table 3. All calculations had been performed with the aid of an IBM 650 computer. The exponential function was calculated by means of a Chebyshev approximation, and all calculations were internally carried out to eight decimal places with the final result being rounded off to seven places, thus giving an error of at most one in the last digit. Attention to the volume in the English-speaking world a year after its publication was drawn to it by a short review which appeared in the *Journal of the Optical Society of America* [420].

As part of Leo Levi's 1968 text *Applied Optics*, a number of tables relating to blackbody radiation were to be found in the appendices of the text [387]. Specially computed for his book, these were included when its author thought sufficiently complete and compact tables were not readily available. It is not that extensive tabulations for the spectral radiant exitance for a wide range of wavelengths and temperatures did not exist by the late 1960s. They did, and Levi was well aware of those by Czerny and Walther (1961), Pivovonsky and Nagel (1961), and Hahn (1964), but found them bulky and awkward to use. He also wished to minimize the chance of his tables becoming obsolete as a change in the value for the second radiation constant was expected, without burdening the user with many subsequent calculations resulting from any possible change.

As a compromise, Levi gave the following five tables. The spectral radiant exitance relative to the maximum value for the spectral radiant exitance (Table 1) and the integrated fractional amount (Table 2) as functions of the wavelength–temperature product from 300 to 9900 μm·K at 10 μm·K intervals to six significant figures, the maximum spectral radiant exitance (Table 3) and the total radiant exitance (Table 4) as functions of temperature from 100 to 9900 K at 10 K intervals and to six significant figures, and the value of the peak wavelength (Table 5) as a function of temperature from 1000 to 50 000 K at 100 K intervals given with four to seven significant figures. Values used for the two radiation constants were 3.7405×10^8 W·μm^2 and 14 387.9 μm·K, respectively. Lacking, however, are any details concerning how the actual calculations were performed.

Levi's five earlier tables were updated and significantly extended some years later in his text *Handbook of Tables of Functions for Applied Optics*. The first table was the same as his earlier Table 1 for a wavelength–temperature product extended from 100 to 9900 μm·K. The second is a similar extension of his earlier Table 2 for a wavelength–temperature product from 100 to 9900 μm·K and now its complement from 10 to 9900 μm·K. The third and the fourth were extensions of Tables 3 and 4 for temperatures from 100 to 99 000 K, while the fifth was an extension of Table 5 for temperatures from 1000 to 99 000 K. The sixth and seventh tables were new additions. In the sixth, the integrated fractional photon exitance amount as a function of the wavelength–temperature product from 100 to 9900 μm·K and its complement from 10 to 9900 μm·K are given, while the seventh contains a tabulation for the total photon exitance as a function of temperature from 100 to 99 000 K. Levi used an updated value for the first radiation constant of 3.7415×10^8 W·μm^2 while the value for the second radiation constant remained unchanged.

Details concerning how Levi performed his computations are now given. He writes that the evaluation for the spectral quantities were straightforward, while the integrals appearing in each fractional amount were evaluated using the exponential and Bernoulli infinite series forms, depending on the size of its argument. Comparing his first and second tables to others, Levi recognized that previously published tables by Hahn (1964) and by Pivovonsky and Nagel (1961), which gave values as functions of both wavelength and temperature instead of as a function of the wavelength–temperature product, may have been more convenient to use; it was however, bought at the expense of much larger interval sizes and far greater page volume. They were also based on a value for the second radiation constant of 1.4380 cm·K, which was no longer the best available value for this constant. And while the tables of Czerny and Walther (1961) were almost identical in form to his first and second tables, except for being expressed in a completely dimensionless form, Levi hoped his more compact tables would be considerably more convenient to use in practice. For the two new tables Levi gave, while the tables of Lowan and Blanch of 1941 covered similar ground their tables were based on a value for the second radiation constant which by 1974 was 0.2% below the presently accepted value. The more recent tables of Apanasowich and Aizenshtadt, while they did consider some actinometric quantities, the total photon exitance was not one of these quantities.

Another set of tables for the spectral radiance appeared in book form in 1972 [254]. It was the product of the Metrological Section of the *Deutsches Amt für Meßwesen und Warenprüfung* (German Office for Metrology and Product Testing) of the former German Democratic Republic in connection with their work on the realization of a radiation–optical temperature scale. It was a trilingual text (German, Russian, and English)[14] whose form was largely influenced by suggestions received from experts in the former Soviet and Eastern Block states. The values for the spectral radiance were given as a function of wavelength and temperature. The table itself was divided up into six separate sections with all of the values given to seven significant figures. In the first section, temperatures from 2000 to 16 000 K at wavelengths 170(10)370 nm were given. In the second section, temperatures from 1000 to 19 000 K at six different interval sizes at wavelengths of 380(5)780 nm were given. In the third section, temperatures from 750 to 21 000 K at six different interval sizes for wavelengths 790(10)990 nm were given. The fourth section contained temperatures from 50 to 10 000 K at three different interval sizes for wavelengths 1.00(0.10)3.50(0.25)7.75 μm. The fifth and final section contained temperatures from 50 to 13 500 K at four different interval width sizes at wavelengths 8.00(0.50)15.00(1.00)30.00(2.00)50.00 μm.

The final section for wavelengths of 320(5)780 nm was at 12 different reference point temperatures used in the 1968 Recommendations for the International Practical Temperature Scale. Values used for the two radiation constants were $c_1 = 3.741844 \times 10^{-16}$ W·m^2 and $c_2 = 1.4388 \times 10^{-2}$ m·K. The value chosen for c_2 conformed with the value recommended in the 1968 Recommendations for the International Practical Temperature Scale, while the value for c_1 was taken from the value given by Barry N. Taylor, William H. Parker and Donald N. Langenberg in 1969, as it was considered to be the most reliable at the time [641]. Values for the spectral radiance had been calculated on a "Robotron 300" computer with the help of a so-called EX 10 subroutine.[15] The Robotron 300 was a medium-sized digital computer in widespread use in the former German Democratic Republic at the time. It was modelled on the IBM 1401 computer and the 300 in its name stood for the desired performance in its attached punch card reader one hoped to achieve; in this case an ability to read 300 punch cards per minute. The error due to rounding in the last listed digit was at most 10 units, but was usually less than 3 units and if interpolation was required, to maintain 10-unit accuracy in the last digit it had to be at least quadratic.

As the 1970s gave way to the 1980s, the decade saw two final, large scale tabulation events for quantities related to blackbody radiation, with each making it into book form. The first of these appeared in China in 1984 [735]. Entitled *Table of Values for Blackbody Radiation* it consisted of eight separate tables and ran to 267 pages. In Table 1 the total radiant exitance and its temperature derivative, and the total photon exitance, as functions of temperature from 50 to 20 000 K for five different interval sizes were given. In Table 2 the integrated fractional amount and the integrated fractional photon exitance amount as functions of the wavelength–temperature product from 200 to 150 000 μm·K at four different interval sizes were tabulated. Table 3 gave the fractional amount of the second kind as a function of the wavelength–temperature product from 500 to 8000 μm·K at 10 μm·K intervals. Table 4 was a simple table for the conversion between wavelength in micrometers and wavenumber per centimeters.

Table 5 was a bit different in that it was divided into two parts, A and B with Part A containing the bulk of the tabulations found in the book. Running for a total of 151 pages, tabulations for the spectral radiant exitance and the integrated amount were given as a function of wavelength ranging from 0.76 to 25.00 μm for seventy-eight different temperatures between 77.36 to 20 000 K. At each different temperature, corresponding values for the peak wavelength, spectral radiant exitance at the peak wavelength, and total radiant exitance were also given. In Part B quantities relating to luminosity were found. Table 6 gave the temperature derivative of the integrated amount as a function of wavelength from 1.00 to 25.00 μm at 0.25 μm intervals, and temperature from 240 to 340 K at 2 K intervals. Table 7 gave the spectral photon exitance and its corresponding integrated amount for wavelengths ranging from 0.25 to 50.00 μm at three different interval sizes for eight different temperatures ranging from 273 to 6000 K. As was found in Table 5A, values for the corresponding peak wavelength, spectral radiant exitance at the peak, and total photon exitance at each different temperature were also given. Table 8, the last of the tables, contained a short tabulation for the spectral photon exitance and its corresponding integrated amount in the linear wavenumber representation as a function of wavenumber from 10 to 250 cm^{-1} at 10 cm^{-1} intervals for eight different temperatures. Values in each table were given to five significant figures. Values used for the two radiation constants of $3.741\,832 \times 10^4$ W·μm^3·cm^{-1}

and $1.438\,786 \times 10^4\,\mu\text{m·K}$ were the 1973 least-squares adjusted values of the fundamental constants given by E. Richard Cohen and Barry N. Taylor [143]. No details concerning how the calculations were made are given. In the West the text appears to have been largely unknown, while within China itself it was modestly influential. It was reviewed in the Chinese literature [725] and referred to by a number of groups working in the infrared community within the country [166, 718, 310, 165]. One suspects tabulations from the West were simply not available in China at the time, other than the tables prepared possibly by Bramson in their original Russian form, as these were occasionally referred to by Chinese authors [166].

The second, and what would turn out to be the last of all the books to be devoted entirely to the tabulation of quantities related to blackbody radiation, came from the Polish engineer Aleksander Sala [581]. Sala worked at the *Instytutu Mechaniki Precyzyjnej* (Institute of Precision Mechanics) in Warsaw, Poland, and his text *Radiant Properties of Materials: Tables of Radiant Values for Black Body and Real Materials* represented the culmination of more than twenty years of work on the radiative properties of materials. Up until that time Sala felt that the literature, both in Poland and elsewhere, lacked a single publication containing a collection of tables devoted entirely to the radiative properties of materials. The book was divided into three parts with the third part constituting the bulk of the book. Before Sala could present tables related to the radiative properties of real materials in the final part, the first and second parts contained introductory material. Part 1 summarised a number of concepts and definitions needed in order to deal with radiating objects and their thermal emission, while Part 2 presented two tables relating to the tabulation of various quantities relating to blackbody radiation.

In the first of Sala's tables, found in the second part of his text, three quantities are given as a function of temperature. They are the spectral radiant exitance in the linear wavelength representation evaluated at its peak value in units of watts per square meter per micrometer $[\text{W·m}^{-2}\text{·}\mu\text{m}^{-1}]$, total radiant exitance in units of watts per square meter $[\text{W·m}^{-2}]$, and values for the wavelength at the spectral peak in units of micrometers $[\mu\text{m}]$. Temperatures ranging from 300 to $5290\,\text{K}$ at $10\,\text{K}$ steps are selected while each tabulated quantity is given, correct to five significant figures.

In Sala's second, and by far larger table, the spectral radiant exitance in units of watts per square meter per micrometer $[\text{W·m}^{-2}\text{·}\mu\text{m}^{-1}]$ and its integrated amount from zero up to some arbitrary wavelength in units of watts per square meter $[\text{W·m}^{-2}]$ are tabulated, correct to five significant figures. Each quantity is tabulated as a function of wavelength ranging from 0.4 to $15.4\,\mu\text{m}$ at steps of $0.1\,\mu\text{m}$ for fifty-six values of the temperature: $250(50)1000(100)5000\,\text{K}$. At each selected temperature the tabulations filled two tight pages. Fifty-six different values for the temperature meant that the second of Sala's tables ran to an impressive 112 pages.

It is unclear what values for the first and second radiation constants and the Stefan–Boltzmann constant Sala actually used in his tabulations. Values one finds in his text for these constants are 3.74×10^{-16} W·m^2, 1.4388×10^{-2} m·K, and 5.6703×10^{-8} W·cm^{-2}·K^{-4}, respectively. A value of 2887.8 µm·K for Wien's displacement constant also appears, so one is lead to suspect these values were used. Neither is any information concerning how the tabulations were performed given, but as the work was the product of the mid-1980s, one suspects a digital computer of some sort was used.

Rather than being professionally typeset Sala's text was typewritten, a common recourse used at the time for inexpensive publications. But Sala's text was hardly inexpensive. As a hardcover reference book that sold for well over $100 at the time [608], it caused one reviewer to remark that besides a number of Greek letters and mathematical symbols appearing deformed and unclean, material presented in this way never quite seemed trustworthy [566]. The text was mainly cited in the literature for the tables it contained in its third part rather than for those found in the second part relating to radiation from a blackbody, most likely a reflection of the widespread availability of digital computers by the mid-1980s, meaning that the need for such things was by then largely in retreat.

7.5 TABLES THEN AND NOW

Of all the tables produced, only two attempts seem to have been made to compare the uncertainties in the various tabular entries. As we saw earlier (see page 284), the first to make any such comparison was Roberts in 1971 [564]. The second was made by Y. Göğüş and J. Kestin in 1976 [259]. Although rarely acknowledged by authors responsible for the tabulations, uncertainties in the tabulated values arose from uncertainties in the values used for the fundamental constants. Göğüş and Kestin, using the recently re-assessed values for the fundamental constants of 1969 [641], limited their attention to a comparison between just three tabulations — those made by Pivovonsky and Nagel (1961), Apanasevich and Aizenshtadt (1964), and those to be discussed shortly by Siegel and Howell (1968). By comparing the tabulated values given for the spectral radiant exitance, divided by the fifth power of the temperature, Göğüş and Kestin concluded the tables of Pivovonsky and Nagel should not be used for very accurate work, while the tables of Apanasevich and Aizenshtadt, and Siegel and Howell could be used. Göğüş and Kestin proceeded to give concise tables for the spectral radiant exitance divided by the fifth power of the temperature and its corresponding integrated fractional amount as functions of the wavelength–temperature product, the spectral radiant exitance in the linear frequency scale divided by the third power of temperature and its corresponding integrated fractional amount as functions of the ratio of frequency to temperature, and values for the maximum wavelength in the linear wavelength representation and the maximum frequency in the linear frequency representation as functions of temperature to very high accuracy. Next to each tabulated value in brackets appeared the associated uncertainty that was imposed by the uncertainties in the values for the fundamental constants used.

By and large, authors making use of tables in their work tended to use a single source only. Occasionally, no one table was sufficient to meet the needs of a particular task, notably so before the appearance of the great tabulations of the 1960s. An example where multiple tables were consulted can be found in the work by William F. Meggers, Charles H. Corliss and Bourdon F. Scribner of 1961 [438]. As part of their work on compiling tables of spectral line intensities, they required values for the spectral radiant exitance of a blackbody at temperatures of approximately 2800 K for wavelengths between 2000 and 9000. As their work came at the moment just before major tabulations for this function over a broad range of wavelengths started to appear, they instead resorted to three different sets of tabulations. Those by Stair and Smith (1943) were used for wavelengths in the 2300 to 3500 range, by Skogland (1929) in the 3200 to 7600 range, and by Lowan and Blanch (1940) in the 7200 to 10 000 range.

The advent and spread of high speed computers at first did not completely remove the need for tables. Many smaller compilations continued to appear throughout this period, often being found in more general handbooks of tables or accompanying specialized texts where the need for such quantities arose. As examples of the former, very modest tables for a range of quantities related to blackbody radiation can be found in Manfred von Ardenne's *Tabellen der Elektronenphysik, Ionenphysik und Übermikroskopie* of 1956 [658], Milton Abramowitz and Irrene A. Stegun's widely acclaimed *Handbook of Mathematical Functions* of 1964 [36], Hans-Werner Drawin and Paul Felenbok's *Data for Plasmas in Local Thermodynamic Equilibrium* of 1965 [191], and George W. C. Kaye and T. H. Laby's *Tables of Physical and Chemical Constants and Some Mathematical Functions* of 1973 [344]. Interestingly the last text, which by 1973 had been in print for over 60 years, even though a section on blackbody radiation had appeared in the very first edition of 1911 [343], it was not until the fourteenth edition of 1973 tabulations relating to blackbody radiation first appeared [257].

In more specialized texts, tables were given to support the material under consideration. Relatively extensive tables for the spectral radiant exitance as a function of wavelength and temperature are to be found in Werner Pepperhoff's *Temperaturstrahlung* of 1956 [495] and in Günter Wyszecki and Walter S. Stiles' *Color Science* of 1967 [717], a very important source of information in its field which, through a second edition, continues to remain in print to this very day. Another extensive tabulation in dimensionless form of the ratio of the spectral radiant exitance relative to its maximum value as a function of the ratio of wavelength relative to its peak value can be found in Joachim Euler and Rudolf Ludwig's *Arbeitsmethoden der optischen Pyrometrie* of 1960 [217].

Nor did the advent and widespread availability of unparalleled computing power see an end to the table making enterprise. As late as 1983, a short set of tables for the spectral radiant exitance and integrated fractional amount as a function of the wavelength–temperature product appeared in the *Journal of the Illuminating Engineering Institute of Japan* [317]. Similar sets of tables appeared in China. Prepared by Deng Jie Zhang, at first the ratio of the spectral radiant exitance to its value at its spectral peak and the integrated fractional amount as a function of the wavelength–temperature product were given to five significant figures [732], then six significant figures [733], using the CODATA 1973 recommended values for the two radiation constants, before updating these some years later [400] using the CODATA 1986 recommended values.

More recently a short table of what amounts to the fractional function of the first kind for the actinometric case as a function of the wavelength–temperature product to six significant figures was given by Wei Jing and Tiegang Fang in the *Journal of Heat Transfer* [333]. The authors claimed to be calculating a new blackbody fractional function $\mathfrak{F}_{0\to\lambda}^*$ defined by

$$
\mathfrak{F}_{0\to\lambda}^* = \frac{\displaystyle\int_0^\lambda \lambda M(\lambda, T)d\lambda}{\displaystyle\int_0^\infty \lambda M(\lambda, T)d\lambda}, \tag{7.33}
$$

but seem to have been unaware their "new" fractional function reduces to $\mathfrak{F}_{q,0\to\lambda}$, the fractional amount for the actinometric case, a fact duly pointed out by one of the authors [632].

Today the greatest number of tabulations for blackbody radiation remain those that continue to be found in texts on radiative heat transfer. Extensive tabulations for the spectral radiant exitance divided by the fifth power of temperature and the integrated fractional amount as a function of the wavelength–temperature product are to be found in the earlier editions of Robert Siegel and John R. Howell's *Thermal Radiation Heat Transfer* [600, 601, 602]. Found in the book's appendices, in the earlier editions of the text seven to eight pages were typically devoted to tabulations of this type. Starting with the fourth edition and continuing into the fifth, these tabulations were sadly wilted down to a mere one and a half pages [603, 309]. Extensive tabulations for spectral radiant exitance in the linear wavelength scale divided by the fifth power of temperature and the integrated fractional amount as functions of the wavelength–temperature product, and the spectral radiant exitance in the linear wavenumber scale divided by the third power of temperature as a function of the ratio of wavenumber to temperature are also to be found in the three editions of Michael F. Modest's text *Radiative Heat Transfer* [451, 452, 453]. Examples of other smaller tabulations, mainly of the spectral radiant exitance and its corresponding integrated fractional amount as functions of the wavelength–temperature product, can be found in a number of recent texts [499, 101, 449, 326, 57, 79, 362, 541, 315].

The enduring inclusion of tables like these in modern day texts led Robert J. Ribando and Edward A. Weller to question why authors continued to feel there was something sacrosanct about blackbody radiation functions [558]. Rather disparagingly, they wrote that despite the fact someone probably received a higher degree 50 or 60 years ago for creating such tables, and indeed they did in at least two instances [266, 140], replicating tables of this sort had become trivial at the time of writing in 1999. Against such criticisms, tables have proved to be incredibly versatile and endlessly adaptable in an age of limitless computing power.

For example, a completely new table of a type not before seen appeared in K. G. Terry Hollands' 2004 text *Thermal Radiation* [295]. The first table Hollands presents had been given many times before, namely the integrated fractional amount as a function of the wavelength–temperature product. His second table, however, is completely new. Given is the wavelength–temperature product as a function of the integrated fractional amount — the inverse of the first. Hollands writes that knowing the inverse is just as important and useful for certain calculations as knowing the function itself. As an example of its use, the so-called median wavelength $\lambda_{\bar{p}}$, which was introduced on page 80 and is the wavelength which divides the total radiant exitance into two equal halves, can be readily found from the second of Hollands' tables. For the median wavelength, since $\mathfrak{F}_{e,0\to\lambda} = 0.5$, from the table $\lambda_{\bar{p}}T = 4.1072 \times 10^{-3}\,\text{m·K}$ so that for any temperature T

$$\lambda_{\bar{p}} = \frac{4.1072 \times 10^{-3}}{T}. \tag{7.34}$$

And so the life of the table no longer flickers as brightly as it once had. A once important calculational aid, while its role today may have diminished to all but a few niche areas, the table is yet to be completely forgotten.

Notes

[1] Neither Rubens nor Kurlbaum actually gave the talk. Instead it was delivered by Friedrich Kohlrausch, then President of the Physikalisch–Technische Reichsanstalt in Charlottenburg where Kurlbaum was working at the time.

[2] The 2014 adjusted value recommended by CODATA for the first radiation constant for spectral radiance is $c_{1L} = 1.191\,042\,953 \times 10^{-16}\,\text{W·m}^2\text{·sr}^{-1}$.

[3] As done in a manner considered in Section 3.2.2. See Eq. (3.74) on page 118.

[4] Yamauti and Okamatu used the symbol E for what today would be M.

[5] Today we would identify the number β as being equal to $5 + W_0(-5e^{-5})$ where $W_0(x)$ is the principal branch of the Lambert W function. See page 43 where this is discussed.

[6] In closed form the second of Yamauti and Okamatu's mathematical constant's K, in terms of the Lambert W function, has a value given by

$$K = \beta^{-5}(e^{\beta} - 1) = -\frac{[5 + W_0(-5e^{-5})]}{W_0(-5e^{-5})} = 0.047\,167\ldots \tag{7.35}$$

[7] The Laplace–Everett formula is an interpolation formula used for estimating the value of a function at an intermediate value of the independent variable when its value is known at a series of equally spaced points by employing central differences of the function of even

order and coefficients which are polynomial functions of the independent variable. It is so named after their independent co-discoverers, the French mathematician Pierre-Simon Laplace (1749–1827) in 1782 and the English physicist Joseph David Everett (1831–1904) in 1900 [391].

[8]In the astronomical literature, the derivative of the spectral radiant exitance of a blackbody with respect to temperature is related to the *Rosseland function*. Named after the Norwegian astrophysicist Svein Rosseland (1894–1985) who first introduced and considered it in 1924 [568], it is given by

$$\frac{\partial M_{e,\nu}^{b}}{\partial T} = 4\sigma T^3 \cdot \frac{15}{4\pi^4} \frac{x^4 e^x}{(e^x - 1)^2} = 4\sigma T^3 R(x).$$

Here $R(x)$ is the Rosseland function while x corresponds to the dimensionless reduced frequency of $h\nu/(k_B T)$.

[9]This logarithmic corrected approximation to Wien's law was discussed in Section 3.1.3.

[10]A Langley [ly] is a unit of energy per unit area commonly employed by meteorologists in the measurement of solar radiation. One Langley corresponds to 11.622 watt-hours per square meter, so that $1\,\text{ly/min} = 697.32\,\text{W·m}^{-2}$ and $1\,\text{ly/day} = 1.0041408 \times 10^6\,\text{W·m}^{-2}$.

[11]Blunt considered the visible portion of the electromagnetic spectrum to lie between the wavelength range of 0.40 to 0.75 μm.

[12]In German it was entitled *Tabellen der Bruchteilfunktionen zum Planckschen Strahlungsgesetz.*

[13]In German its title was: *Siebenstellige Tabellen der Planck-Funktion für den sichtbaren Spektralbereich.*

[14]In German it went by the title: *Tabellen der spektralen Strahldichte des schwarzen Körpers* while in English: *Tables of Black Body Spectral Radiance.*

[15]We have been unable to determine what an EX 10 subroutine was.

8 Beyond the analogue aids of old

What started out in the early 1950s with a handful of large and extremely expensive electronic computers which only a small number of universities and institutions could afford to purchase and run, led ultimately to the digital revolution, and with it the demise in the nomogram, the slide rule, and tables as the computational aids of choice among practitioners. Shortly after its arrival in the mid-1940s, due to its ability to perform error-free computations at high speed, the significance of the digital computer as an important research tool was quickly recognised [271]. While cardboard slide charts made by Perrygraf for Teledyne Judson Technologies and BAE Systems, together with brief sets of tables in a number of texts have managed to survive to the present day, today it is the digital computer or similar electronic devices that are almost universally employed when it comes to calculations related to thermal radiation.

The spread of large mainframe computers in the 1960s allowed for the computation of tables for large, well resourced, and often well funded projects, but their prohibitively high cost meant that for a time individual access to the machines was extremely rare. For many people, it would not be until the mid to late 1970s that access to hand-held programmable calculators or digital computers became, for the very first time, an affordable reality. The dawning of the new digital age heralded the beginning of fast, accurate, computation on demand. Accompanying this change, a number of programs for several of the more widely used computer programming languages and devices began to appear. With functionality, accuracy, and speed far beyond anything an analogue aid could provide, the latter's fate was all but sealed as the emerging digital revolution swept all before it.

An early example of a computer program for calculating quantities associated with blackbody radiation was given by Donald C. Todd in 1968. As part of his report for a project sponsored by the Air Force Avionics Laboratory at Wright-Patterson Air Force Base in Ohio, Todd gave a subroutine for the evaluation of the integrated fractional amount in both energetic and photonic units [647]. The program was written in FORTRAN IV and calculated the fractional amounts to five significant figures. While Todd recognized that several useful aids such as slide rules and tables, capable of performing calculations of this type already existed, he felt his computer program would also be useful to others. It was, of course, true only for those who had access to the necessary computing resources needed, something many in the late 1960s still did not have.

By the early 1970s the situation had started to change. For the very first time affordable digital computing power became within reach for many with the arrival of hand-held scientific and programmable calculators. From the beginning, the intention of the scientific calculator was clear. Its immediate goal was to displace the more advanced slide rules available at the time. To do this, it had to be capable of performing all of the mathematical operations and function evaluations of the former. This intention was often reflected in the name chosen by some manufacturers for their first generation scientific calculators, namely that of a "slide rule calculator." It was not long before manufacturers of these new digital pocket calculators realized they could be made programmable, allowing users to calculate their own, user defined functions. These devices proved to be a real boon to those working in the infrared. With all this new-found computational power at one's fingertips, it was inevitable that these calculators would quickly supersede the nomograms, slide rules, and tables of old.

Some of the earliest hand-held programmable calculators pressed into service for calculating quantities related to blackbody radiation were the Hewlett-Packard 25 [697] and 65 [125, 306] and the Texas Instruments SR-52 [126, 697, 306] and SR-56 [697]. Each of the published programs for these programmable calculators were capable of calculating all quantities found on the blackbody radiation slide rule made for Electro Optical Industries, Inc., but to a much improved five, or sometimes seven, significant figure accuracy. No doubt many other programs for these early programmable calculators were written, but sadly it was all too easy for no permanent account of their existence to be left behind in the literature.

As access and availability to computers increased, one way that former users of the various analogue aids could now satisfy their computational needs was by writing simple programs of their own. In the early 1980s computer programs written in BASIC started to appear. In 1983, George W. Hopkins described what he called a "radiation slide rule program" written in BASIC. It too was capable of calculating all quantities found on the radiation slide rule made for Electro Optical Industries, Inc. [298]. Wolfe, a decade later, gave eleven programs written in BASIC that could not only calculate a number of different radiometric and actinometric quantities related to a blackbody, but also graph many of these quantities as well [700, 701]. The growing use of spreadsheets on desktop computers also become a popular and simple choice [700, 381, 382]. In time, descriptions for the calculation of various quantities relating to a blackbody using one of the more advanced mathematical software packages that had become available by the late 1980s, such as Mathematica [323] and MATLAB [427, 670, 308], were also given.

Michael F. Modest in the later editions of his text *Radiative Heat Transfer* provided two computer programs that could be used to calculate several quantities associated with blackbody radiation [452, 453]. The first of these programs calculated the fractional function of the first kind when the argument is $x = n\lambda T$ (here n corresponds to the refractive index of the medium). The second, for a given temperature and wavelength or wavenumber, calculated the fractional function of the first kind and the ratio of the spectral radiant exitance in either the linear wavelength representation to the fifth power of the absolute temperature, or the linear wavenumber representation to the third power of the absolute temperature. In the second edition of his text, programs written in FORTRAN and in C++ are given, while in the third edition programs written in MATLAB are also provided. Modest writes that all programs started out life as FORTRAN programs, but as a gesture towards the growing C++ and MATLAB communities, the simplest codes were first converted to C++ and later MATLAB. All programs could be downloaded free of charge from the book's dedicated website hosted by the publisher.

Mikron Infrared, Inc., a small company engaged in the development and manufacture of equipment used in non-contact temperature measurements, released a Windows-based program in 1997 called the Mikron Blackbody Radiation Calculator. Once installed, the calculator could compute and graph up to five spectral radiant exitance curves as a function of either wavelength, frequency, or wavenumber at once for a given temperature and emissivity. At each selected temperature, values for the peak wavelength, spectral radiant exitance at the peak wavelength, total radiant exitance, and total radiance were given to three significant figures — a number significantly less than the number that could be found using tables produced two decades earlier. All selected temperatures had to lie between 1×10^{-12} and 1×10^{15} K. The calculator could also calculate the amount of energy radiated by a blackbody or greybody into a given spectral band, and it was also possible to plot the normalized integrated fractional amount. These in turn could be either saved or printed out. The program could be obtained free of charge on request and was sent to interested users as an e-mail attachment.

Of a more ephemeral nature are those calculators which have appeared from time to time online. The *Blackbody Emission Calculator* found on the Electro Optical Industries, Inc. webpage [210], is one such calculator. It is a very simple calculator. For a given temperature in Kelvin, the spectral radiant exitance as a function of wavelength in micrometers is plotted and values for the peak wavelength and the spectral radiant exitance at the peak are given. Another simple calculator can be found at the website CalcTool [597]. The calculator, called the *Black Body Emission Max*, calculates the peak wavelength found in the spectral distribution curve within the linear wavelength representation at a given temperature. The temperature needed to produce a given peak value in wavelength can also be found.

An *Infrared Radiance Calculator* is hosted on the Military Sensing and Information Analysis Center (SENSIAC) webpage [594]. The calculator was created by Anthony LaRocca [378]. After selecting a temperature in Kelvin and upper and lower values for the wavelength in micrometers, on hitting "calculate" the following six quantities are calculated: total radiance, spectral radiance at the lower and upper values for the two wavelengths entered, peak wavelength, band radiance, and temperature derivative of the in-band radiance between the selected wavelength limits. All values are given to either three or four significant figures.

The most complete calculator currently found online is without doubt the *Blackbody Calculator* by SpectralCalc.com [245]. For a selected temperature in either units of Kelvin [K], Celsius [°C], or Fahrenheit [°F], the peak wavelength, spectral radiant exitance at the peak wavelength, total radiant exitance, and total radiance are all displayed in one panel. In a second panel below the first, for a selected wavelength, a value for the spectral radiance is given. In the third and final window, the in-band radiance between selected lower and upper wavelength limits is given. On pressing the "calculate" button, all values for those quantities listed above are given to six significant figures. A plot for the spectral radiance between the wavelength limits chosen also appears. All values can be given in either energetic or photonic units using one of the three linear spectral represents of wavelength, frequency, or wavenumber. The calculation for the in-band radiance is implemented in C++ using the two infinite series expansion forms given by Eqs (3.49) and (3.51) [615].

A number of calculator programs have been written for some of the more popular hand-held electronic devices which have appeared since the turn of the millennium. One such example was Black Body v1.0 for the Palm Pilot which was released as freeware on 5 August 2001 by David S. Flynn [226]. Flynn, who was a PhD student working in the infrared at the time, had been asked by his government sponsors, the Air Force Research Laboratory at Eglin Air Force Base in Florida, to write a blackbody program for them using CplxCalPro — the first programmable graphics calculator for the Palm platform. At a selected temperature in Kelvin the calculator gave the total radiant exitance, the maximum wavelength at the spectral peak, and the corresponding value for the spectral radiance at the peak value. The spectral radiance as a function of wavelength was also plotted and the in-band radiance could be calculated between two selected wavelengths. The program also gave an average for the in-band radiance corresponding to a rectangle waveband at the lower and upper wavelengths selected.

Most recently, with the widespread availability of smart-phones and their ability to run the now ubiquitous "app," a number of applications specific to the calculations of various quantities associated with blackbody radiation have been released for the two most dominant smart-phone platforms — Android and the iPhone. *CalRadiance* for the iPhone by Frederic Alves was released on 10 March 2013 [46]. On release it was listed as version 1.2.0, was 1.43MB in size, required iOS 6.1 or greater to run, was claimed to be the first app of its type for the iPhone, and was free. In appearance it takes the form of a very simple calculator, yet is an incredibly powerful little app. In addition to the ten numeric keys, the decimal point key, and the equal sign key, three orange keys running down the left side appear. These are, from top to bottom, "Temp" for temperature, λ_1 for lower wavelength, and λ_2 for upper wavelength. On entering values for each of these three variables the following quantities are displayed in a panel above the key pad: total radiance, in-band radiance, spectral radiance at λ_1, spectral radiance at λ_2, and the maximum wavelength at the spectral peak. All values are given to six significant figures. In the settings menu for the calculator, one can select between energetic or photonic units, normal mode or derivative mode where values for each quantity are given with respect to their temperature derivative, the unit for wavelength in either micrometers or nanometers, and the unit for temperature in either Kelvin or Celsius. Only quantities in the linear wavelength representation are given. Sometime later, version 1.3.1 appeared, which fixed a small number of bugs and changed the color scheme used for the graphical user interface of the calculator, but since early 2014 it appears to be no longer available.

A second app for Android-based smart-phones going by the name *Blackbody Radiation Calculator* was published on 15 December 2013 [496]. Written by Steve Peters, the app is free, is 895kB in size and requires Android 2.12 or higher in order to run. It is a simple calculator which attempts to implement some of the features of the Infrared Radiation Calculator slide charts of old. How the app came about is interesting. In 2012, after having picked up a free Infrared Radiation Calculator from BAE Systems who was an exhibitor at the *Military Sensing Symposium Specialty Group on Electro-Optical and Infrared Countermeasures* he was attending, Peters thought no more of the rule and placed it in the bottom of his desk drawer where it stayed for the best part of a year [497].

Interested in developing a calculator app, and looking for something unique, the idea occurred to Peters of implementing the Infrared Radiation Calculator in a modern digital guise as a smart-phone application. Peters reasoned some people out in the field may occasionally have a need for such an application, particularly when one does not have ready access to a computer. Being unsure of the level of interest for such a thing, he begun by keeping it simple and decided not to implement all the features of the Infrared Radiation Calculator.

On launching the app one is presented with five options. The first provides information on the symbols and nomenclature used, while the second provides background material on the theory of a blackbody. The third is the first of the calculators. Selecting a temperature (in either units of Kelvin, Celsius, or Fahrenheit), and wavelength in micrometers, the following five quantities are calculated: spectral radiance, spectral radiant exitance, spectral photon radiance, spectral photon exitance, and the peak wavelength at the spectral maximum. All quantities are calculated to five significant figure accuracy. The fourth option opens a second calculator. On entering a value for the temperature the total radiant exitance and the total photon exitance to five significant figures is calculated. The fifth option gives a listing of emissivities for a number of materials at various different temperatures. While Peters had originally planned to add the capability to perform in-band spectral radiant exitance calculations, it is currently lacking. Features such as this plus several others may be added at a future date if demand from users justifies their addition.

A final app for iOS based systems was released on 16 June 2014 by Burton S. Chambers III from Two Tori LLC [122]. Known as a *Blackbody Tool* it is 700kB in size and sells for $2.99. Like Peters, Chambers was familiar with the Infrared Radiation Calculator and reasoned that developing such a tool for one's smart-phone would occasionally be helpful to the working engineer or scientist who needs to perform a quick calculation but does not necessarily have access to a computer or the internet [121]. Unlike Peters, Chambers' app is fully featured.

In operation, four different values need to be entered. These are the temperature (in units of either Kelvin, Celsius, Fahrenheit, or Rankine), the wavelength (in either micrometers or nanometers), and lower and upper wavelength limits for the selected spectral band of interest. Upon pressing the "calculate" button, the following six quantities are calculated: total radiance, total radiant exitance, spectral radiance, wavelength at the spectral peak, the value for the spectral radiance at the spectral peak, and the in-band radiance. The equivalent six actinometric quantities in photonic units can also be calculated. In addition to the linear wavelength representation, linear wavenumber and linear frequency spectral representations can be selected. All calculated quantities are given to four significant figures, a minor drawback of the application. A plot of the spectral radiance for the spectral band selected can also be made. The app has been updated twice since the initial release of version 1.0. Version 1.1 appeared on 6 July 2014 while the most recent version at the time of writing, version 1.2, appeared on 10 November 2014.

In a recent book by Cornelius J. Willers, he presented from a radiometric perspective the design and analysis of a variety of electro-optical system [737]. The utilization of blackbody radiation in such systems is discussed in detail and in *Appendix D* of Willers' book, a variety of computer programs are included in Matlab® and Python® that support radiometric calculations.[1,2]

It is almost certain that these latest digital incarnations will not be the last. For as long as the blackbody remains of central importance to the study of thermal radiation, there will always be a need to calculate quantities associated with it. As we head into the future, how these calculations will continue to be made, and on what devices, only time will tell.

Notes

[1] https://github.com/NelisW/pyradi/blob/master/pyradi/ryplanck.py

[2] http://nbviewer.jupyter.org/github/Nelisw/ComputationalRadiometry/blob/master/07-Optical-Sources.ipynb#Planck-law-technical-implementional-detail

A Miscellaneous mathematical results

Proofs to several of the more important mathematical results stated without proof in the text are given here.

A.1   SERIES EXPANSION FOR THE LAMBERT W FUNCTION

In this appendix a series expansion for the Lambert W function about the origin will be developed using the Lagrange inversion theorem. This theorem allows for the compositional inverse of a formal power series to be found. Informally, if the series expansion for a function $y(x)$ is known, the series for x in terms of y can be found by inverting the series in a process referred to as series reversion.

For the case of the Lambert W function, as the inverse for this function is relatively simple, the Lagrange inversion theorem provides a particularly simple and elegant way of finding a series expansion for the principal branch of the Lambert W function about the origin. But first we begin with a short statement of the inversion theorem.

Theorem A.1: The Lagrange inversion theorem

Suppose the dependence between the variables w and z is implicitly defined by an equation of the form

$$f(w) = z,$$

where f is analytic at the point $w = a$ and $f'(a) \neq 0$. Then it is possible to invert the equation for w

$$w = g(z),$$

in the neighborhood of $b = f(a)$, where g is analytic at the point $f(a)$. The series expansion of g about the point $w = a$ is given by

$$g(z) = a + \sum_{n=1}^{\infty} \left[\frac{d^{n-1}}{dw^{n-1}} \left(\frac{w - a}{f(w) - f(a)} \right)^n \right]_{w=a} \frac{(z - f(a))^n}{n!}. \qquad (A.1)$$

■

When applied to the Lambert W function, a slightly simpler form resulting from a special case for the Lagrange inversion theorem can be applied. If

$z = f(w) = w/\phi(w)$ and $\phi(0) \neq 0$, taking $a = 0$ gives $b = f(0) = 0$ and Lagrange's inversion theorem reduces to

$$g(z) = \sum_{n=1}^{\infty} \left[\frac{d^{n-1}}{dw^{n-1}} \phi(w)^n \right]_{w=0} \frac{z^n}{n!}. \tag{A.2}$$

As the Lambert W function is the inverse of the function $f(w) = w e^w = w/e^{-w}$, we observe that $\phi(w) = e^{-w}$. Noting that $\phi(0) = 1 \neq 0$ and

$$\left[\frac{d^{n-1}}{dw^{n-1}} \phi(w)^n \right]_{w=0} = \left[\frac{d^{n-1}}{dw^{n-1}} e^{-nw} \right]_{w=0} = (-n)^{n-1},$$

from the special case of Lagrange's inversion theorem the required power series expansion for the principal branch of the Lambert W function about the origin follows

$$W_0(z) = \sum_{n=1}^{\infty} \frac{(-n)^{n-1}}{n!} z^n. \tag{A.3}$$

The ratio test establishes that the series converges if $|z| < 1/e$.

A.2   BERNOULLI NUMBERS AND THE RIEMANN ZETA FUNCTION

When the argument of the Riemann zeta function is a positive even integer it can be written in closed form. The result is

$$\zeta(2n) = \frac{(-1)^{n+1}(2\pi)^{2n} B_{2n}}{2(2n)!}, \quad n \geqslant 1. \tag{A.4}$$

Here B_{2n} are the Bernoulli numbers. These numbers are among one of the most interesting and important number sequences in mathematics. They first appeared in the posthumous work *Ars Conjectandi* by the Swiss mathematician Jakob Bernoulli (1654–1705) in connection with sums of powers of consecutive integers. Published in 1713, the numbers are named after their discoverer. Several different proofs for Eq. (A.4) are known to exist. The proof we choose to give here makes use of a result coming from the theory of the Fourier series.

The *Bernoulli numbers* B_n are defined by the generating function [331]

$$\frac{t}{e^t - 1} = \sum_{n=0}^{\infty} B_n \frac{t^n}{n!}, \quad |t| < 2\pi. \tag{A.5}$$

They can be found by multiplying Eq. (A.5) by the Maclaurin series representation for the function $e^t - 1$, expanding the right-hand side, and equating equal powers of t on either side of the identity.

Performing the series multiplication, after collecting equal powers of t, yields

$$t = B_0 t + \left(B_1 + \frac{1}{2}B_0\right) t^2 + \left(\frac{1}{2}B_2 + \frac{1}{2}B_1 + \frac{1}{6}B_0\right) t^3$$
$$+ \left(\frac{1}{6}B_3 + \frac{1}{4}B_2 + \frac{1}{6}B_1 + \frac{1}{24}B_0\right) t^4 + \cdots .$$

Equating equal powers of t leads to the system of equations

$$B_0 = 1$$

$$B_1 + \frac{1}{2}B_0 = 0$$

$$\frac{1}{2}B_2 + \frac{1}{2}B_1 + \frac{1}{6}B_0 = 0$$

$$\frac{1}{6}B_3 + \frac{1}{4}B_2 + \frac{1}{6}B_1 + \frac{1}{24}B_0 = 0$$

$$\vdots$$

Starting with $B_0 = 1$, solving recursively, these equations generate all higher order Bernoulli numbers. The first few are:

$$B_0 = 1, \; B_1 = -\frac{1}{2}, \; B_2 = \frac{1}{6}, \; B_3 = 0, \; B_4 = -\frac{1}{30}, \; B_5 = 0, \; B_6 = \frac{1}{42}, \; \ldots$$

The Bernoulli numbers form a sequence of rational numbers. Apart from B_1, all other odd order Bernoulli numbers are zero. To show this is the case, for all odd integer orders greater than or equal to $2n+1$ where $n = 1, 2, 3, \ldots$, if t is replaced with $-t$ in Eq. (A.5) and the result of the latter subtracted from the former, one finds

$$\frac{t}{e^t - 1} - \frac{-t}{e^{-t} - 1} = -t = \sum_{n=0}^{\infty} [1 - (-1)^n] B_n \frac{t^n}{n!}.$$

On equating equal powers for t, we see that

$$B_1 = -\frac{1}{2}$$

$$B_{2n+1} = 0, \; n = 1, 2, 3, \ldots$$

If on the other hand, after replacing t with $-t$ in Eq. (A.5) the result of the latter is added to the former, one finds

$$\frac{t}{e^t - 1} + \frac{-t}{e^{-t} - 1} = t \coth \frac{t}{2} = \sum_{n=0}^{\infty} [(1 + (-1)^n] B_n \frac{t^n}{n!}. \tag{A.6}$$

As

$$1 + (-1)^n = \begin{cases} 2 & : \quad n \text{ even} \\ 0 & : \quad n \text{ odd} \end{cases}$$

shifting the summation index from n to $2n$ and setting $z = t/2$, one has

$$\coth z = \frac{1}{z} + \sum_{n=1}^{\infty} \frac{2^{2n} B_{2n}}{(2n)!} z^{2n-1}, \quad 0 < |z| < \pi, \tag{A.7}$$

the Maclaurin series for the hyperbolic cotangent function. Recognising

$$\cot z = i \coth iz,$$

where i is the imaginary unit, the Maclaurin series for the cotangent function quickly follows. Replacing z with iz in Eq. (A.7), after taking care of the imaginary term $i^{2n-1} = (-1)^{n-1}i$ which arises, one finds

$$\cot z = \frac{1}{z} + \sum_{n=1}^{\infty} \frac{(-1)^n 2^{2n} B_{2n}}{(2n)!} z^{2n-1}, \quad 0 < |z| < \pi. \tag{A.8}$$

We will return to Eq. (A.8) in a moment. For now we find an alternative series expansion for the cotangent function which will be obtained from a Fourier series expansion for a particular periodic function.

Let $\alpha \in \mathbb{R} \backslash \mathbb{Z}$ and consider the periodic function

$$f(x) = \cos \alpha x, \quad -\pi \leqslant x \leqslant \pi$$
$$f(x) = f(x + 2\pi), \quad x \in \mathbb{R}$$

We note the function $f(x)$ will be continuous at π as $\cos \alpha x$ is even but will not be differentiable at this point. Expanding $f(x)$ as a Fourier series on the interval $[\pi, \pi]$, as the function is even, the sine coefficient will obviously be equal to zero. For the cosine coefficient

$$a_n = \frac{1}{\pi} \int_{-\pi}^{\pi} \cos \alpha \cos nx \, dx. \tag{A.9}$$

Using the prosthaphaeresis formula which relates products of two trigonometric functions to the sum or difference between them, the integral appearing in Eq. (A.9) can be readily evaluated. The result is

$$a_n = \frac{1}{\pi} \int_{-\pi}^{\pi} \frac{1}{2} [\cos(\alpha - n)x + \cos(\alpha + n)x] dx = \frac{(-1)^n}{\pi} \frac{2\alpha}{\alpha^2 - n^2} \sin \pi \alpha, \ n \geqslant 0. \tag{A.10}$$

From the general Fourier series expansion for a periodic function with period 2π

$$f(x) \sim \frac{a_0}{2} + \sum_{n=1}^{\infty} [a_n \cos(nx) + b_n \sin(nx)], \quad -\pi \leqslant x \leqslant \pi. \tag{A.11}$$

The tilde allows for equality between the left and right-hand sides, in addition to points where the function is continuous, at those points in the function on the interval where finite jump discontinuities occur in accordance with the Dirichlet condition. As the cosine function is everywhere continuous on $[-\pi, \pi]$ one has

$$\cos \alpha x = \frac{\alpha \sin \pi \alpha}{\pi} \left(\frac{1}{\alpha^2} + \sum_{n=1}^{\infty} \frac{2(-1)^n}{\alpha^2 - n^2} \cos nx \right), \quad -\pi \leqslant x \leqslant \pi. \quad \text{(A.12)}$$

Setting $x = \pi$ and $\alpha = z$ in Eq. (A.12), after rearranging we are led to the extraordinary formula

$$\pi z \cot \pi z = 1 + 2 \sum_{n=1}^{\infty} \frac{z^2}{z^2 - n^2}. \quad \text{(A.13)}$$

If the summand appearing in Eq. (A.13) is rewritten as

$$\frac{z^2}{z^2 - n^2} = -\frac{z^2}{n^2} \frac{1}{1 - z^2/n^2},$$

the infinite geometric series

$$\frac{1}{1 - u} = \sum_{n=0}^{\infty} u^n, \quad |u| < 1,$$

allows one to write, after a shift of the summation index

$$\frac{z^2}{z^2 - n^2} = -\sum_{k=1}^{\infty} \frac{z^{2k}}{n^{2k}}, \quad |z| < n. \quad \text{(A.14)}$$

The series in Eq. (A.13) can now be expressed as

$$\pi z \cot \pi z = 1 - 2 \sum_{n=1}^{\infty} \sum_{k=1}^{\infty} \frac{z^{2k}}{n^{2k}} = 1 - 2 \sum_{k=1}^{\infty} \left(\sum_{n=1}^{\infty} \frac{1}{n^{2k}} \right) z^{2k}, \quad \text{(A.15)}$$

after interchanging the order of the summations. Recognizing the summation term in brackets in Eq. (A.15) as the Riemann zeta function, the series becomes

$$\pi z \cot \pi z = 1 - 2 \sum_{n=1}^{\infty} \zeta(2n) z^{2n}, \quad \text{(A.16)}$$

where we have changed the summation index from k back to n again.

The change of variable from z to πz made in the Maclaurin series for the cotangent function as given by Eq. (A.8) produces

$$\pi z \cot \pi z = 1 + \sum_{n=1}^{\infty} \frac{(-1)^n (2\pi)^{2n} B_{2n}}{(2n)!} z^{2n}, \quad |z| < 1. \quad \text{(A.17)}$$

Comparing Eqs (A.16) and (A.17), on equating equal powers of z^{2n} for $n \geqslant 1$ yields

$$\zeta(2n) = \frac{(-1)^{n+1}(2\pi)^{2n}B_{2n}}{2(2n)!}, \quad n \geqslant 1,$$

and proves the result. For the particular case of $\zeta(4)$ used in the text, setting $n = 2$ in the above equation we find

$$\zeta(4) = -\frac{\pi^4}{3} \cdot B_4 = -\frac{\pi^4}{3} \cdot -\frac{1}{30} = \frac{\pi^4}{90}.$$

A.3   NUMBER OF REAL ROOTS TO AN EQUATION

In this appendix we show the number of real roots to the equation

$$x^\alpha - \alpha^4 x + \alpha^4 - 1 = 0, \quad x \geqslant 1 \tag{A.18}$$

is exactly two for all $\alpha > 1$.

Before proceeding with its proof we begin with a few preliminaries first. A well-known inequality for the logarithmic function is

$$\ln x < x - 1, \quad x > 1, \tag{A.19}$$

and follows trivially from the concavity of the log function. If x is replaced with $u+1$ in the above inequality, a second inequality involving the logarithmic function is

$$u > \ln(u + 1), \quad u > 0. \tag{A.20}$$

A third inequality which will be needed, and follows directly from Eq. (A.19), is

$$e^{nx} > e^n x^n, \quad n > 0, x > 1. \tag{A.21}$$

To see this, from Eq. (A.19) one has

$$\ln < x - 1$$
$$\ln x + 1 < x$$
$$\ln x + \ln e < x$$
$$\ln(ex) < x,$$

for $x > 1$. Multiplying the above result by n, where $n > 0$, gives $n \ln(ex) < nx$ or $\ln(ex)^n < nx$. As the exponential function is a monotonically increasing function of its argument, on exponentiating both sides of the inequality, the result immediately follows.

A bound for $x^{\frac{3}{x-1}}$ where $x > 1$ will now be found. From Eq. (A.19), as $\ln(x) > 0$ for all $x > 1$, we have

$$0 < \ln x < x - 1, \quad x > 1.$$

or after rearranging

$$0 < \frac{1}{x-1} \ln x < 1,$$

since the term $x - 1$ is positive for all $x > 1$. On multiplying both sides by three gives

$$0 < \ln\left(x^{\frac{3}{x-1}}\right) < 3, \quad x > 1. \tag{A.22}$$

Since the exponential function is a monotonic increasing function of its argument, on exponentiating both sides of the inequality given by Eq. (A.22) one finds

$$1 < x^{\frac{3}{x-1}} < e^3, \quad x > 1, \tag{A.23}$$

a bound for $x^{\frac{3}{x-1}}$ for all $x > 1$.

A second bound we require is

$$1 < 1 + \frac{1}{x} + \frac{1}{x^2} + \frac{1}{x^3} < 4, \quad x > 1. \tag{A.24}$$

To prove this result, let

$$h(x) = 1 + \frac{1}{x} + \frac{1}{x^2} + \frac{1}{x^3}, \quad x > 1.$$

As

$$h'(x) = -\left(\frac{1}{x^2} + \frac{2}{x^3} + \frac{3}{x^4}\right) < 0,$$

for all $x > 1$ we see that h is a decreasing function on its domain. Since $h(1) = 4$ and $\lim_{x \to \infty} h(x) = 1$ the bound as given by Eq. (A.24) immediately follows.

The last preliminary result we require is the following inequality

$$x^{\frac{3}{x-1}} > 1 + \frac{1}{x} + \frac{1}{x^2} + \frac{1}{x^3}, \quad x > 1. \tag{A.25}$$

A proof for this inequality is now given. As the natural logarithmic function is a monotonic increasing function of its argument, this is equivalent to showing

$$\ln x > \frac{1}{3}(x-1)\ln\left(1 + \frac{1}{x} + \frac{1}{x^2} + \frac{1}{x^3}\right), \quad x > 1 \tag{A.26}$$

From the second of the natural logarithmic inequalities given, namely Eq. (A.20), if we set $u = 1/x + 1/x^2 + 1/x^3$, from the bound given by Eq. (A.24) u is clearly greater than zero. Thus

$$\frac{1}{x} + \frac{1}{x^2} + \frac{1}{x^3} > \ln\left(1 + \frac{1}{x} + \frac{1}{x^2} + \frac{1}{x^3}\right), \quad x > 1. \tag{A.27}$$

So in proving Eq. (A.26) it suffices to show

$$\ln x > \frac{1}{3}(x-1)\left(\frac{1}{x} + \frac{1}{x^2} + \frac{1}{x^3}\right), \quad x > 1. \tag{A.28}$$

Now let $f(x) = \ln x$ and

$$g(x) = \frac{1}{3}(x-1)\left(\frac{1}{x} + \frac{1}{x^2} + \frac{1}{x^3}\right).$$

Observing that $f'(x) = 1/x$ and $g'(x) = 1/x^4$ such that

$$f'(x) - g'(x) = \frac{1}{x} - \frac{1}{x^4} = \frac{(x-1)(x^2+x+1)}{x^4} > 0,$$

for $x > 1$ we see that $f'(x) > g'(x)$ for all $x > 1$. Since f, g are differentiable on $\mathbb{R}$, $f(1) = g(1) = 0$ and $f'(x) > g'(x)$ for all $x > 1$, one can conclude $f(x) > g(x)$ for all $x > 1$ (this result is given, for example, in [68]) and hence the inequality given by Eq. (A.28) follows thereby proving Eq. (A.26)

We are now in a position to prove the main result. Let

$$f(x) = x^\alpha - \alpha^4 x + \alpha^4 - 1, \quad x \geqslant 1.$$

f is differentiable (and hence continuous) for all $x \geqslant 1$. Clearly, $f(1) = 0$ for all $\alpha > 1$, so $x = 1$ is a root to the equation. From the First Derivative Test for extrema, stationary points in f occur when $f'(x) = 0$. So

$$f'(x) = \alpha x^{\alpha-1} - \alpha^4 = 0 \tag{A.29}$$

yielding $x = \alpha^{\frac{3}{\alpha-1}}$ when x is real. The nature of the stationary point can be determined from the Second Derivative Test. Since $f''(x) = \alpha(\alpha-1)x^{\alpha-2}$, at the stationary point

$$f''\left(\alpha^{\frac{3}{\alpha-1}}\right) = \alpha(\alpha-1)\alpha^{\frac{3(\alpha-2)}{\alpha-1}} > 0, \quad \text{for } \alpha > 1, \tag{A.30}$$

and shows $x_{\min} = \alpha^{\frac{3}{\alpha-1}}$ is a local minimum. At the local minimum the value of the function f is

$$f\left(\alpha^{\frac{3}{\alpha-1}}\right) = -\frac{(\alpha-1)}{\alpha^3}\left[\alpha^{\frac{3}{\alpha-1}} - \left(1 + \frac{1}{\alpha} + \frac{1}{\alpha^2} + \frac{1}{\alpha^3}\right)\right] < 0, \quad \alpha > 1, \tag{A.31}$$

where use of the inequality given by Eq. (A.25) has been made. So the value of the function f at the local minimum is negative for all $\alpha > 1$.

Now as $f''(x) = \alpha(\alpha-1)x^{\alpha-2} > 0$ for all $x \geqslant 1$ and $\alpha > 1$, f is convexed. So $f(x)$ is increasing for $x > x_{\min}$ and therefore cannot have more than one root on the interval $[x_{\min}, \infty)$. From the bound given by Eq. (A.23) we see $\alpha^{\frac{3}{\alpha-1}} \in (1, e^3)$ for all $\alpha > 1$. Choosing a value for x larger than the largest possible value for $x_{\min}$, say $x = e^4$ for example, we have

$$f(e^4) = (e^{4\alpha} - e^4\alpha^4) + (\alpha-1)(\alpha+1)(\alpha^2+1) > 0,$$

for all $\alpha > 1$ where the inequality given by Eq. (A.21) with $n = 4$ has been used.

As $f(x_{\min}) < 0$ and $f(e^4) > 0$, by the Intermediate Value Theorem, there is at least one $c \in [x_{\min}, e^4]$ such that $f(c) = 0$. But it was just shown f as an increasing function can not have more than one root on the interval $[x_{\min}, \infty)$. Hence f has exactly one root on the interval $[x_{\min}, \infty)$ for $\alpha > 1$ and shows the equation $f(x) = 0$ for $x \geqslant 1$ has exactly two real roots for all $\alpha > 1$, as required to prove.

B Computer program for the fractional function of the first kind

The GNU Octave (version 3.2.4) program listed below calculates the fractional function of the first kind for the radiometric, actinometric, and eta cases. The program acts as a simple calculator. Here the user is prompted to enter the case type s (either 2, 3, or 4), a wavelength (in meters), and a temperature (in Kelvin) with a value for the corresponding fractional function of the first kind outputted. Values found are accurate to nine significant figures. A value of $1.438\,777\,36 \times 10^{-2}\,\text{m}\cdot\text{K}$ for the second radiation constant has been used.

```
function [c] = fnFractional(s,L,T),
format long
%
% This program calculates the blackbody fractional function
% of the first kind for either the radiometric, actinometric,
% or eta cases accurate to 9 significant figures.
%
% The octave-specfun package needs to be loaded before running
% the program. The octave-specfun package provides special
% mathematical functions for octave, including Riemann's zeta
% function and the Lambert W function. It is loaded using the
% command "pkg load specfun" once Octave is up and running.
%
% syntax
% s : Case type (radiometric, actinometric, eta)
% L : wavelength (in metres)
% T : temperature (in Kelvin)
%
%
% User input of case type (s), wavelength (L), and
% temperature (T) is required.
%
% For the three different cases considered we have:
% Radiometric: s = 4
% Actinometric: s = 3
% Eta: s = 2
```

```
% Ensuring the case type s is only one of three integer
% values between 2 and 4.
s = input('Case type, s: ');
if s~=floor(s)
    error(sprintf(['For case-type, only integer values\n'...
                    'between 2 and 4 allowed. Please\n'...
                    're-enter a correct value for s.']))
end
if s < 2
    error(sprintf(['For case-type, only integer values\n'...
                    'between 2 and 4 allowed. Please\n'...
                    're-enter a correct value for s.']));
elseif s > 4
    error(sprintf(['For case-type, only integer values\n'...
                    'between 2 and 4 allowed. Please\n'...
                    're-enter a correct value for s.']));
end
%
% User selected values for the wavelength (L) and temperature
  (T).
%
% The value for the wavelength (L) must be positive.
L = input('Wavelength (in metres): ');
if L < 0
    error(sprintf(['The wavelength must be positive.\n'...
                    'Please enter a positive value\n'...
                    'for the wavelength.']))
end
%
% The value for the temperature (T) must be positive.
T = input('Temperature (in Kelvin): ');
if T < 0
    error(sprintf(['The temperature must be positive.\n'...
                    'Please enter a positive value\n'...
                    'for the temperature.']))
end
%
% For 9 significant figure accuracy only the first 10 terms
% in either summation need to be summed. The division point
% between the two sums is taken at z = 2.0. As no more than
% the first 20 Bernoulli numbers are needed, these can be
% written out explicitly.
%
% First 20 Bernoulli numbers
```

```
%
BernoulliN(2) = 1/6;
BernoulliN(4) = -1/30;
BernoulliN(6) = 1/42;
BernoulliN(8) = -1/30;
BernoulliN(10) = 5/66;
BernoulliN(12) = -691/2730;
BernoulliN(14) = 7/6;
BernoulliN(16) = -3617/510;
BernoulliN(18) = 43867/798;
BernoulliN(20) = -174611/330;
%
% Second radiation constant C2.
C2 = 1.43877736e-2;
% z is a dimensionless parameter equal to
% C2/(wavelength*temperature).
z = C2./(L.*T.);
%
% Calculating the generalised fractional function of the
% first kind.
n = 10;
b = 0.;
for m = 1:length(z),
if z(m)>=2.0
  for i = 1:n
     for k = 0:s-1
        b = b + exp(-i.*z(m))./i.^s.*factorial(s-1)./...
           factorial(k).*(i.*z(m)).^k;
        end
     end
  c(m) = b./(factorial(s-1)*zeta(s));
else
  for i = 1:n
     b = b + BernoulliN(2*i)./((2.*i+3)*factorial(2.*i))...
        .*z(m).^(2*i+3);
     end
  c(m) = 1-(z(m).^(s-1)./(s-1)-z(m).^s./(2.*s)+b)./...
        (factorial(s-1).*zeta(s));
end
b = 0;
printf("F_{0-LT} = %10.9g\n",c(m));
end
```

C Chronological table of events

Chronological table of events in the development of blackbody radiation and its associated computational aids. The table lists when the various computational aids described in the text appeared. Major scientific developments within the field of thermal radiation are also given.

Year	Event
1859,1860	G. R. Kirchhoff postulates the existence of a perfect blackbody and announces the law that now bears his name.
1875	W. Crookes invents the radiometer.
1879	J. Štefan empirically deduces the law that now bears his name.
1880	A.-P.-P. Crova produces the first three-dimensional plot for the spectral radiant exitance of a blackbody.
1884	L. E. Boltzmann deduces theoretically Štefan's law of 1879.
1893,1894	W. Wien deduces his homologous law for thermal radiation.
1896	W. Wien deduces his approximate law for thermal radiation valid in the low temperature, short wavelength limit.
1899	O. Lummer and E. Pringsheim are the first to suspect the possible failure of Wien's law based on their experimental work at high temperatures and long wavelengths.
1900	H. Rubens and F. Kurlbaum experimentally confirm the failure of Wien's law at high temperatures and long wavelengths.
	M. Planck deduces his law for thermal radiation.
	H. Rubens makes the very first calculations for the spectral radiant exitance using Planck's new radiation law.
	Lord Rayleigh deduces his approximate law for thermal radiation valid in the high temperature, long wavelength limit.
1905	Lord Rayleigh re-derives his law from 1900 except this time gives an explicit value for the constant of proportionality.
	J. H. Jeans corrects a minor error in Rayleigh's deduction of 1905.
1906	E. Hertzsprung produces what is thought to be one of the first sets of tables for the spectral radiant exitance.
1910	F. E. Fowle gives a short table for the total and spectral radiant exitance.
1912	P. Debye gives the Bernoulli infinite series form for the integrated fractional amount for the radiometric case.
1913	E. G. Warburg, G. Leithäuser, E. Hupka, and C. Müller give a short table of values for the ratio of the spectral radiant exitance.
1914	F. E. Fowle updates his tables of 1910.
1917	A. Einstein derives Planck's law using classical statistical mechanics and parts of Bohr's theory.
1920	F. E. Fowle updates his tables of 1914.
	W. W. Coblentz tabulates a number of quantities useful for evaluating the spectral radiant exitance.
	W. E. Forsythe gives a one page table of values of interest to

(*Cont.*)

(*Continued*)

Year	Event
	optical pyrometrists.
1925	M. K. Frehafer and C. L. Snow produce a number of graphs and tables relating to the spectral radiant exitance.
1926	J. W. T. Walsh gives a nomogram for the spectral radiant exitance. It is found in one of the appendices of his text *Photometry*.
	G. N. Lewis coins the term "photon" to represent the smallest unit of radiant energy.
1927	M. P. A. C. Fabry tabulates the integrated fractional amount in dimensionless form.
1928	L. L. Holladay gives tables for the integrated fractional amount.
1929	J. F. Skogland produces an extensive set of tables while working for the US Bureau of Standards.
	F. E. Fowle revises and extends his tables of 1920.
	E. Lax and M. Pirani give a simple nomogram for the total radiant exitance.
1931	R. Davis and K. S. Gibson extend the 1925 tables of Frehafer and Snow.
	H. Zanstra gives the exponential and Bernoulli infinite series forms for the integrated fractional amount for both the radiometric and actinometric cases.
	W. de Groot gives an unusual table for finding the spectral radiance.
1932	A. Brill presents a number of tables for various thermal quantities.
1933	F. E. Fowle revises and extends his tables for one last time.
	Planck's function in dimensionless form appears in the first edition of E. Jahnke and F. Emde's *Funktionentafeln*.
1936	P. H. Moon describes a sliding template to find the spectral radiant exitance in his text *The Scientific Basis of Illuminating Engineering* and gives a set of tables for the spectral radiant exitance similar to those given by Skogland in 1929.
	Y. Omoto gives an unusual graph and a table for the Planck function in dimensionless form.
1936,1937	Z. Yamauti and M. Okamatu give a number of tables related to the spectral radiant exitance.
1937,1938	Moon gives the first half of his extensive set of tables for the spectral radiant exitance.
1938	Z. Miduno presents series solutions for the integrated fractional amount and gives a nomogram for the spectral radiant exitance.
1939	K. Terada gives several tables and nomograms for the spectral radiant exitance, its ratio at the spectral peak, and the integrated fractional amount.
1940	M. F. Béhar gives a simple alignment chart to find the spectral radiant exitance.
1940,1941	A. N. Lowan and G. Blanch give a number of tables for quantities related to thermal radiation.
1941	H. G. Kienle gives a nomogram for the relative gradient of a blackbody used in finding color temperatures for stars.
1942	W. M. Elsasser produces a set of tables for the integrated spectral radiant exitance and its temperature derivative.
1943	E. Q. Adams and W. E. Forsythe give tables for the spectral radiant exitance in the temperature gap between the highest temperature used by Skogland and the lowest temperature used by Moon.

(*Cont.*)

(*Continued*)

Year	Event
	R. Stair and W. O. Smith extend tabulations of the relative spectral radiant exitance to shorter wavelengths and higher temperatures.
	F. H. Sears and M. C. Joyner produce a brief set of tables.
1944	M. Czerny describes and builds the first slide rule for calculating quantities associated with blackbody radiation.
1946	A nomogram in manuscript form is described by A.-W. Kron.
1947	H. G. W. Harding and R. B. Sisson produce a set of tables for the spectral radiant exitance to seven significant figures.
	The Admiralty slide rule is first described by M. W. Makowski.
	M. Boll includes a number of tables and graphs related to blackbody radiation in his general volume of tables.
1947?	The ARISTO Nr. 10048 slide rule is released.
1947, 1948	P. H. Moon gives the second half of his extensive set of tables for the spectral radiant exitance.
1948	The General Electric GEN-15 slide chart designed and described by A. H. Canada is released.
1948?	The F5100 Radiation Slide Rule designed by Makowski is released. It later becomes known as simply the "Admiralty" rule. by A. G. Thornton, Ltd.
1951	W. Brügel gives a number of tables in the appendix of his text *Physik und Technik der Ultrarotstrahlung*.
1952	The revised General Electric GEN-15A slide chart is released.
1953	H. G. W. Harding and T. Vickers produce the first of the computer-generated tables. Their tables for the spectral radiant exitance are given to nine significant figures.
	E. Vandekerkhove produces a number of tables and graphs related to blackbody radiation for astronomical use.
	J. C. de Vos produces a moderately extensive set of tables for the spectral radiant exitance as part of his thesis work.
	R. W. Kavanagh, E. K. Björnerud, and S. S. Penner devise two detailed compound nomograms for the analysis of radiation spectra.
1954	M. Czerny defines what he refers to as the fractional functions of the first and second kinds.
	R. V. Dunkle updates the tables of Lowan and Blanch by converting them to engineering units.
	W. E. Forsythe updates Fowle's tables of 1933.
1955	R. L. LaFara, E. L. Miller, W. E. Pearson, and J. F. Peoples produce an extensive set of computer-generated tables in dimensionless form.
	C. W. Allen gives a number of short tables in the first edition of his text *Astrophysical Quantities*.
	J. K. McDonald gives an extensive set of computer-generated tables for the spectral radiance as a function of frequency.
	C. W. Allen includes some brief tables on blackbody radiation in his text *Astrophysical Quantities*.
	A. M. Crooker and W. L. Ross refine Moon's sliding template of 1936.
1956	The updated General Electric GEN-15B slide chart is released.
	F. R. Gilmore produces an extensive set of tables for the ratio of the spectral radiant exitance and the integrated fractional amount in dimensionless form.

(*Cont.*)

(*Continued*)

Year	Event
	W. Pepperhoff gives an extensive set of tables for the spectral radiant exitance in his text *Temperaturstrahlung*.
	R. S. Knox further refines Moon's sliding template of 1936.
1957	J. P. Chernoch devises a nomogram to simplify many infrared calculations.
	K.-H. Böhm and B. Schlender produce computer-generated tables for the integrated fractional amounts while at a computing workshop held by IBM Germany.
	M. M. Reynolds, R. J. Corruccini, M. M. Fulk, and R. M. Burley produce a modest set of tables for the *American Institute of Physics Handbook*.
	T. P. Gill gives an alignment chart for calculating the temperature derivative of the spectral radiant exitance.
	Tables by M. Czerny and A. Walther, which will later appear in book form, are published as part of a technical report for the *Deutsche Glastechnische Gesellschaft*.
1957?	The ARISTO Nr. 922 slide rule is released.
1958	A. LaRocca and G. J. Zissis describe a slide rule for radiance calculations.
1958–1962?	The various Infrared Slide Rules made for Infrared Industries, Inc. appear.
1959	C. B. Tanner and S. M. Robinson produce an extensive set of computer-generated tables for the total radiant exitance in a number of different units commonly used by meteorologists.
	E. B. Mayfield updates the 1957 tables of Reynolds *et al.* as part of his thesis work.
	A. G. DeBell produces a computer-generated set of tables for his work on flame temperatures.
1960	T. R. Harrison gives a number of tables in his text *Radiation Pyrometry and its Underlying Principles of Radiant Heat Transfer*.
	H. L. Hackforth gives a very simple nomogram in his text *Infrared Radiation*.
	C. D. Reid describes a do-it-yourself type sliding template.
	S. A. Golden produces two sets of tables while at the Rockerdyne division of North American Aviation.
	W. M. Elsasser updates his tables of 1942 and includes a tabulation for the temperature derivative of the spectral radiant exitance.
	An extensive set of tables in dimensionless form are given by J. Euler and R. Ludwig in their text *Arbeitsmethen der optischen Pyrometrie*.
1961	A book of tables by M. Czerny and A. Walther is published in the United States and Germany.
	A book of tables by M. Pivovonsky and M. R. Nagel is published in the US.
	A book of tables by P. A. Apanasevich and V. S. Aizenshtadt is published in Russia.
	G. T. Stevenson extends the 1957 tables of Reynolds *et al.*
1962	E. J. Pisa extends Stevenson's tables of 1961 to other quantities related to a blackbody.
	R. G. Walker produces an extensive set of computer-generated tables for the spectral radiance as a function of wavenumber.
	C. C. Ferriso produces a 650 page computer-generated tabulation for the spectral radiance as a function of wavenumber for her work on

(*Cont.*)

(*Continued*)

Year	Event
	flame temperatures.
	The Radiation Calculator slide chart from Block Associates, Inc. becomes available.
1963	The Photon Detector Slide Rule made for Vahlo becomes available.
	S. A. Twomey produces a computer-generated tabulation for the spectral radiance as a function of wavenumber for meteorological use.
	T. R. Bowen produces an extensive set of tables for the spectral radiance.
1964	A compendium volume of tables by M. A. Bramson is published in Russia
	A book of tables to seven significant figures is published in Germany by D. Hahn, J. Metzdorf, U. Schley and J. Verch.
	R. M. Goody includes a modest set of tabulations at temperatures of terrestrial interest in his book *Atmospheric Radiation*.
	A brief set of tables for Planck's function in dimensionless form appears in M. Abramowitz and I. A. Stegun's widely acclaimed *Handbook of Mathematical Functions*.
1965	The revised General Electric GEN-15C is released.
	L. H. Byrne produces an extensive set of computer-generated tables for the spectral radiance as a function of wavenumber.
1965?	Updated plastic version of the Admiralty rule made by Blundell Harling Limited is released.
1967	R. M. Blunt produces a number of computer-generated tables.
	A set of tables are given by G. Wyszecki and W. S. Stiles in their text *Color Science*.
	D. E. Erminy gives a number of Laurent polynomial and non-rational approximations to Planck's equation.
1968	The Autonetics Photon Calculator, a circular slide rule designed by H. J. Eckweiler, appears.
	T. E. Michels gives a detailed account outlining how to evaluate the integral appearing in the fractional amount together with a number of computer-generated tables.
	A number of tables appear in the appendices of L. Levi's text *Applied Optics*.
	Modest sets of tables appear in the appendices of the first edition of R. Siegel and J. R. Howell's text *Thermal Radiation Heat Transfer*.
	D. C. Todd gives a FORTRAN IV subroutine for the evaluation of the integrated fractional amount in both energetic and photonic units.
1968,1969	R. K. H. Gebel gives many computer-generated tables for a range of quantities needed in his work on detector sensitivities.
1969?	The elusive DENEM Nuclear Radiation Calculator slide chart is produced.
1970	The Cussen rule made by Pickett for Electro Optical Industries, Inc., is released.
	M. Suzuki, M. Habu, and T. Nagasaka produce a number of tables as a by-product of their work on a new spectral radiometric standard.
1971	J. Roberts issues a *Departmental Bulletin* consisting of tabulations for various thermal quantities.
1972	S. L. Sargent produces a compact set of tables for the integrated fractional amount.

(*Cont.*)

(*Continued*)

Year	Event
	A book of tables for the spectral radiance by E. Geyer, H. König, E. Wahls is published in Germany.
1973	M. Hatch gives a method for fitting Planck's equation using Chebyshev polynomials.
1974	L. Levi extends his previous set of tables from 1968 in his text *Handbook of Tables of Functions for Applied Optics*.
	R. B. Johnson and E. E. Branstetter propose an approximation to the integrated fractional amount based on Gauss–Laguerre quadrature.
1975	R. J. Chandos gives a number of programs for the Hewlett-Packard 65 and the Texas Instruments SR-52 programmable calculators.
1976	The Infrared Radiation Calculator slide chart from Sensors, Inc. is released.
	Y. Göğüş and J. Kestin compare the accuracy of a number of widely used tables.
1978	A. Zanker gives a ladder nomogram for the spectral radiant exitance.
	W. L. Wolfe gives a number of programs for the Hewlett-Packard 25 and the Texas Instruments SR-52 and SR-56 programmable calculators.
1980	R. Pavelle shows the Planck integral cannot be evaluated in terms of a finite series of elementary functions.
	W. A. Feibelman produces forty computer-generated graphs for the spectral radiant exitance to aid astronomers working with data collected by the International Ultraviolet Explorer satellite.
1982	T. B. Andersen gives Wien's displacement constant in terms of what would later become known as the Lambert W function.
1983	T. Ishikawa gives a short set of tables for the spectral radiant exitance and the integrated fractional amount.
	G. W. Hopkins gives a radiation slide rule program written in BASIC.
1984	A book of tables by H. Zhu, X. Liu, Q. Zheng, C. Yang, and F. Yu is published in China.
	M. Janes extends the 1974 approximation of Johnson and Branstetter by using generalised Gauss–Laguerre quadrature.
1985	V. I. Matveev gives a simple graph for estimating the fractional amount in two spectral bands in the infrared.
1986	The last book of tables is published by A. Sala in Poland.
	The Infrared Radiation Calculator slide charts from EG&G Judson, Infrared Information Analysis Center, and the Engineering Summer Conferences become available.
	D. J. Zhang publishes the first and second of a set of three tables.
1986,1987	B. A. Clark shows the Planck integral can be written in closed form in terms of polylogarithms.
1991	The Infrared Radiation Calculator from EG&G Judson is re-released.
1993	W. L. Wolfe gives a large number of computer programs written in BASIC useful for evaluating blackbody radiation functions.
	D. J. Zhang publishes the third of his set of tables, updating those he published in 1986.
	M. F. Modest gives two computer programs written in FORTRAN and C++ for calculating several quantities associated with blackbody radiation.
1995	The Infrared Radiation Calculator from EG&G Judson is released for a third and final time.

(*Cont.*)

(*Continued*)

Year	Event
1997	The Mikron Blackbody Radiation Calculator becomes available.
1998,1999, 2002	G. Páez and M. S. Scholl develop their truncated series approach to evaluating the integrated fractional amount.
2000	S. R. Valluri, D. J. Jeffrey and R. M. Corless show the constant appearing in Wien's displacement law can be written in closed form in terms of the Lambert W function.
2001	D. S. Flynn releases Black Body v1.0 for the Palm Pilot.
2003	M. F. Modest extends his computer programs from 1993 to include code for MATLAB.
2004	K. G. T. Hollands gives a new table for the inverse of the integrated fractional amount in his text *Thermal Radiation*.
2008	Y. F. Wang and J. X. Mao describe a fast and efficient way to evaluate Planck's integral using MATLAB.
2008,2009	The Infrared Radiation Calculator slide chart for Teledyne Judson Technologies, LLC becomes available.
2010	The online Blackbody Calculator by `SpectralCalc.com` is released.
	The online Infrared Radiance Calculator created by A. LaRocca is released.
2012	The Infrared Radiation Calculator slide chart for BAE Systems becomes available.
	A polylogarithmic reformulation of the blackbody problem is given by S. M. Stewart.
2013	The CalRadiance app developed by F. Alves for the iPhone is released.
	The Blackbody Radiation Calculator app for Android is released by S. Peters.
	W. Jing and T. Fang give a short set of tables for the actinometric integrated fractional amount.
2014	Commemorative Electro Optical Industries, Inc., slide rule marking 50 years since the foundation of the company is released.
	B. S. Chambers III releases Blackbody Tool app for the iPhone.

References

1. ————. Special slide rule invented by Mr Makowski, Admiralty Research Laboratory, for showing energy and wavelength distribution required for investigation of thermal receivers: question of Crown copyright and manufacture. The National Archives (TNA): Public Record Office (PRO) ADM 1/22042, 1945.

2. ————. A slide rule for black body radiation calculations. *European Scientific Notes*, 2(16):234–236, 1948.

3. ————. Radiation slide rule. *The Michigan Technic*, 67(6):26, 1949.

4. ————. Radiation slide rule. *Electronic Medical Digest*, 33:42, 1949.

5. ————. Slide rule measures invisible heat rays. *Science News Letter*, 55(6):89, 1949.

6. ————. M. W. Makowski. *A slide rule for radiation calculations*. Rev. Scient. Instr. **20**, 876–884, *1949*, nr. 12. (Dez.). *Physikalische Berichte*, 34(4):956, 1953.

7. ————. We hear that ... Arthur J. Cussen. *Physics Today*, 11(3):50, 1958.

8. ————. Fundamentals of infrared technology. *University of Michigan Engineering Summer Conferences*, 61(52):8–9, 1959.

9. ————. Infrared. *Industrial Research*, 2:41, 1960.

10. ————. Radiation slide rule. *Air Engineering*, 3:77, 1961.

11. ————. Radiation slide rule permits one step calculations. *The Microwave Journal*, 4(1):90, 98, 1961.

12. ————. Names in the news. *Missiles and Rockets*, 10:42, 1962.

13. ————. Radiation slide rule. *Electronic Design*, 16(14):168, 1962.

14. ————. New values for the physical constants: recommended by NAS-NRC. *National Bureau of Standards Technical News Bulletin*, 47(10):175–177, 1963.

15. ————. Notes and news. *Infrared Physics*, 3(3):183, 1963.

16. ————. Photon detector slide rule. *Electronic Design*, 11:128, 1963.

17. ————. *Recommended unit prefixes; Defined values and conversion factors; General physical constants*. Number 253 in National Bureau of Standard Miscellaneous Publications. US Government Print Office, Washington, DC, 1963.

18. ————. Infrared physicists and engineers. *Journal of the Optical Society of America*, 54(2):v, 1964.

19. ————. Concerned with infrared detectors? *Journal of the Optical Society of America*, 54(11):iv, 1964.

20. ————. Of optics and opticists. *Applied Optics*, 4(6):722,763, 1965.

21. ————. Three intensive one-week courses for: engineers, scientists. *University of Michigan Engineering Summer Conferences*, 67(115):1–14, 1966.

22. ————. A new black body radiation slide rule. *Physics Today*, 24(10):83, 1971.

23. ————. Calculating blackbody radiation. *Physics Bulletin*, 23(6):364, 1972.

24. ————. A new blackbody radiation sliderule. *Materials Research and Standards*, 12(5):80, 1972.

25. ———. Radiation calculation aid. *Optics and Laser Technology*, 4(4):190, 1972.

26. ———. New radiation sliderule. *Laser+Elektro-Optik*, 5:56, 1973.

27. ———. Michigan Engineering Summer Conferences. *Journal of the Optical Society of America*, 64(4):577, 1974.

28. ———. Infrared-radiation calculator. *Analytical Chemistry*, 48(13):1076A, 1976.

29. ———. IR calculator. *Electrical Design News*, 21:148, 1976.

30. ———. IR calculator. *Optical Spectra*, 10:69, 1976.

31. ———. IR slide rule. *Instrumentation Technology*, 23:75, 1976.

32. ———. Infrared-radiation calculator. *Experimental Mechanics*, 17(2):12N, 1977.

33. ———. University of Michigan announces two summer conferences. *Bulletin of the American Meteorological Society*, 65(5):456, 1984.

34. ———. EG&G Judson offers free IR radiation calculator. *Photonics Spectra*, 25(9):58, 1991.

35. ———. Infrared radation. *Lasers & Optronics*, 10:62, 1991.

36. M. Abramowitz and I. A. Stegun, editors. *Handbook of Mathematical Functions with Formulas, Graphs, and Mathematical Tables*. National Bureau of Standards, Washington, DC, 1964.

37. Abstract 11-33673. Makowski (M. W.) [Polish U. C., London]. Règle à curseur pour les calculs de rayonnement. (A slide rule for radiation calculations). *Rev. sci. Instrum.*, U.S.A. (1949), 20, no. 12, 876–84. *Bulletin Analytique du Centre National de la Recherche Scientifique*, 11(8):2510, 1950.

38. Abstract 1740. A slide rule for radiation calculations. M. W. Makowski. Rev. Sci. Instruments 20, 876–84 (1949) Dec. *Nuclear Science Abstracts*, 4(6):278, 1950.

39. D. P. Adams. *An Index of Nomograms*. The Technology Press of Massachusetts Institute of Technology and John Wiley & Sons, Inc., New York, 1950.

40. E. Q. Adams. A table of logarithmic corrections to the Wien radiation law. *Journal of the Optical Society of America*, 37(9):695–697, 1947.

41. E. Q. Adams and W. E. Forsythe. Radiometric and colorimetric characteristics of the blackbody between $2800°K$ and $3800°K$. *Bulletin of the Scientific Laboratories of Denison University*, 38:52–68, 1943.

42. A. G. Agnese, M. La Camera, and E. Recami. Black-body laws derived from a minimum knowledge of physics. *Il Nuovo Cimento B*, 114(12):1367–1374, 1999.

43. J. Alcott. Radiation slide rule. *Oklahoma State Engineer*, 14(3):32, 1949.

44. C. W. Allen. *Astrophysical Quantities*. The Athlone Press, London, 1955.

45. C. W. Allen. *Astrophysical Quantities*. The Athlone Press, London, third edition, 1973.

46. F. Alves. CalRadiance. http://www.calradiance.com, 2013.

47. T. B. Andersen. An exact expression for the Wien displacement constant. *Journal of Quantitative Spectroscopy and Radiative Transfer*, 27(2):663–664, 1982.

48. A. Anderson-Quinn, A. G. De Bell, and S. Simmons. Blackbody Source. US Patent 3,205,343, 1962.

49. G. E. Andrews, R. Askey, and R. Roy. *Special Functions.* Cambridge University Press, Cambridge, 1999.

50. H. W. Andrews. A slide rule for radiation calculations. *Journal of the Oughtred Society,* 11(1):32–35, 2002.

51. L. C. Andrews. *Special Functions of Mathematics for Engineers.* Oxford University Press/SPIE Optical Engineering Press, Oxford, second edition, 1998.

52. K. Ångström. Energy in the visible spectrum of the Hefner standard. *Physical Review* [Series 1], 17(4):302–314, 1903.

53. P. A. Apanasevich and V. S. Aizenshtadt. *Tablitsy Raspredeleniya Energii i Fotonov v Spektre Ravnovesnogo Izlucheniya.* Akademiya Nauk Belorusskoĭ SSR, Minsk, 1961.

54. P. A. Apanasevich and V. S. Aizenshtadt. *Tables for the Energy and Photon Distribution in Equilibrium Radiation Spectra.* The MacMillan Company, New York, 1965.

55. R. C. Archibald. Review of 'Parry Moon, **A**. "A Table of Planck's Function from 3500 to 8000°K," *Jn. Math. Phys.,* v. 16, 1937, p. 133–157. 17.5 × 25.4 cm. Also M.I.T., E.E. Dept., *Contribution* no. 131, 1938. **B**. *A Table of Planck's Function 2000 to 3500°K,* Cambridge, Mass. Inst. Techn., Dept. Electr. Engineering, 1947, 80 p. 15.2 × 22.8 cm. **C**. "A Table of Planckian Radiation," Opt. Soc. Amer., *Jn.* v. 38, 1948, p. 291–294.'. *Mathematical Tables and Other Aids to Computation,* 3(23):175–176, 1948.

56. Autonetics. *Black body photon calculator.* North American Aviation, Inc., Anaheim, California, 1967.

57. H. D. Baehr and K. Stephan. *Heat and Mass Transfer.* Spring-Verlag, Berlin, second edition, 2006.

58. E. Baisch. Versuche zur Prüfung des Wien–Planckschen Strahlungsgesetzes im Bereich kurzer Wellenlängen. *Annalen der Physik* [Vierte Folge], 35(8):543–590, 1911.

59. D. J. Baker and W. L. Brown. Presentation of Spectra. *Applied Optics,* 5(8):1331–1333, 1966.

60. D. W. Ball. Wien's displacement law as a function of frequency. *Journal of Chemical Education,* 90(9):1250–1252, 2013.

61. R. E. Ball. *The Fundamentals of Aircraft Combat Survivability Analysis and Design.* American Institute of Aeronautics and Astronautics, Inc., Reston, VA, second edition, 2003.

62. S. S. Ballard. Optical activities in the universities. *Applied Optics,* 4(2):219–220, 1965.

63. C. R. Barber. The International Practical Temperature Scale of 1968. *Metrologia,* 5(2):35–44, 1969.

64. G. I. Barenblatt. *Dimensional Analysis.* Gordon and Breach Science Publishers, New York, 1987.

65. E. S. Barr, B. B. Barrow, R. West, W. B. Fowler, and W. A. Shurcliff. Frequency scale for spectra (Letters to the Editor). *Science,* 151(3709):400–404, 1966.

66. W. E. Barrows. *Light, Photometry and Illumination: A Throughly Revised Edition of "Electrical Illuminating Engineering".* McGraw-Hill Book Company, New York, 1912.

67. F. O. Bartell. Projected solid angle and blackbody simulators. *Applied Optics*, 28(6):1055–1057, 1989.

68. R. G. Bartle and D. R. Sherbert. *Introduction to Real Analysis*. John Wiley & Sons, Inc., New York, 2000.

69. H. Bateman. Review of 'Project for Computation of Mathematical Tables (A. N. Lowan, technical director), *Miscellaneous Physical Tables. Planck's Radiation Functions, and Electronic Functions*. Prepared by the Federal Works Agency, Works Projects Aministration for the City of New York, conducted under the sponsorship of the National Bureau of Standards. New York, 1941, vii, 58 p. 20.2 × 26.3 cm. Reproduced by a photo offset process. Sold by the U.S. Bureau of Standards, Washington, D.C. The work was not distributed until 1943. $1.50; foreign price $1.75.'. *Mathematical Tables and Other Aids to Computation*, 1(3):75–76, 1943.

70. E. Bauer. Sur la loi du rayonnement noir et la théorie des quanta. Remarques sur un travail de M. J. de Boissoudy. *Journal de Physique Théorique et Appliquée*, 3(1):641–649, 1913.

71. J. A. Bearden and J. S. Thomsen. A survey of atomic constants. *Nuovo Cimento. Supplemento*, 5(2):267–360, 1957.

72. M. F. Béhar. Industrial Pyrometry. *Instruments*, 13(12):383–396, 1940.

73. M. F. Béhar, editor. *The Handbook of Measurement and Control*. Instruments Publishing Company, Pittsburgh, PA, 1951.

74. F. Benford. The law of anomalous numbers. *Proceedings of the American Philosophical Society*, 78(4):551–572, 1938.

75. F. Benford. Laws and corollaries of the black body. *Journal of the Optical Society of America*, 29(2):92–96, 1939.

76. F. Benford. The Blackbody. Part I. Its Physical Theory and Its Spectral Distribution and other Radiation Characteristics. *General Electric Review*, 46(7):377–382, 1943.

77. F. Benford. The Blackbody. Part II. Summation of Radiation, Brightness and Color Temperatures, Pressure of Radiation. Blackbody as Quanta Generator, and Efficiency of Radiation. *General Electric Review*, 55(8):433–440, 1943.

78. F. Benford. The Blackbody. *Journal of the American Society of Naval Engineers*, 55(4):718–738, 1943.

79. T. L. Bergman, A. S. Lavine, F. P. Incropera, and D. P. DeWitt. *Fundamentals of Heat and Mass Transfer*. John Wiley & Sons, Inc., Hoboken, New Jersey, seventh edition, 2011.

80. J. Bertrand. Sur l'homogénéité dans les formules de physique. *Comptes rendus hebdomadaires des séances de l'Académie des sciences*, 86(15):916–920, 1878.

81. U. Besson. Paradoxes of thermal radiation. *European Journal of Physics*, 30(5):995–1007, 2009.

82. M. L. Bhaumik and M. A. Levine. Infrared Physics. In K. Seyrafi, editor, *Engineering Design Handbook on Infrared Military Systems, Part 1*, pages 2-13–2-17. Army Materiel Command, AD-763 495, 1971.

83. R. T. Birge. Probable values of the general physical constants. *The Physical Review Supplement*, 1(1):1–73, 1929.

84. R. T. Birge. A new table of values of the general physical constants. *Reviews of Modern Physics*, 13(4):233–239, 1941.

85. G. Birkhoff. *Hydrodynamics, a study in logic, fact and similitude*. Princeton University Press, Princeton, 1960.

86. M. J. Block. A personal memoir of the first commercial FTIR for analytical chemistry and its critical path. *The Spectrum*, 20(2):12–14, 2001.

87. R. M. Blunt. *Black Body Functions for Pyrotechnicists*. US Naval Ammunition Depot, Crane, Indiana, March 1967.

88. A. H. Boerdijk. The value of the constant in Wien's displacement law. *Philips Research Reports*, 8(4):291–303, 1953.

89. H. Bohle. *Electrical Photometry and Illumination: A Treatise on Light and Its Distribution, Photometric Apparatus, and Illuminating Engineering*. Charles Griffin & Company, London, 1912.

90. K.-H. Böhm and B. Schlender. Tabelle von Integralen über die Kirchhoff–Planck–Funktion. *Zeitschrift für Astrophysik*, 43(2):95–97, 1957.

91. M. Boll. *Tables Numériques Universelles des Laboratoires et Bureaux D'étude*. Dunod, Paris, 1947.

92. L. Boltzmann. Ableitung des Stefan'schen Gesetzes, betreffend die Abhängigkeit der Wärmestrahlung von der Temperatur aus der electromagnetischen Lichttheorie. *Annalen der Physik und Chemie* [Neue Folge], 22(6):291–294, 1884.

93. D. Bornemeier. A review of blackbody radiation laws. In G. H. Suits, J. J. Cook, N. Smith, and I. Sattinger, editors, *The University of Michigan Notes for a Program of Study in Remote Sensing of Earth Resources*, pages I-64–I-65. The University of Michigan, Ann Arbor, Michigan, 1968.

94. T. R. Bowen. *Blackbody Radiation Tables*. Advanced Research Projects Agency, US Army Missile Command, Redstone Arsenal, Huntsville, Alabama, May 1963.

95. R. W. Boyd. *Radiometry and the Detection of Optical Radiation*. Wiley, New York, 1983.

96. R. N. Bracewell. The maximum of the Planck energy spectrum. *Nature*, 174(4429):563–564, 1954.

97. M. A. Bramson. *Infrakrasnoe Izluchenie Nagretykh Tel (Infrared radiation from heated bodies)*. Nauka Press, Moscow, 1964.

98. M. A. Bramson. *Spravochnye Tablitsy po Infrakrasnomu Izlucheniyu Nagretykh Tel (Reference tables on infrared radiation from heated bodies)*. Nauka Press, Moscow, 1964.

99. M. A. Bramson. *Infrared Radiation: A Handbook for Applications*. Plenum Press, New York, 1968.

100. K. Brecher. Why don't we see in the infrared? *Bulletin of the American Astronomical Society*, 30(4):1293, 1998.

101. M. Q. Brewster. *Thermal Radiative Transfer and Properties*. John Wiley & Sons, Inc., New York, 1992.

102. P. W. Bridgman. *Dimensional Analysis*. Yale University Press, New Haven, 1932.

103. A. Brill. Die Temperaturen der Fixsterne. In G. Eberhard, A. Kohlschütter, and H. Ludendorff, editors, *Handbuck der Astrophysik. Band V, Erste Hälfer. Das Sternsystem. Erster Teil*, pages 128–209. Verlag von Julius Springer, Berlin, 1932.

104. M. Brillouin. Rayonnements et Thermodynamique. *Annales de Physique*, 1(1):163–170, 1914.

105. P. B. Brito, F. Fabião, and A. Staubyn. Euler, Lambert, and the Lambert W-function today. *The Mathematical Scientist*, 33(2):127–133, 2008.

106. J. M. Brown. *Tables and Conversions for Microclimatology.* USDA Forest Service, General Technical Report NC-8, North Central Forest Experiment Station, St. Paul, Minnesota, 1973.

107. H. A. Brück. Review of 'Tables of the Fractional Functions for the Planck Radiation Law' by M. Czerny and A. Walther. Springer, Berlin, 1961. vi + 55 pp., DM 28.00. *Journal of Atmospheric and Terrestrial Physics*, 24(11):999, 1962.

108. W. Brügel. *Physik und Technik der Ultrarotstrahlung.* Curt R. Vincentz Verlag, Hannover, 1951.

109. W. Brügel. *Physik und Technik der Ultrarotstrahlung.* Curt R. Vincentz Verlag, Hannover, second edition, 1961.

110. H. E. Buchanan. On a certain integral arising in quantum mechanics. *National Mathematics Magazine*, 10(7):247–248, 1936.

111. E. Buckingham. On physically similar systems; illustrations of the use of dimensional equations. *Physical Review*, 4(4):345–376, 1914.

112. M. Bukshtab. *Applied Photometry, Radiometry, and Measurement of Optical Losses.* Springer, New York, 2012.

113. E. G. Bylander. Components for electrooptics. In C. A. Harper, editor, *Handbook of Components for Electronics*, pages 5–21. McGraw-Hill Book Company, New York, 1977.

114. L. H. Byrne. *Tables of the Blackbody Radiation Function. Volume 1* (For wavenumbers from 250 to 2745 cm^{-1} and temperatures from 150 to 350°K). Goddard Space Flight Center, Greenbelt, Maryland, January 1965.

115. F. E. Cady and H. B. Dates. *Illuminating Engineering.* Wiley, New London, 1925.

116. P. S. Callahan. Nondestructive temperature and radiance measurements on night flying moths. *Applied Optics*, 7(9):1811–1817, 1968.

117. H. L. Callendar. Note on radiation and specific heat. *The London, Edinburgh, and Dublin Philosophical Magazine and Journal of Science*, [Sixth Series], 26(154):787–791, 1913.

118. H. L. Callendar. Thermodynamics of radiation. *The London, Edinburgh, and Dublin Philosophical Magazine and Journal of Science*, [Sixth Series], 27(161):870–880, 1914.

119. A. H. Canada. *Infrared: Its Military and Peacetime Uses.* General Electric Company, Utica, New York, 1947.

120. A. H. Canada. Simplified calculation of black-body radiation. *General Electric Review*, 51(12):50–54, 1948.

121. B. S. Chambers, III. private communication, 2014.

122. B. S. Chambers, III. Blackbody Tool. `http://support.twotori.com/products/blackbody-tool-or-bbtool`, 2014.

123. R. J. Chandos. *Interim Manual Blackbody Radiation Sliderule.* Electro Optical Industries Inc., Santa Barbara, California, 1970.

124. R. J. Chandos. *Blackbody Radiation Sliderule: Manual of Instruction and Information.* Electro Optical Industries Inc., Santa Barbara, California, December 1971.

125. R. J. Chandos. *A Program for Blackbody Radiation Calculations: Planck's Law Programs for the Hewlett Packard Model 65 Programmable Calculator.* Electro Optical Industries Inc., Santa Barbara, California, 1975.

126. R. J. Chandos. *A Program for Blackbody Radiation Calculations II: Planck's law programs for the Texas Instruments model SR-52 programmable calculator.* Electro Optical Industries Inc., Santa Barbara, California, 1975.

127. R. J. Chandos. private communication, 2012.

128. S. L. Chang and R. T. Rhee. Blackbody radiation functions. *International Communications in Heat and Mass Transfer*, 11(5):451–455, 1984.

129. A. J. Chapman. *Heat Transfer.* The Macmillian Company, New York, second edition, 1967.

130. A. Chappell, editor. *Optoelectronics: theory and practice.* McGraw-Hill Book Company, New York, 1978.

131. C. V. L. Charlier. Das Strahlungsgesetz. *Arkiv för matematik, astronomi och fysik*, 7(31):8 pp., 1912.

132. C. V. L. Charlier. Das Strahlungsgesetz. Zweite Mitteilung. *Arkiv för matematik, astronomi och fysik*, 9(11):12 pp., 1913.

133. J. J. Chen and J. D. Lin. A theoretical model for the nongray radiation drying of polyvinyl alcohol/water solutions. *Drying Technology: An International Journal*, 22(4):853–875, 2004.

134. J. P. Chernoch. Infrared calcations made simple. *Aviation Age*, 28(1):116–117, 1957.

135. W.-C. Chiu. On the interpretation of the energy spectrum. *American Journal of Physics*, 35(7):642–648, 1967.

136. W.-C. Chiu. The Wien displacement law. *Bulletin of the American Meteorological Society*, 49(12):1142, 1968.

137. J. W. Christy and J. R. Rochester. Low-temperature blackbody radiation source. US Patent 6,232,614, 1998.

138. B. A. Clark. Using MACSYMA to derive numerical methods to compute radiation integrals. *Transaction of the American Nuclear Society*, 53:230–231, 1986.

139. B. A. Clark. Computing multigroup radiation integrals using polylogarithm-based methods. *Journal of Computational Physics*, 70(2):311–329, 1987.

140. L. F. Cleveland. *Tables of Spectral Energy Distribution based on Planck's Radiation Formula.* MSc thesis, Massachusetts Institute of Technology, Cambridge, Massachusetts, 1935.

141. W. W. Coblentz. Methods for computing and intercomparing radiation data. *Scientific Papers of the Bureau of Standards*, 15:617–624, 1920.

142. E. R. Cohen, J. W. M. DuMond, T. W. Layton, and J. S. Rollett. Analysis of variance of the 1952 data on the atomic constants and a new adjustment, 1955. *Review of Modern Physics*, 27(4):363–380, 1955.

143. E. R. Cohen and B. N. Taylor. The 1973 least-squares adjustment of the fundamental constants. *Journal of Physical and Chemical Reference Data*, 2(4):663–734, 1973.

144. R. M. Corless, G. H. Gonnet, D. E. G. Hare, and D. J. Jeffrey. Lambert's W function in Maple. *Maple Technical Newsletter (MapleTech)*, 9:12–22, 1993.

145. R. M. Corless, G. H. Gonnet, D. E. G. Hare, D. J. Jeffrey, and D. E. Knuth. On the Lambert W function. *Advances in Computational Mathematics*, 5(4):329–359, 1996.

146. R. M. Corless and D. J. Jeffrey. The Lambert W function. In N. J. Higham, M. R. Dennis, P. Glendinning, P. A. Martin, F. Santosa, and J. Tanner,

editors, *The Princeton Companion to Applied Mathemtics*. Princeton University Press, New Jersey, 2015.

147. S. Corrsin. A simple geometrical proof of Buckingham's π-theorem. *American Journal of Physics*, 19(3):180–181, 1951.

148. A. N. Cox, editor. *Allen's Astrophysical Quantities*. AIP Press/Springer, New York, fourth edition, 2000.

149. A. M. Crooker and W. L. Ross. A note on black body radiation. *Canadian Journal of Physics*, 33(5):257–260, 1955.

150. A. Crova. Étude des radiations émises par les corps incandescents. Mesure optique des hautes températures. *Annales de Chimie et de Physique, Série 5*, 19:472–550, 1880.

151. E. Császár. Die experimentelle Prüfung der Planckschen Strahlungsformel auf lichtelektrischem Wege. *Zeitschrift für Physik*, 14(1):220–225, 1923.

152. E. Császár. Die Verallgemeinerung der Formel der schwarzen Strahlung. *Zeitschrift für Physik*, 90(9-10):667–673, 1934.

153. E. Császár. Eine Bemerkung zu den Gesetzen der schwarzen Strahlung. *Die Naturwissenschaften*, 30(17-18):265–266, 1942.

154. W. D. Curtis, J. D. Logan, and W. A. Parker. Dimensional analysis and the Pi theorem. *Linear Algebra and its Applications*, 47:117–126, 1982.

155. A. J. Cussen. Overview of Blackbody Radiation Sources. In R. D. Ennulat, editor, *Infrared Sensor Technology*, volume 344 of *Proceedings of SPIE*, pages 2–15, 1982.

156. M. Cussen. private communication, 2012.

157. D. Cvijović. New integral representations of the polylogarithm function. *Proceedings of the Royal Society A*, 463:897–905, 2007.

158. M. Czerny. Ein Hilfsmittel zur Integration des Planckschen Strahlungsgesetzes. *Physikalische Zeitschrift*, 45(9/12):205–206, 1944.

159. M. Czerny. Einige Bemerkungen über die Intensitätsverteilung der schwarzen Strahlung. *Physikalische Zeitschrift*, 45(9/12):207–208, 1944.

160. M. Czerny. Zur Integration des Planckschen Strahlungsgesetzes. *Zeitschrift für Physik*, 139(3):302–308, 1955.

161. M. Czerny, L. Genzel, and G. Heilmann. Über den Strahlungsstrom im Inneren von Glaswannen. *Glastechnische Berichte*, 28(5):185–190, 1955.

162. M. Czerny and L. Genzel. Zur Berechnung des Strahlungsstromes in Glasbad von Schmelzwannen. *Glastechnische Berichte*, 30(1):1–7, 1957.

163. M. Czerny and A. Walther. *Bruchteilfunktions-Tabellen zum Rechnen mit dem Planckschen Strahlungsgesetz*. Deutsche Glastechnische Gesellschaft, Frankfurt am Main, 1957.

164. M. Czerny and A. Walther. *Tables of the Fractional Functions for the Planck Radiation Law*. Springer-Verlag, Berlin, 1961.

165. J. Dai, Z. Mi, and X. Dai. Peak laws, scaling laws and some related physical constants in black-body radiation. *Chinese Journal of Infrared Research*, 8(4):277–284, 1989.

166. X. Dai, X. Xu, and X. Wang. The analytic expression of some physical quantities in black-body radiation. *Chinese Journal of Infrared Research*, 5(4):247–256, 1986.

167. J. D'Ans and E. Lax. *Taschenbuch für Chemiker und Physiker*. Springer-Verlag, Berlin, 1943.

168. J. D'Ans and E. Lax. *Taschenbuch für Chemiker und Physiker.* Springer-Verlag, Berlin, second edition, 1949.

169. C. G. Darwin. Radiation theory. In R. Glazebrook, editor, *A Dictionary of Applied Physics. Volume IV,* pages 566–572. MacMillan and Co., London, 1923.

170. R. Das. Wavelength- and frequency-dependent formulations of Wien's displacement law. *Journal of Chemical Education,* 92(6):1130–1134, 2015.

171. C. N. Davies. Tyndall and Stefan's Radiation Law. *Nature,* 157(4000):879, 1946.

172. R. Davis and K. S. Gibson. Fliters for the reproduction of sunlight and daylight and the determination of color temperature. *National Bureau of Standards. Miscellaneous Publications,* 114, 1931.

173. A. L. Day and C. E. Van Orstrand. The black body and the measurement of extreme temperatures. *The Astrophysical Journal,* 19(1):1–40, 1904.

174. R. de A. Martins. The origin of dimensional analysis. *Journal of the Franklin Institute,* 311(5):331–337, 1981.

175. W. de Beauclair. Alwin Walther, IPM, and the development of calculator/computer technology in Germany, 1930–1945. *Annals of the History of Computing,* 8(4):334–350, 1986.

176. J. de Boissoudy. Sur la constante de la loi du rayonnement. *Comptes rendus hebdomadaires des séances de l'Académie des sciences,* 156:1364–1366, 1913.

177. J. de Boissoudy. Sur une nouvelle forme de la loi du rayonnement noir et de l'hypothèse des quanta. *Journal de Physique Théorique et Appliquée,* 3(1):385–396, 1913.

178. J. de Boissoudy. Sur la loi du rayonnement noir. Réponse a M. E. Bauer. *Journal de Physique Théorique et Appliquée,* 3(1):649–651, 1913.

179. N. G. de Bruijn. *Asymptotic Methods in Analysis.* North-Holland Publishing Co., Amsterdam, 1958.

180. W. de Groot. Tabellen voor de stralingsformule van Planck. *Physica. Nederlands Tijdschrift voor Natuurkunde,* 11(8):265–274, 1931.

181. J. C. de Vos. *The Emissivity of Tungsten Ribbon: the Tungsten Striplamp as a Standard Source of Radiation.* Proefschrift, Vrije Universiteit, Amsterdam, 1953.

182. W. de Wiveslie Abney and E. R. Festing. The relation between electric energy and radiation in the spectrum of incandescence lamps. *Proceedings of the Royal Society of London,* 37(232-234):157–173, 1884.

183. A. G. DeBell. *Tables for the spectral radiant intensity of a blackbody and of a tungsten ribbon.* Research Report 59-32, Rocketdyne, A Division of North American Aviation, Inc., Canoga Park, California, 1959.

184. P. Debye. Zur Theorie der spezifischen Wärme. *Annalen der Physik* [Vierte Folge], 39(4):789–839, 1912.

185. E. L. Dereniak and D. G. Crowe. *Optical Radiation Detectors.* Wiley, New York, 1984.

186. S. Dey and S. Gupta, editors. *Numerical Methods.* McGraw Hill Education (India), New Delhi, 2013.

187. D. P. Doane and L. E. Seward. Measuring skewness: a forgotten statistic? *Jouranl of Statistics Education,* 19(2):1–18, 2011.

188. J. G. Dorfman. Golitsyn, Boris Borisovich. In C. C. Gillispie, editor, *Dic-*

tionary of Scientific Biography. Volume V, pages 461–462. Charles Scribner's Sons, New York, 1972.

189. R. C. Dougal. The centenary of the fourth-power law. *Physics Education*, 14(4):234–238, 1979.

190. J. W. Draper. On the production of light by heat. *London, Edinburgh, and Dublin Philosophical Magazine and Journal of Science* [Third Series], 30(202):345–360, 1847.

191. H.-W. Drawin and P. Felenbok, editors. *Data for Plasmas in Local Thermodynamic Equilibrium*. Gauthier-Villars, Paris, 1965.

192. L. A. Drew. Tabulated values of the Stefan–Boltzman function. *Minnesota Forestry Research Notes*, 249, April 1974.

193. B. K. Driver and R. D. Driver. Simplicity of solutions of $x'(t) = bx(t-1)$. *Journal of Mathematical Analysis and Applications*, 157(2):591–608, 1991.

194. P. Drude. *The Theory of Optics*. Longmans, Green and Co., London, 1902.

195. C. V. Drysdale. Note on the luminous efficiency of a black body. *Proceedings of the Physical Society of London*, 21(1):573–580, 1907.

196. J. W. M. DuMond and E. R. Cohen. Our knowledge of the atomic constants F, N, m, and h in 1947, and of other constants derivable therefrom. *Review of Modern Physics*, 20(1):82–108, 1948.

197. J. W. M. DuMond and E. R. Cohen. Least-squares adjustment of the atomic constants, 1952. *Review of Modern Physics*, 25(3):691–708, 1953.

198. R. V. Dunkle. Thermal-radiation tables and applications. *Transactions of the American Society of Mechanical Engineers*, 76:549–552, 1954.

199. A. L. Dunklee. *An experiment to measure secondary ion-electron emission on dielectrics*. MSc thesis, Cornell University, Ithaca, New York, 1968.

200. J. W. Eaton, D. Bateman, S. Hauberg, and R. Wehbring, editors. *The GNU Octave 3.8 reference manual - Part 1/2*. Samurai Media Limited, London, 2014.

201. J. W. Eaton, D. Bateman, S. Hauberg, and R. Wehbring, editors. *The GNU Octave 3.8 reference manual - Part 2/2*. Samurai Media Limited, London, 2014.

202. H. J. Eckweiler. A Blackbody Photon Calculator. *Applied Optics*, 7(7):1409–1411, 1968.

203. D. K. Edwards. Radiative transfer characteristics of materials. *Transactions of the American Society of Mechanical Engineers: Journal of Heat Transfer*, 91(1):1–15, 1969.

204. EG&G Judson. *Infrared Radiation Calculator and User Guide*. EG&G Judson, Montgomeryville, Pennsylvania, 1986.

205. P. Ehrenfest. Bemerkung zu einer neuen Ableitung des Wienschen Verschiebungsgesetzes. *Physikalische Zeitschrift*, 7(15):527–528, 1906.

206. P. Ehrenfest. Bemerkung zu einer neuen Ableitung des Wienschen Verschiebungsgesetzes. Antwort auf Herrn Jeans' Entgegnung. *Physikalische Zeitschrift*, 7(23):850–852, 1906.

207. P. Ehrenfest. Welche Züge der Lichtquanetenhypothese spielen in der Theorie der Wärmestrahlung eine wesentliche Rolle? *Annalen der Physik* [Vierte Folge], 36(11):91–118, 1911.

208. A. Einstein. Zum gegenwärtigen Stand des Strahlungsproblems. *Physikalische Zeitschrift*, 10(6):185–193, 1909.

209. L. Eisner. Instrumentation teaching equipment. Part three: miscellaneous. *Chemical Education*, 41(9):A636, 1964.

210. Electro Optical Industries, Inc. Blackbody Energy Calculator. `http://www.electro-optical.com`, 2013.

211. W. M. Elsasser. Heat transfer by infrared radiation in the atmosphere. *Harvard Meteorological Studies*, 6, 1942.

212. W. M. Elsasser and M. F. Culbertson. Atmospheric radiation tables. *Meteorological Monographs*, 4(23), 1960.

213. R. B. Emmons. Efficient computations of blackbody functions. *Optical Engineering*, 19(2):SR-038,SR-040,SR-042, 1980.

214. P. S. Epstein. On Planck's quantum of action. *American Journal of Physics*, 22(6):402–405, 1954.

215. D. E. Erminy. Some approximations to the Planck function in the intermediate region with applications in optical pyrometry. *Applied Optics*, 6(1):107–117, 1967.

216. R. S. Estey. The correlation of color temperatures based on the Wien and Planck radiation formulas. *Journal of the Optical Society of America*, 28(8):293–295, 1938.

217. J. Euler and R. Ludwig. *Arbeitsmethod der optischen Pyrometrie*. Verlag G. Braun, Karlsruhe, 1960.

218. L. Euler. *Institutiones Calculi Integralis*, volume I. Impensis Academiae Imperialis Scientiarum, Petropoli, 1768.

219. L. Euler. De serie lambertina plurimisque eius insignibus proprietatibus. *Acta Academiae Scientarum Imperialis Petropolitinae*, pages 29–51, 1779,1783.

220. J. H. Evans. Dimensional analysis and the Buckingham Pi theorem. *American Journal of Physics*, 40(12):1815–1822, 1972.

221. C. Fabry. *Introduction Générale à la Photométrie*. Éditions de la Revue d'Optique théorique et instrumentale, Paris, 1927.

222. W. G. Fastie. Ambient temperature independent thermopilies for radiation pyrometry. *Journal of the Optical Society of America*, 41(11):823–829, 1951.

223. A. K. Federman. Some general methods of integration of partial differential equations of first order. *Annales de i'Instit Polytechnique Pierre le Grand à St. Pétersbourg. Mathématique, physique, sciences naturelles et appliquées*, 16(1):97–155, 1911.

224. W. A. Feibelman. *Blackbody Curves for the IUE Spectral Range $\lambda 1150$ to $\lambda 3200$ from 6,000 to 200,000 K*. NASA Technical Memorandum 81997, August 1980.

225. C. C. Ferriso. *Blackbody Radiation Tables*. General Dynamics/Astronautics, General Dynamics Corporation, San Diego, California, September 1962.

226. D. S. Flynn. Black Body v1.0. `http://www.freewarepalm.com/calculator/blackbody.shtml`, August 2001.

227. W. Foerst, editor. *Ullmanns Encyklopädie der technischen Chemie*, volume 11. Urban & Schwarzenberg, München, 1960.

228. L. Foitzik. Über die Darstellung der spektralen Energieverteilung von Strahlungsquellen. *Exprimentelle Technik der Physik*, 1(4-5):209–213, 1953.

229. P. D. Foote. A new relation derived from Planck's law. *Bulletin of the Bureau of Standards (US)*, 12(3):479–482, 1916.

230. P. D. Foote. A new relation derived from Planck's law. *Physical Review*, 7(2):224–225, 1916.

231. W. E. Forsythe. 1919 report of Standards Committee on pyrometry. *Journal of the Optical Society of America*, 4(5):305–332, 1920.

232. W. E. Forsythe. *Smithsonian Physical Tables*. The Smithsonian Institution, Washington, DC, ninth revised edition, 1954.

233. J. B. J. Fourier. *Théorie analytique de la chaleur*. Gauthier-Villars, Paris, 1822.

234. F. E. Fowle. *Smithsonian Physical Tables*. The Smithsonian Institution, Washington, DC, fifth revised edition, 1910.

235. F. E. Fowle. *Smithsonian Physical Tables*. The Smithsonian Institution, Washington, DC, sixth revised edition, 1914.

236. F. E. Fowle. *Smithsonian Physical Tables*. The Smithsonian Institution, Washington, DC, seventh revised edition, 1920.

237. F. E. Fowle. Radiation from a perfect (black-body) radiator. In E. W. Washburn, editor, *International Critical Tables of Numerical Data, Physics, Chemistry and Technology*, pages 238–242. McGraw-Hill Book Company, Inc., New York, 1929.

238. F. E. Fowle. *Smithsonian Physical Tables*. The Smithsonian Institution, Washington, DC, eighth revised edition, 1933.

239. A. Franklin. *The Neglect of Experiment*. Cambridge University Press, Cambridge, 1986.

240. M. K. Frehafer. New tables and graphs for facilitating computation of spectral energy distribution by Planck's formula. *Journal of the Optical Society of America*, 7(1):74–75, 1923.

241. M. K. Frehafer and C. L. Snow. Tables and graphs for facilitating computation of spectral energy distribution by Planck's formula. *National Bureau of Standards. Miscellaneous Publications*, 56, 1925.

242. F. N. Fritsch, R. E. Shafer, and W. P. Crowley. Algorithm 443: Solution of the transcendental equation $we^w = x$. *Communication of the ACM*, 16(2):123–124, 1973.

243. I. C. Gardner. Validity of the Cosine-Fourth-Power Law of Illumination. *Journal of Research. National Bureau of Standards*, 39:213–219, 1947.

244. P. R. Gast. Blackbody Radiation. In S. L. Valley, editor, *Handbook of Geophysics and Space Environments*, pages B-1–B-9. Office of Aerospace Research, Air Force Cambridge Research Laboratories, Bedford, Massachusetts, 1965.

245. GATS, Inc. Blackbody Calculator. `http://www.spectralcalc.com/blackbody_calculator`, 2013.

246. R. K. H. Gebel. Normalized cumulative blackbody functions, and their applications in thermal radiation calculations. *Bulletin of the American Physical Society, Series 2*, 13(7):959–960, 1968.

247. R. K. H. Gebel. The normalized cumulative blackbody functions, their applications in thermal radiation calculations, and related subjects. Aerospace Research Laboratories, Wright-Patterson Air Force Base, Ohio, January 1969.

248. R. K. H. Gebel. The fundamental infra-red threshold in thermal image detection as affected by detector cooling and related problems. In J. D. McGee, D. McMullan, E. Kahan, and B. L. Morgan, editors, *Photo-Electronic Image Devices*, pages 685–704. Academic Press, London, 1969.

249. R. W. Gelinas and R. H. Genoud. *A broad look at the performance of infrared detectors.* RAND Corporation Report P-1697, May 1959.

250. General Electric. *Radiation Calculator. Example problems showing use of the General Electric Radiation Calculator.* General Electric, Schenectady, New York, 1956.

251. L. Genzel. Review of 'Tables of blackbody radiation functions.' Mark Pivovonsky and Max R. Nagel, Macmillan, New York, 1961. 481 pages. $12.50. *Infrared Physics*, 2(2):121, 1962.

252. L. Genzel, W. Martienssen, and H. A. Mueser. Marianus Czerny. *Physics Today*, 39(7):83, 1986.

253. A. A. Gershun. On the spectral density of radiation. *Uspekhi Fizicheskikh Nauk*, 46(3):388–395, 1952.

254. E. Geyer, H. König, and E. Wahls. *Tables of black body spectral radiance.* VEB Deutscher Verlag der Wissenschaften, Berlin, 1972.

255. S. R. Ghorpade and B. V. Limaye. *A course in calculus and real analysis.* Springer Science+Business Media, LLC, New York, 2006.

256. T. P. Gill. Some problems in low-temperature pyrometry. *Journal of the Optical Society of America*, 47(11):1000–1005, 1957.

257. E. J. Gillham. Thermal radiation. In A. E. Bailey, E. J. Gillham, E. F. G. Herington, and B. Rose, editors, *Tables of physical and chemical constants and some mathematical functions*, pages 76–77. Longman Group Limited, London, 1973.

258. F. R. Gilmore. *A table of the Planck radiation function and its integral.* Project RAND Research Memorandum RM-1743, July 1956.

259. Y. Göğüş and J. Kestin. Black-body radiation functions. *Journal of Pure and Applied Sciences* (Middle East Technical University), 9(1):113–132, 1976.

260. M. Golay. *Introduction to astronomical photometry.* D. Reidel Publishing Company, Dordrecht, 1974.

261. S. A. Golden. *Spectral and integrated blackbody radiation functions.* Research Report 60-23, Rocketdyne, A Division of North American Aviation, Inc., Canoga Park, California, 1960.

262. R. L. Goodstein. On the evaluation of Planck's integral. *Edinburgh Mathematical Notes*, 37:17–20, 1949.

263. R. M. Goody. *Atmospheric radiation. I Theoretical basis.* Oxford, London, 1964.

264. R. M. Goody and Y. L. Yung. *Atmospheric radiation. Theoretical basis.* Oxford University Press, Oxford, second edition, 1989.

265. H. Görtler. Zur Geschichte des π-Theorems. *Zeitschrift für Angewandte Mathematik und Mechanik*, 55(1):3–8, 1975.

266. R. L. Gouchoe. *Computation of spectral energy distribution for incandescent sources.* BSc thesis, Massachusetts Institute of Technology, Cambridge, Massachusetts, 1934.

267. R. L. Graham, D. E. Knuth, and O. Patashnik. *Concrete mathematics.* Addison-Wesley Publishing Company, Reading, Massachusetts, second edition, 1994.

268. S. G. Grenishen and L. N. Kaporskiĭ. Optical radiation. Part 1. *Journal of Optical Technology*, 67(1):79–89, 2000.

269. D. A. Grier. Table making for the relief of labour. In M. Campbell-Kelly,

M. Croarken, R. Flood, and E. Robson, editors, *The History of Mathematical Tables. From Sumer to Spreadsheets*, pages 264–292. Oxford University Press, Oxford, 2003.

270. P. R. Griffiths. Forty years of FT-IR spectrometry: Strong-men, Connesmen, and Block-busters or how Mertz raised the Hertz. *Analytical Chemistry*, 64(18):868–875, 1992.

271. H. R. J. Grosch. High speed arithmetic: the digital computer as a research tool. *Journal of the Optical Society of America*, 43(4):306–310, 1953.

272. M. M. Gurevich. On the spectral distribution of radiant energy. *Uspekhi Fizicheskikh Nauk*, 56(3):417–424, 1955.

273. M. M. Gurevich. Spectral distribution of radiant energy. *Soviet Physics Uspekhi*, 5(6):908–912, 1963.

274. A. E. Haas. Note on the photon emission of a black body. *Journal of the Optical Society of America*, 28(5):167, 1938.

275. A. E. Haas and E. Guth. The relation between Stefan's radiation law and Nernst's heat theorem. *Physical Review*, 53(4):324, 1938.

276. H. L. Hackforth. *Infrared Radiation*. McGraw-Hill Book Company, Inc., New York, 1960.

277. D. Hahn, J. Metzdorf, U. Schley, and J. Verch. *Seven-Place Tables of the Planck Function for the Visible Spectrum*. Friedr. Vieweg & Sohn, Braunschweig, 1964.

278. J. A. Hall and C. R. Barber. The International Temperature Scale—1948 revision. *British Journal of Applied Physics*, 1(4):81–86, 1950.

279. H. G. W. Harding. The colour temperature of light sources. *Proceedings of the Physical Society. Section B*, 63(9):685–699, 1950.

280. H. G. W. Harding and R. B. Sisson. Distribution coefficients for the calculation of colours on the C.I.E. trichromatic system for total radiators at $1500-250-3500°$K., and $2360°$K. ($C_2 = 14\,350$). *Proceedings of the Physical Society*, 59(5):814–827, 1947.

281. H. G. W. Harding and T. Vickers. *The Spectral Distribution of Power from a Planckian Radiator ($C_2 = 1.438$ cm.deg)*. National Physical Laboratory, Teddington, Middlesex, September 1953.

282. T. R. Harrison. *Radiation Pyrometry and Its Underlying Principles of Radiant Heat Transfer*. John Wiley & Sons, Inc., New York, 1960.

283. M. Hatch. Approximations for the Planck function. *Applied Optics*, 12(3):617–619, 1973.

284. B. Hayes. Why W? *American Scientist*, 93(2):104–108, 2005.

285. N. D. Hayes. Roots of the transcendental equation associated with a certain difference–differential equation. *Journal of the London Mathematical Society*, 25(3):226–232, 1950.

286. M. Hazewinkel, editor. *Encyclopeadia of Mathematics. Volume 1*. Kluwer Academic Publishers, Dordrecht, 1995.

287. M. A. Heald. Where is the 'Wien peak'? *American Journal of Physics*, 71(12):1322–1323, 2003.

288. E. O. Hercus. *Elements of Thermodynamics and Statical Mechanics*. Melbourne University Press, Carlton, 1950.

289. E. Hertzsprung. Über die optische Stärke der Strahlung des schwarzen Körpers und das minimale Lichtäquivalent. *Zeitschrift für wissenschaftliche Photographie, Photophysik und Photochemie*, 4:43–54, 1906.

290. G. Hettner. Die Bedeutung von Rubens Arbeiten für die Plancksche Strahlungsformel. *Die Naturwissenschaften*, 10(48):1033–1038, 1922.

291. A. Hobson. Solar blackbody spectrum and the eye's sensitivity. *American Journal of Physics*, 71(4):295, 2003.

292. G. Hofmann. Zur Darstellung der spektralen Verteilung der Strahlungsenergie. *Arch Meteorol Geophys Bioklimatol B*, 6(3):274–279, 1955.

293. B. R. Holeman. Review of 'Infrared System Engineering' by Richard D. Hudson, Jr. pp. xxvi+642; John Wiley & Sons, 1969. price 185s. *Journal of the Royal Naval Scientific Services*, 26(4):279, 1971.

294. L. L. Holladay. Proportion of energy radiated by incandescent solids in various spectral regions. *Journal of the Optical Society of America and Review of Scientific Instruments*, 17(5):329–342, 1928.

295. K. G. T. Hollands. *Thermal Radiation Fundamentals*. Begell House, Inc., New York, 2004.

296. M. R. Holter, S. Nudelman, G. H. Suits, W. L. Wolfe, and G. J. Zissis. *Fundamental of Infrared Technology*. The MacMillan Company, New York, 1962.

297. L. Hopf. Über Modellregeln und Dimensionsbetrachtungen. *Die Naturwissenschaften*, 8(6):107–111, 1920.

298. G. W. Hopkins. Basic Algorithms for Optical Engineering. In R. R. Shannon and J. C. Wyant, editors, *Applied Optics and Optical Engineering*, volume 9, pages 1–32. Academic Press, New York, 1983.

299. H. C. Hottel. Radiant-Heat Transmission. In W. H. McAdams, editor, *Heat Transmission*, pages 55–125. McGraw-Hill Book Company, New York, third edition, 1954.

300. H. C. Hottel and A. F. Sarofim. *Radiative Transfer*. McGraw-Hill Book Company, New York, 1967.

301. J. N. Howard. Review of 'Tables of the Fractional Function for the Planck Radiation Law' by M. Czerny and A. Walther, Spring-Verlag, Berlin, 1961. *Applied Optics*, 1(3):342, 358, 1962.

302. J. N. Howard. From the Editor. *Applied Optics*, 4(6):676, 1965.

303. J. N. Howard. From the Editor. *Applied Optics*, 7(4):693–694, 1968.

304. J. N. Howard. Radiation slide rule. *Applied Optics*, 11(9):2107, 1972.

305. J. N. Howard. Foreword – Dr. John N. Howard. In J. Fox, editor, *Proceedings of the Symposium on Optical and Acoustical Micro-electronics*, volume 23, page xiii. Polytechnic Press, Brooklyn, New York, 1975.

306. J. N. Howard. From the Editor. *Applied Optics*, 18(8):a85,1232, 1979.

307. J. N. Howard. Advertisements in JOSA. *Optics and Photonics News*, 13(11):10–11, 2002.

308. John N. Howard. *Thermophotovoltaic energy conversion in submarine nuclear power plants*. MSc thesis, Naval Postgraduate School, Monterey, California, 2011.

309. J. R. Howell, R. Siegel, and M. P. Mengüç. *Thermal Radiation Heat Transfer*. CRC Press, Boca Raton, fifth edition, 2011.

310. J. Huang, X. Xu, W. Zhong, H. Li, X. Dai, X. Wang, and Z. Mi. Mathematical and physical constants in displacement laws of black-body radiation. *Chinese Journal of Infrared Research*, 7A(4):281–289, 1988.

311. R. D. Hudson, Jr. *Infrared System Engineering*. John Wiley & Sons, Hoboken, New Jersey, 1969.

312. H. E. Huntley. *Dimensional Analysis*. MacDonald, London, 1952.

313. W. G. Hyzer. Slide-rule-type calculators can save you valuable time, whatever your mathematical background. *Photomethods*, 17(11):24,26,28,108,109, 1974.

314. W. G. Hyzer. Computational aids – slide rule and electronic – can aid the scientific and industrial photographer. *Photomethods*, 18(1):20, 1975.

315. F. P. Incropera, D. P. DeWitt, T. L. Bergman, and A. S. Lavine. *Foundations of Heat Transfer*. John Wiley & Sons, Inc., Hoboken, New Jersey, sixth edition, 2013.

316. Infrared Information Analysis Center. *Infrared Information Analysis Center: User's Guide Supplement*. Environmental Research Institute of Michigan, Ann Arbor, Michigan, 1993.

317. T. Ishikawa. Normalized spectral-distribution function of blackbody radiation. *Journal of the Illuminating Engineering Institute of Japan*, 67(1):13–14, 1983.

318. A. Ivanov. Izlucheniye (Radiation). In P. K. Martens, editor, *Tekhnicheskaya Entsiklopediya (Technical Encyclopedia)*, volume 8, pages 756–763. Aktsionernoye Obshchestvo, Moscow, 1929.

319. H. E. Ives. Luminous Efficiency. *Transactions of the Illuminating Engineering Society*, 5(2):113–137, 1910.

320. E. Jahnke and F. Emde. *Funktionentafeln mit Formeln und Kurven*. B. G. Teubner, Leipzig, 1933.

321. E. Jahnke, O. Lummer, and E. Pringsheim. Kritisches zur Herleitung der Wien'schen Spectralgleichung. *Annalen der Physik* [Vierte Folge], 4(1):225–230, 1901.

322. P. K. Jain. On blackbody radiation. *Physics Education*, 26(3):190–194, 1991.

323. P. K. Jain. IR, visible and UV components in the spectral distribution of blackbody radiation. *Physics Education*, 31(3):149–155, 1996.

324. J. A. Jamieson, R. H McFee, G. N. Plass, R. H. Grube, and R. G. Richards. *Infrared Physics and Engineering*. McGraw-Hill Book Company, Inc., New York, 1963.

325. M. Janes. The Gauss–Laguerre approximation method for the evaluation of integrals in thermal radiation theory. *Infrared Physics*, 24(1):49–56, 1984.

326. W. S. Janna. *Engineering Heat Transfer*. CRC Press, Boca Raton, second edition, 2000.

327. J. H. Jeans. On the Laws of Radiation. *Proceedings of the Royal Society of London. Series A, Containing Papers of a Mathematical and Physical Character*, 76(513):545–552, 1905.

328. J. H. Jeans. On the partition of energy between matter and æther. *The London, Edinburgh, and Dublin Philosophical Magazine and Journal of Science*, [Sixth Series], 10(55):91–98, 1905.

329. J. H. Jeans. Bemerkung zu einer neuen Ableitung des Wienschen Verschiebungsgesetzes. Erwiderung auf Herrn P. Ehrenfests Abhandlung. *Physikalische Zeitschrift*, 7(19):667, 1906.

330. J. H. Jeans. Bemerkung zu einer neuen Ableitung des Wienschen Verschiebungsgesetzes. Erwiderung auf die Kritik des Herrn Ehrenfests. *Physikalische Zeitschrift*, 8(3):91–92, 1907.

331. A. Jeffrey and H.-H. Dai. *Handbook of Mathematical Formulas and Integrals*. Academic Press, Amsterdam, fourth edition, 2008.

332. K. Jellinek. *Physikalische Chemie; der homogenen und heterogenen Gasreaktionen, unter besonderer Berücksichtigung der Strahlungs- und Quantenlehre, sowie des Nernstschen Theorems.* Verlag von S. Hirzel, Leipzig, 1913.

333. W. Jing and T. Fang. Note on a new blackbody fraction function used for surfaces with linear emissivity in a wavelength interval. *Journal of Heat Transfer*, 135(5):054506, 2013.

334. R. B. Johnson. Correctly making panoramic imagery and the meaning of optical center. In P. Z. Mouroulis, W. J. Smith, and R. B. Johnson, editors, *Current Developments in Lens Design and Optical Engineering IX*, volume 7060 of *Proceeding of SPIE*, page 70600F, 2008.

335. R. B. Johnson and E. E. Branstetter. Integration of Planck's equation by the Laguerre–Gauss quadrature method. *Journal of the Optical Society of America*, 64(11):1445–1449, 1974.

336. R. B. Johnson, C. Feng, and J. D. Fehribach. On the validity and techniques of temperature and emissivity measurements. In R. D. Lucier, editor, *Thermosense X: Thermal Infrared Sensing for Diagnostics and Control*, volume 934 of *Proceeding of SPIE*, pages 202–206, 1988.

337. R. B. Johnson and S. M. Stewart. A history of slide rules for blackbody radiation computations. In M. G. Turner, editor, *Tribute to William Wolfe*, volume 8483 of *Proceeding of SPIE*, page 848302, SPIE, Bellingham, WA, 2012.

338. H. Kangro. *Early History of Planck's Radiation Law.* Taylor & Francis Ltd, London, 1976.

339. J. Kaspar. Radiometry. In D. E. Gray, editor, *American Institute of Physics Handbook*, pages 6-153–6-156. McGraw-Hill Book Company, New York, third edition, 1972.

340. S. Katz. Review of 'Tables of Blackbody Radiation Functions' by M. Pivovonsky and M. R. Nagel, Macmillan, New York, 1961. 481 pp. $12.50. *Applied Optics*, 1(3):334, 342, 1962.

341. R. W. Kavanagh, E. K. Björnerud, and S. S. Penner. Nomogram for the Evaluation of Blackbody Radiancy and of Peak and Total Intensities for Spectral Lines with Doppler Contour. *Journal of the Optical Society of America*, 43(5):380–382, 1953.

342. R. W. Kavanagh and S. S. Penner. Nomogram for the Evaluation of Blackbody Radiancy and of Peak and Total Intensities for Spectral Lines with Lorentz Contour. *Journal of the Optical Society of America*, 43(5):383–384, 1953.

343. G. W. C. Kaye and T. H. Laby. *Tables of Physical and Chemical Constants and some Mathematical Functions.* Longman's, Green, and Co., London, 1911.

344. G. W. C. Kaye and T. H. Laby. *Tables of Physical and Chemical Constants and some Mathematical Functions.* Longman Group Limited, London, fourteenth edition, 1973.

345. H. Kayser. *Handbuch der Spectroscopie. Zweiter Band.* Verlag von S. Hirzel, Leipzig, 1902.

346. P. G. Kendall. Black body radiation source. US Patent 2,952,762, 1960.

347. R. L. Kenyon. The FIAT Review of German science. first in a series on science and technology in Germany. *Chemical and Engineering News*, 25(14):962–963, 1947.

348. R. J. Keyes and T. M. Quist. Low-level coherent and incoherent detection in the infrared. In R. K. Willardson and A. C. Beer, editors, *Semiconductors and Semimetals. Volume 5: Infrared Detectors*, pages 321–359. Academic Press, New York, 1970.

349. A. Kh. Khrgian. B. B. Golitsyn and his work in physics and geophysics. *Izvestiya, Atmospheric and Oceanic Physics*, 27(5):409–411, 1991.

350. H. Kienle. Zur Berechnung von Farbtemperaturen. *Zeitschrift für Astrophysik*, 20(4):239–245, 1941.

351. H. Kim, S. C. Lim, and Y. H. Lee. Size effect of two-dimensional thermal radiation. *Physics Letters A*, 375(27):2661–2664, 2011.

352. R. Kingslake. *Applied Optics and Optical Engineering, Volume II, The Detection of Light and Infrared Radiation*. Academic Press, New York, 1969.

353. R. Kingslake and R. B. Johnson. *Lens Design Fundamentals*. Academic Press, Burlington, second edition, 2010.

354. G. Kirchhoff. Ueber das Verhältniss zwischen dem Emissionsvermögen und dem Absorptionsvermögen der Körper für Wärme and Licht. *Annalen der Physik und Chemie* [Zweite Folge], 109(2):275–301, 1860.

355. G. Kirchhoff. On the relation between the radiating and absorbing powers of different bodies for light and heat. *The London, Edinburgh, and Dublin Philosophical Magazine and Journal of Science*, [Fourth Series], 20(130):1–21, 1860.

356. M. J. Klein. Max Planck and the beginning of the quantum theory. *Archive for History of Exact Sciences*, 1(5):459–479, 1961.

357. R. S. Knox. Direct-reading Planck distribution slide rule. *Journal of the Optical Society of America*, 46(10):879–881, 1956.

358. W. Kofink. Kleine Bemerkung zum Planckschen Strahlungsgesetz. *Naturwissenschaften*, 38(10):234, 1951.

359. W. H. Kohl. *Materials Technology for Electron Tubes*. Reinhold Publishing Corporation, New York, 1951.

360. N. S. Kopeika. *A System Engineering Approach to Imaging*. SPIE Press, Bellingham, 1998.

361. L. J. Kozlowski and W. F. Kosonocky. *Infrared Detector Arrays*, volume II of *Handbook of Optics*, chapter 33. McGraw Hill, New York, third edition, 2010.

362. F. Kreith, R. M. Manglik, and M. S. Bohn. *Principles of Heat Transfer*. Cengage Learning, Stamford, CT, seventh edition, 2011.

363. A.-W. Kron. Nomographische Darstellung der absoluten und relativen Strahlung nach Planck im sichtbaren Wellenlängenbereich. Manuscript. Institut für Praktische Mathematik der Technischen Hochschule Darmstadt, 1946.

364. P. W. Kruse, L. D. McGlauchlin, and R. B. McQuistant. *Elements of Infrared Technology: Generation, Transmission, and Detection*. John Wiley & Sons, Inc., New York, 1962.

365. F. Kurlbaum. Ueber eine Methode zur Bestimmung der Strahlung in absolutem Maass und die Strahlung des schwarzen Körpers zwischen 0 und 100 Grad. *Annalen der Physik und Chemie* [Neue Folge], 65(8):746–760, 1898.

366. F. Kurlbaum and H. Rubens. Über die Emission langer Wellen durch den schwarzen Körper. *Verhandlungen der Deutschen Physikalischen Gesellschaft*, 2(13):181, 1900.

367. R. L. LaFara, E. L. Miller, W. E. Pearson, and J. F. Peoples. *Tables of Black Body Radiation and the Transmission Factor for Radiation through Water Vapor.* US Naval Ordnance Plant, Indianapolis, Indiana, November 1955.

368. J. H. Lambert. Observationes variae in mathesin puram. *Acta Helveticae physico-mathematico-anatomico-botanico-medic,* 3:128–168, 1758.

369. V. Lampret, J. Peternelj, and A. Krainer. Luminous flux and luminous efficacy of black-body radiation: an analytical approximation. *Solar Energy,* 73(5):319–326, 2002.

370. J. Landen. A new method of computing the sums of certain series. *Philosophical Tansactions of the Royal Society of London,* 51:553–565, 1760.

371. S. P. Langley. The bolometer and radiant energy. *Proceedings of the American Academy of Arts and Sciences,* 16:342–358, 1881.

372. S. P. Langley. The spectrum of an Argand burner. *Science,* 1(17):481–484, 1883.

373. S. P. Langley. Experimental determination of wave-lengths in the invisible prismatic spectrum. *American Journal of Science* [Third Series], 27(159):169–188, 1884.

374. S. P. Langley and F. W. Very. On the cheapest form of light from studies at the Allegheny Observatory. *The London, Edinburgh and Dublin Philosophical Magazine and Journal of Science* [Fifth Series], 30(184):260–280, 1890.

375. J. Larmor. On the relations of radiation to temperature. In *Report of the Sevenieth Meeting of the British Association for the Advancement of Science,* pages 657–659. John Murray, London, 1900.

376. J. Larmor. On the relations of radiation to temperature. *Nature,* 63(1626):216–218, 1900.

377. J. Larmor. Theory of Radiation. In *The Encyclopædia Britannica,* volume 22, pages 785–793. Cambridge University Press, Cambridge, 11th edition, 1911.

378. A. LaRocca. Artifical sources. In M. Bass, editor, *Handbook of Optics. Volume II. Design, Fabrication and Testing; Sources and Detectors; Radiometry and Photometry,* pages 15.1–15.54. McGraw-Hill Book Company, Inc., New York, third edition, 2010.

379. A. LaRocca and G. J. Zissis. *A Slide Rule for Radiance Calculations.* University of Michigan, Report No. 2144-247-T, May 1958.

380. S. H. Lavington. *Early British Computers: the story of vintage computers and the people who built them.* Manchester University Press, Manchester, 1980.

381. D. Lawson. A closer look at Planck's blackbody equation. *Physics Education,* 32(5):321–326, 1997.

382. D. Lawson. The blackbody fraction, infinite series and spreadsheets. *International Journal of Engineering Education,* 20(6):984–990, 2004.

383. E. Lax and M. Pirani. *Temperaturstrahlung Fester Körper.* Verlag von Julius Springer, Berlin, 1929.

384. C. H. Lees. Physics at the British Association. *Nature,* 62(1614):562–565, 1900.

385. E.-M. Lémeray. Sur les racines de l'equation $x = a^x$. *Nouvelles Annales de Mathématiques,* 15:548–556, 1896.

386. E.-M. Lémeray. Sur les racines de l'equation $x = a^x$. Racines imaginaires. *Nouvelles Annales de Mathématiques,* 16:54–61, 1897.

387. L. Levi. *Applied Optics.* John Wiley & Sons, New York, 1968.

388. L. Levi. *Handbook of Tables of Functions for Applied Optics*. CRC Press, Inc., Cleveland, Ohio, 1974.

389. G. N. Lewis and E. Q. Adams. Notes on quantum energy. *Physical Review*, 4(4):331–344, 1914.

390. R. A. Lewis. Let's talk terahertz! *American Journal of Physics*, 79(4):341, 2011.

391. G. J. Lidstone. Notes on Everett's interpolation formula. *Proceedings of the Edinburgh Mathematical Society*, 40:21–26, 1921.

392. R. J. List. *Smithsonian Meteorological Tables*. The Smithsonian Institution, Washington, DC, sixth revised edition, 1949.

393. Z. Liu. The discussion of Wien's displacement law's formula. *Journal of Baoji Teacher College (Natural Science)*, 2(1):52–54, 1993.

394. E. Lommel. Theorie der Absorption unf Fluorescenz. *Annalen der Physik und Chemie* [Neue Folge], 3(2):251–283, 1878.

395. H. A. Lorentz. De theorie der strahling en de tweede wet der thermodynamica. In *Koninklijke Akademie van Wetenschappen te Amsterdam*, volume 9, pages 418–434, 1900.

396. H. A. Lorentz. The theory of radiation and the second law of thermodynamics. In *Koninklijke Akademie van Wetenschappen te Amsterdam*, volume 3, pages 436–450, 1901.

397. A. N. Lowan. *Tables of the Exponential Function e^x*. National Bureau of Standards, New York, 1939.

398. A. N. Lowan. *Miscellaneous Physical Tables. Planck's radiation functions and electronic functions*. National Bureau of Standards, New York, 1941.

399. A. N. Lowan and G. Blanch. Tables of Planck's radiation and photon functions. *Journal of the Optical Society of America*, 30(2):70–81, 1940.

400. J. Lu, Z. L. Xue, and D. J. Zhang. Some new constants in infrared physics and new function table of $f(\lambda T)$, $F(\lambda T)$ of blackbody (III). *Journal of Harbin University of Science and Technology*, 17(1):40–48, 1993.

401. R. Luck and J. W. Stevens. Explicit solutions for transcendental equations. *SIAM Review*, 44(2):227–233, 2002.

402. O. Lummer. Ueber die Strahlung des absolut schwarzen Körpers und seine Verwirklichung. *Naturwissenschaftliche Rundschau*, 11(6):65–68, 1896.

403. O. Lummer. Ueber die Strahlung des absolut schwarzen Körpers und seine Verwirklichung (Fortsetzung). *Naturwissenschaftliche Rundschau*, 11(7):81–83, 1896.

404. O. Lummer. Ueber die Strahlung des absolut schwarzen Körpers und seine Verwirklichung (Schluss). *Naturwissenschaftliche Rundschau*, 11(8):93–95, 1896.

405. O. Lummer. Le rayonnement des corps noirs. In Ch.-Éd. Guillaume and L. Poin caré, editors, *Rapports présentés au Congrès International de Physique réuni a Paris en 1900. Tome II*, pages 41–99. Gauthier-Villars, Paris, 1900.

406. O. Lummer and E. Jahnke. Ueber die Spectralgleichung des schwarzen Körpers und des blanken Platins. *Annalen der Physik* [Vierte Folge], 3(10):283–297, 1900.

407. O. Lummer and F. Kurlbaum. Bolometrische untersuchungen. *Annalen der Physik und Chemie* [Neue Folge], 46(6):204–224, 1892.

408. O. Lummer and F. Kurlbaum. Der electrisch geglühte „absolut schwarz" Körper und seine Temperaturemessung. *Verhandlungen der Physikalischen Gesellschaft zu Berlin*, 17(9):106–111, 1898.

409. O. Lummer and E. Pringsheim. Die Strahlung eines „schwarzen" Körpers zwischen 100 und 1300°C. *Annalen der Physik und Chemie* [Neue Folge], 63(13):395–410, 1897.

410. O. Lummer and E. Pringsheim. „Die Vertheilung der Energie im Spectrum des schwarzen Körpers". *Verhandlungen der Deutschen Physikalischen Gesellschaft*, 1(1):23–41, 1899.

411. O. Lummer and E. Pringsheim. 1. Die Vertheilung der Energie im Spectrum des schwarzen Körpers und des blanken Platins; 2. Temperaturbestimmung fester glühender Körper. *Verhandlungen der Deutschen Physikalischen Gesellschaft*, 1(12):215–235, 1899.

412. O. Lummer and E. Pringsheim. Ueber die Strahlung des schwarzen Körpers für lange Wellen. *Annalen der Physik* [Vierte Folge], 2(9):163–180, 1900.

413. O. Lummer and E. Pringsheim. Notiz zu unserer Arbeit: Ueber die Strahlung eines „schwarzen" Körpers zwischen 100° und 1300°C. *Annalen der Physik* [Vierte Folge], 3(9):159–160, 1900.

414. O. Lummer and E. Pringsheim. Der elektrisch geglühte „schwarze" Körper. *Annalen der Physik* [Vierte Folge], 5(8):829–836, 1901.

415. O. Lummer and E. Pringsheim. Kritisches zur schwarzen Strahlung. *Annalen der Physik* [Vierte Folge], 6(9):192–210, 1901.

416. O. Lummer and E. Pringsheim. Die strahlungstheoretische temperaturskala und ihre verwirklichung bis 2300° abs. *Berichte der Deutschen Physikalischen Gesellschaft*, 5(1):3–13, 1903.

417. Q. Luo, Z. Wang, and J. Han. A Padé approximant approach to two kinds of transcendental equations with applications in physics. *European Journal of Physics*, 36(3):035030, 2015.

418. D. K. Lynch and B. H. Soffer. On the solar spectrum and the color sensitivity of the eye. *Optics and Photons News*, 10(3):28–30, 1999.

419. L. Ma, J. Yang, and J. Nie. Two forms of Wien's displacement law. *Latin American Journal of Physics Education*, 3(3):566–568, 2009.

420. D. L. Macadam. Review of 'Seven-Place Tables of the Planck Function for the Visible Spectrum.' Dietrich Hahn, Joachim Metzdorf, Ulrich Schley, and Jochim Verch. Friedrich Vieweg und Sohn, Braunschweig, Germany, and Academic Press Inc., New York, 1964, Pp. xxi+135. Price \$5.50. *Journal of the Optical Society of America*, 55(1):112, 1965.

421. E. O. Macagno. Historico-critical review of dimensional analysis. *Journal of the Franklin Institute*, 292(6):391–402, 1971.

422. W. Macke. Review of 'M. Czerny–A. Walther, Tabellen der Bruchteilfunktionzum Planckschen Strahlungsgesetz.' VIII+59 S. m. 8 Abb. Berlin/Göttingen/Heidelberg 1961. Springer-Verlag. Preis geb. DM 28,–. *Zeitscgrift für Angewandte Mathematik und Mechanik*, 42(4-5):218, 1962.

423. R. P. Madden. A. Francis Turner. Frederic Ives Medalist for 1971. *Journal of the Optical Society of America*, 62(8):927–930, 1972.

424. M. Makowski and L. A. J. Verra. A slide rule for radiation calculations. The National Archives (TNA): Public Record Office (PRO) ADM 213/438, September 1947.

425. M. W. Makowski. A slide rule for radiation calculations. *Review of Scientific Instruments*, 20(12):876–884, 1949.

426. M. W. Makowski. Erratum: a slide rule for radiation calculations. *Review of Scientific Instruments*, 21(4):336, 1950.

427. X. P. V. Maldague. *Theory and Practice of Infrared Technology for Nondestructive Testing*. John Wiley & Sons, Inc., New York, 2001.

428. J. M. Marr and F. P. Wilkin. A better presentation of Planck's radiation law. *American Journal of Physics*, 80(5):399–405, 2012.

429. G. L. Matloff, T. Taylor, and C. Powell. Phobos/deimos sample return via solar sail. *Annuals of the New York Academy of Sciences*, 1065:429–440, 2005.

430. V. I. Matveev. Estimating relative black-body densities in the spectral ranges of infrared vision devices. *Measurement Techniques*, 28(12):1071–1073, 1985.

431. J. A. Mauro. *Optical Engneering Handbook*. General Electric Company, Syracuse, New York, 1966.

432. J. C. Maxwell. Remarks on the mathematical classification of physical quantities. *Proceedings of the Mathematical Society of London*, 3(34):224–233, 1871.

433. E. B. Mayfield. *Radiometric Temperature Measurements of Short Duration Events*. PhD thesis, The University of Utah, Salt Lake City, Utah, 1959.

434. E. B. Mayfield. *Black Body Radiation Tables using Planck's Radiation Distribution Formula*. US Naval Ordnance Test Station, China Lake, California, 1959.

435. E. B. Mayfield. *Radiometric Temperature Measurements of Short Duration Events*. US Naval Ordnance Test Station, China Lake, California, December 1961.

436. J. K. McDonald. Tables of the Planck radiation function $B_\nu(T)$. *Publications of the Dominion Astrophysical Observatory* (Victoria, BC), 10(5):127–143, 1955.

437. G. B. McIntosh. Camera manufacturers 1, Stefan–Boltzmann 0. *Think Thermally*, Winter:8–9, 2009.

438. W. F. Meggers, C. H. Corliss, and B. F. Scribner. Tables of spectral-line intensities. Part II. Arranged by wavelengths. *National Bureau of Standards Monograph*, 32, 1961.

439. J. Mehra and H. Rechenberg. *The Historical Development of Quantum theory. Volume 1, Part 1. The quantum theory of Planck, Einstein, Bohr and Sommerfeld: its foundation and the rise of its difficulties 1900–1925*. Springer-Verlag, New York, 1982.

440. C. Mendoza. Yet another proof of the π-theorem. *Mechanics Research Communications*, 23(3):299–303, 1996.

441. M. P. Mengüç and R. Viskanta. On the radiative properties of polydispersions: a simplified approach. *Combustion Science and Technology*, 44(3-4):143–159, 1985.

442. G. S. Merrill. Tungsten Lamps. *Transactions of the American Institute of Electrical Engineers*, 29(2):1709–1729, 1910.

443. T. E. Michels. *Planck Functions and Integrals: Methods of Computation*. National Aeronautics and Space Administration, Washington, DC, March 1968.

444. W. Michelson. Essai théorique sur la distribution de l'énergie dans les spectres des solides. *Journal de Physique Théorique et Appliquée*, 6(1):467–479, 1887.

445. W. Michelson. Theoretical essay on the distribution of energy in the spectra of solids. *The London, Edinburgh, and Dublin Philosophical Magazine and Journal of Science* [Fifth Series], 25(156):425–435, 1888.

446. Z. Miduno. Table and graph for the calculations of black body radiations. *Proceedings of the Physico-Mathematical Society of Japan*, 20(11):951–961, 1938.

447. Z. Miduno. Errata:–Table and graph for the calculations of black body radiations. *Proceedings of the Physico-Mathematical Society of Japan*, 21(2):89–90, 1939.

448. A. A. Mikhailov, editor. *Physics of Stars and Stellar Systems*. Israel Program for Scientific Translations Press, Jerusalem, 1969.

449. A. F. Mills. *Basic Heat and Mass Transfer*. Prentice Hall, Upper Saddle River, New Jersey, 1999.

450. G. Moddel. Fractional bandwidth normalization for optical spectra with applications to the solar blackbody spectrum. *Applied Optics*, 40(3):413–416, 2001.

451. M. F. Modest. *Radiative Heat Transfer*. McGraw-Hill, New York, 1993.

452. M. F. Modest. *Radiative Heat Transfer*. Academic Press, San Diego, second edition, 2003.

453. M. F. Modest. *Radiative Heat Transfer*. Academic Press, San Diego, third edition, 2013.

454. P. J. Mohr, B. N. Taylor, and D. B. Newell. CODATA recommended values of the fundamental physical constants: 2010. *Reviews of Modern Physics*, 84(4):1527–1605, 2012.

455. P. Moon. *The Scientific Basis of Illuminating Engineering*. McGraw-Hill Book Company, New York, 1936.

456. P. Moon. A Table of Planck's Function from 3500 to 8000°K. *Journal of Mathematics and Physics*, 16:133–157, 1937.

457. P. Moon. *A Table of Planck's Function from 3500 to 8000°K*. Contribution from the Department of Electrical Engineering, No. 131. Publications from the Massachusetts Institute of Technology, 1938.

458. P. Moon. *A Table of Planck's Function: 2000 to 3500° K*. Contribution from the Department of Electrical Engineering, No. 316. Publications from the Massachusetts Institute of Technology, 1947.

459. P. Moon. A Table of Planckian Radiation. *Journal of the Optical Society of America*, 38(3):291–294, 1948.

460. P. Moon and D. E. Spence. Analytic expressions in photometry and colorimetry. *Journal of Mathematics and Physics*, 25:111–190, 1946.

461. P. Moon and D. E. Spence. Approximations to Planckian distributions. *Journal of Applied Physics*, 17(6):506–514, 1946.

462. M. R. Nagel and A.-W. Kron. *Zahlentafeln und graphische Darstellungen zur Planck'schen Strahlungsfunktion*. Technische Hochschule Darmstadt, Institut für praktische Mathematik, 1952.

463. E. L. Nichols. Preliminary note on the efficiency of the acetylene flame. *Physical Review* [Series 1], 11(4):215–229, 1900.

464. E. L. Nichols and M. L. Crehore. Studies of the lime light. *Physical Review* [Series 1], 2(3):161–169, 1894.

465. F. E. Nicodemus. *Self-Study Manual on Optical Radiation Measurements: Part I – Concepts.* Number 910-1 in National Bureau of Standard Technical Note. US Government Print Office, Washington, DC, 1976.

466. NIST. CODATA internationally recommended 2014 values of the fundamental physical constants. `physics.nist.gov/cuu/Constants`, 2015.

467. R. E. Nyswander. The distribution of energy in the spectrum of the tungsten filament. *Physical Review* [Series 1], 28(6):438–445, 1909.

468. F. W. J. Olver, D. W. Lozier, R. F. Boisvert, and C. W. Clark. *NIST Handbook of Mathematical Functions.* Cambridge University Press, Cambridge, 2010.

469. Y. Omoto. Table radiation function. *Journal of the Illuminating Engineering Institute of Japan*, 20(4):139–143, 1936.

470. J. M. Overduin. Eyesight and the solar Wien peak. *American Journal of Physics*, 71(3):216–219, 2003.

471. M. N. Özişik. *Radiative Transfer and Interactions with Conduction and Convection.* John Wiley & Sons, New York, 1973.

472. G. Páez and M. S. Scholl. Integrable and differentiable approximations to Planck's equation. In M. S. Strojnik and B. F. Andresen, editors, *Infrared Spaceborne Remote Sensing VI*, volume 3437 of *Proceedings of SPIE*, pages 371–377, 1998.

473. G. Páez and M. S. Scholl. Integrable and differentiable approximations to the generalized Planck's equations. In G. C. Holst, editor, *Infrared Imaging Systems: Design, Analysis, Modeling, and Testing X*, volume 3701 of *Proceedings of SPIE*, pages 95–105, 1999.

474. G. Páez, M. Strojnik, and T. Kranjc. Error evaluation in the series expansion of the generalized Planck's equation for radiation integrals. In M. Strojnik and B. F. Andresen, editors, *Infrared Spaceborne Remote Sensing IX*, volume 4486 of *Proceedings of SPIE*, pages 501–512, 2002.

475. J. Palacios. *Análisis Dimensional.* Espasa-Calpe, S.A., Madrid, 1956.

476. J. M. Palmer and B. G. Grant. *The Art of Radiometry.* SPIE Press, Bellingham, Washington, 2010.

477. J. R. Partington. Tyndall and Stefan's radiation law. *Nature*, 157(4000):879, 1946.

478. J. R. Partington. *An Advanced Treatise on Physical Chemistry. Volume One.* Longmans, Green and Co., London, 1949.

479. F. Paschen. Ueber Gesetzmäßigkeiten in den Spectren fester körper und über eine neue Bestimmung der Sonneutemperatur. *Nachrichten von der Königlichen Gesellschaft der Wissenschaften zu Göttingen, Mathematisch-Physikalische Klasse*, pages 295–304, 1895.

480. F. Paschen. On the existence of law in the spectra of solids bodies, and on a new determination of the temperature of the sun. *Astrophysical Journal*, 2:202–211, 1895.

481. F. Paschen. Ueber Gesetzmässigkeiten in den Spectren fester Körper. Erste Mittheilung. *Annalen der Physik und Chemie* [Neue Folge], 58(7):455–492, 1896.

482. F. Paschen. Ueber Gesetzmässigkeiten in den Spectren fester Körper. Zweite Mittheilung. *Annalen der Physik und Chemie* [Neue Folge], 60(4):662–723, 1897.

483. F. Paschen. Ueber die Vertheilung der Energie im Spectrum des schwarzen

Körper bei niederen Temperaturen. *Sitzungsberichte der Königlich Preussischen Akademie der Wissenschaften zu Berlin*, (I)(22):405–420, 1899.

484. F. Paschen. Ueber die Vertheilung der Energie im Spectrum des schwarzen Körper bei höheren Temperaturen. *Sitzungsberichte der Königlich Preussischen Akademie der Wissenschaften zu Berlin*, (II)(53):959–976, 1899.

485. F. Paschen. On the distribution of energy in the spectrum of the black body at low temperature. *Astrophysical Journal*, 10:40–57, 1899.

486. F. Paschen. On the distribution of energy in the spectrum of the black body at high temperature. *Astrophysical Journal*, 11:288–306, 1900.

487. F. Paschen. Ueber das Strahlungsgesetz des schwarzen Körpers. *Annalen der Physik* [Vierte Folge], 4(2):277–298, 1901.

488. F. Paschen. Ueber das Strahlungsgesetz des schwarzen Körpers. Entgegnung auf Ausführungen der Herren O. Lummer und E. Pringsheim. *Annalen der Physik* [Vierte Folge], 6(11):646–658, 1901.

489. F. Paschen and H. Wanner. A photometric method for the determination of the exponential constant of the emission function. *Astrophysical Journal*, 9:300–307, 1899.

490. F. Paschen and H. Wanner. Eine photometrische Methode zur Bestimmung der Expoentialconstanten der Emissionsfunction. *Sitzungsberichte der Königlich Preussischen Akademie der Wissenschaften zu Berlin*, (I)(2):5–11, 1899.

491. C. C. Paterson and B. P. Dudding. The estimation of high temperature by the method of colour identity. *Proceedings of the Physical Society of London*, 27(1):230–262, 1914.

492. R. Pavelle. The Planck integral cannot be evaluated in terms of a finite series of elementary functions. *Journal of Mathematical Physics*, 21(1):14, 1980.

493. R. Pavelle. Erratum: The Planck integral cannot be evaluated in terms of a finite series of elementary functions [J. Math. Phys. 21, 14 (1980)]. *Journal of Mathematical Physics*, 21(8):2313, 1980.

494. S. S. Penner, R. W. Kavanagh, and E. K. Björnerud. *Nomogram for the Evaluation of Blackbody Radiancy and of Peak and Total Intensities for Spectral Lines with Various Contours*. US Navy, Office of Naval Research, Contract Nonr-220(03), NR 015 210, Technical Report No. 14, California Institute of Technology, 1953.

495. W. Pepperhoff. *Temperaturstrahlung*. Verlag von Dr. Dietrich Steinkopff, Darmstadt, 1956.

496. S. Peters. Blackbody Radiation Calculator. https://play.google.com/store/apps/details?id=com.peters.blackbodyradiationcalculator, 2013.

497. S. Peters. private communication, 2014.

498. L. E. Picolet. Review of 'Tables and graphs for facilitating the computation of spectral-energy distribution by Planck's formula' by M. Katherine Frehafer, Associate Physicist, and Chester L. Snow, Draftsman, Bureau of Standards, assisted by Harry J. Keegan. Miscellaneous publication of the Bureau of Standards, no. 56, for sale by the Superintendent of Documents, Government Printing Office, Washington, D.C. Price, thirty-five cents. *Journal of the Franklin Institute*, 200(5):703–704, 1925.

499. L. J. Pinson. *Electro-optics*. John Wiley & Sons, New York, 1985.

500. E. J. Pisa. *Tables of Black-Body Radiation Functions and their Derivatives.* US Naval Ordnance Test Station, China Lake, California, December 1964.

501. M. Pivovonsky and M. R. Nagel. *Tables of Blackbody Radiation Functions.* The Macmillan Company, New York, 1961.

502. M. Pivovonsky and M. R. Nagel. *Tables of Blackbody Radiation Functions.* Literary Licensing, LLC, Whitefish, Montana, 2013.

503. M. Planck. Über irreversible Strahlungsvorgänge. Erste Mittheilung. *Sitzungsberichte der Königlich Preussischen Akademie der Wissenschaften zu Berlin* (Erster Halbband), (6):57–68, 1897.

504. M. Planck. Über irreversible Strahlungsvorgänge. Zweite Mittheilung. *Sitzungsberichte der Königlich Preussischen Akademie der Wissenschaften zu Berlin* (Zweiter Halbband), (34):715–717, 1897.

505. M. Planck. Über irreversible Strahlungsvorgänge. Dritte Mittheilung. *Sitzungsberichte der Königlich Preussischen Akademie der Wissenschaften zu Berlin* (Zweiter Halbband), (52):1122–1145, 1897.

506. M. Planck. Über irreversible Strahlungsvorgänge. Vierte Mittheilung. *Sitzungsberichte der Königlich Preussischen Akademie der Wissenschaften zu Berlin* (Zweiter Halbband), (34):449–476, 1898.

507. M. Planck. Über irreversible Strahlungsvorgänge. Fünfte Mittheilung (Schluss). *Sitzungsberichte der Königlich Preussischen Akademie der Wissenschaften zu Berlin* (Erster Halbband), (25):440–480, 1899.

508. M. Planck. Ueber irreversible Strahlungsvorgänge. *Annalen der Physik* [Vierte Folge], 1(1):69–122, 1900.

509. M. Planck. Entropie und Temperatur strahlender Wärme. *Annalen der Physik* [Vierte Folge], 1(4):719–737, 1900.

510. M. Planck. Über eine Verbesserung der Wien'schen Spectralgleichung. *Verhandlungen der Deutschen Physikalischen Gesellschaft*, 2(13):181, 1900.

511. M. Planck. Ueber eine Verbesserung der Wien'schen Spectralgleichung. *Verhandlungen der Deutschen Physikalischen Gesellschaft*, 2(13):202–204, 1900.

512. M. Planck. Zur Theorie des Gesetzes der Energieverteilung im Normalspectrum. *Verhandlungen der Deutschen Physikalischen Gesellschaft*, 2(17):237–245, 1900.

513. M. Planck. Ueber das Gesetz der Energieverteilung im Normalspectrum. *Annalen der Physik* [Vierte Folge], 4(3):553–563, 1901.

514. M. Planck. Über irreversible Strahlungsvorgänge. (Nachtrag). *Annalen der Physik* [Vierte Folge], 6(12):818–831, 1901.

515. M. Planck. Über irreversible Strahlungsvorgänge. (Nachtrag). *Sitzungsberichte der Königlich Preussischen Akademie der Wissenschaften zu Berlin* (Erster Halbband), (25):544–555, 1901.

516. M. Planck. *The Theory of Heat Radiation.* P. Blakiston's Sons & Co., Philadelphia, 1914.

517. M. Planck. *Eight Lectures on Theoretical Physics, Delivered at Columbia University in 1909.* Columbia University Press, Columbia, 1915.

518. M. Planck. On an improvement of Wien's equation for the spectrum. In D. ter Haar, editor, *The Old Quantum Theory*, pages 79–81. Pergamon Press, Oxford, 1967.

519. M. Planck. On the theory of the energy distribution law of the normal spectrum. In D. ter Haar, editor, *The Old Quantum Theory*, pages 82–90. Pergamon Press, Oxford, 1967.

520. M. Planck. The genesis and present state of development of the quantum theory. In *Nobel Lectures, Physics 1901–1921*, pages 407–418. Elsevier Publishing Company, Amsterdam, 1967.

521. J. N. Plendl and H. S. Plendl. Center frequency formulation of the blackbody radiation spectrum. *International Journal of Infrared and Millimeter Waves*, 10(11):1377–1386, 1989.

522. B. E. Pobedrya and D. V. Georgievskii. On the proof of the π-theorem in dimension theory. *Russian Journal of Mathematical Physics*, 13(4):431–437, 2006.

523. H. C. Pocklington. The radiation of a black body on the electromagnetic theory. In *Report of the Seventieth Meeting of the British Association for the Advancement of Science*, pages 654–655. John Murray, London, 1900.

524. G. Pólya and G. Szegö. *Aufgaben und Lehrsätze aus der Analysis I*. J. Springer, Berlin, 1925.

525. S. M. Pompea, D. W. Bergener, and D. F. Shepard. Optically black coating with improved infrared absorption and process of formation. US Patent 4,589,972, 1984.

526. L. S. Pontryagin. On zeros of some transcendental functions. *Izvestia Akademii Nauk SSSR. Seriya Matematicheskaya*, 6(3):115–134, 1942.

527. L. S. Pontryagin. On the zeros of some elementary transcendental functions. *American Mathematical Society Translations Series 2*, 1:95–110, 1955.

528. J. S. Preston. Computer—or table? *Nature*, 193(4812):207, 1962.

529. T. Preston. *The Theory of Heat*. Macmillan and Co., London, third edition, 1919.

530. I. G. Priest. A one-term pure exponential formula for the spectral distribution of radiant energy from a complete radiator. *Journal of the Optical Society of America and Review of Scientific Instruments*, 2(1):18–22, 1919.

531. I. G. Priest. A new formula for the spectral distribution of energy from a complete radiator. *Physical Review*, 13(4):314–317, 1919.

532. I. G. Priest. A new formula for the spectral distribution of energy from a complete radiator. II., Concerning the value of the constant D_2. *Physical Review*, 13(4):314–317, 1919.

533. E. Pringsheim. Ueber die Gesetze der schwarzen Strahlung. *Verhandlungen der Gesellschaft Deutscher Naturforscher und Ärzte. Zweiter Theil, I Hälfte*, pages 27–30, 1901.

534. E. Pringsheim. Ueber Temperaturbestimmungen mit Hülfe der Strahlungsgesetze. *Verhandlungen der Gesellschaft Deutscher Naturforscher und Ärzte. Zweiter Theil, I Hälfte*, pages 31–36, 1902.

535. E. Pringsheim. über die Strahlungsgesetze. II. Anwendungen der Strahlungsgesetze. *Archiv der Mathematik und Physik*, 7(4):296–308, 1904.

536. H. J. Queisser. Slow solar ascent. In B. Kramer, editor, *Advances in solid state physics*, volume 44, pages 3–12. Springer-Verlag, Berlin, 2004.

537. H. J. Queisser. Detailed balance limit for solar cell efficiency. *Materials Science and Engineering B*, 159-160(3):322–328, 2009.

538. H. J. Queisser. private communication, 2012.

539. G. C. Quinn. The infrared thermometer lets you see temperatures. *Factory*, 122(2):80–83, 1964.

540. P. Rabinowitz and G. Weiss. Tables of abscissas and weights for numerical evaluation of integrals of the form $\int_0^\infty e^{-x} x^n f(x)\, dx$. *Mathematical Tables and Other Aids to Computation*, 13(68):285–294, 1959.

541. M. M. Rathore and R. R. A. Kapuno, Jr. *Engineering Heat Transfer*. Jones & Barlett Learning, LLC, Sudbury, Massachusetts, second edition, 2011.

542. Lord Rayleigh. On the light from the sky, its polarization and colour. *The London, Edinburgh, and Dublin Philosophical Magazine and Journal of Science*, [Fourth Series], 41(271):107–120, 1871.

543. Lord Rayleigh. *The Theory of Sound*. MacMillian and Co, London, 1877-1878.

544. Lord Rayleigh. Distribution of energy in the spectrum. *Nature*, 27(702):559–560, 1883.

545. Lord Rayleigh. On the character of the complete radiation at a given temperature. *The London, Edinburgh, and Dublin Philosophical Magazine and Journal of Science*, [Fifth Series], 27(169):460–469, 1889.

546. Lord Rayleigh. On the question of the stability of the flow of fluids. *The London, Edinburgh, and Dublin Philosophical Magazine and Journal of Science*, [Fifth Series], 34(202):59–70, 1892.

547. Lord Rayleigh. Note on the pressure of radiation, showing an apparent failure of the usual electromagnetic equations. *The London, Edinburgh, and Dublin Philosophical Magazine and Journal of Science*, [Fifth Series], 45(277):522–525, 1898.

548. Lord Rayleigh. Remarks upon the law of complete radiation. *The London, Edinburgh, and Dublin Philosophical Magazine and Journal of Science*, [Fifth Series], 49(301):539–540, 1900.

549. Lord Rayleigh. The dynamical theory of gases and of radiation. *Nature*, 72(1855):54–55, 1905.

550. Lord Rayleigh. The constant of radiation as calculated from molecular data. *Nature*, 72(1863):243–244, 1905.

551. Lord Rayleigh. The principle of similitude. *Nature*, 95(2368):66–68, 1915.

552. F. Reiche. *Die Quantentheorie. Ihr Ursrung und ihre Entwicklung*. J. Springer, Berlin, 1921.

553. C. D. Reid. Nomographic-type slide rule for obtaining spectral blackbody radiation directly. *Review of Scientific Instruments*, 31(8):886–890, 1960.

554. M. Reiss. The Cos^4 Law of Illumination. *Journal of the Optical Society of America*, 35(4):283–288, 1945.

555. M. Reiss. Notes on the Cos^4 Law of Illumination. *Journal of the Optical Society of America*, 38(11):980–986, 1948.

556. M. M. Reynolds, R. J. Corruccini, M. M. Fulk, and R. M. Burley. Radiometry. In D. E. Gray, editor, *American Institute of Physics Handbook*, pages 6-64–6-67. McGraw-Hill Book Company, Inc., New York, 1957.

557. D. Riabouchinsky. Méthode des variables de dimensions zéro, et son application en aérodynamique. *L'Aérophile*, 1(Septembre):407–408, 1911.

558. R. J. Ribanda and E. A. Weller. The verification of an analytical solution: an important engineering lesson. *Journal of Engineering Education*, 88(3):281–283, 1999.

559. O. W. Richardson. *The Electron Theory of Matter*. Cambridge University Press, Cambridge, 1914.

560. M. J. Riedl. private communication, 2012.

561. B. Riemann. Ueber die Anzahl der Primzahlen unter einer gegebenen Grösse. *Monatsberichte der Königlichen Preußische Akademie der Wissenschaften zu Berlin*, pages 671–680, 1859.

562. K. F. Riley, M. P. Hobson, and S. J. Bence. *Mathematical Methods for Physics and Engineering*. Cambridge University Press, Cambridge, second edition, 2002.

563. R. H. Risch. The problem of integration in finite terms. *Transactions of the American Mathematical Society*, 139:167–189, 1969.

564. J. Roberts. *Black Body Radiation Functions. For monochromatic emissive power (spectral radiance). Relative spectral radiance to maximum fractional black emission (relative cumulative spectral radiance). The auxiliary error function.* Department of Chemical Engineering, University of Newcastle, January 1971.

565. S. M. Robinson. private communication, 2012.

566. D. M. Roessler. Review of 'Radiant Properties of Materials: Tables of Radiant Values for Black Body and Real Materials'. By A. Sala, Elsevier, 1986. 479 pp, $125. *Optics News*, 13(8):48–49, 1987.

567. W. D. Ross. Methods of representing radiation formulas. *Journal of the Optical Society of America*, 44(10):968–969, 1954.

568. S. Rosseland. Note on the absorption of radiation within a star. *Monthly Notices of the Royal Astronomical Society*, 84(7):525–528, 1924.

569. G.-C. Rota. Book reviews. L. Lewin, *Polylogarithms and Associated Funtions*, North-Holland, 1981, 359 pp. *Advances in Mathematics*, 59(2):184, 1986.

570. H. Rubens. Vérification de la formule du rayonnement de Planck dans le domaine des grandes longueurs d'onde. In P. Langevin and M. de Broglie, editors, *La théorie du Rayonnement et les Quanta*, pages 87–92. Gauthier-Villars, Paris, 1912.

571. H. Rubens and E. Aschkinass. Über die Eigenschaften der Reststrahlen des Steinsalzes. *Verhandlungen der Deutschen Physikalischen Gesellschaft zu Berlin*, 17(5):42–45, 1898.

572. H. Rubens and F. Kurlbaum. Über die Emission langwelliger Wärmestrahlen durch den schwarzen Körper bei verschiedenen Temperaturen. *Sitzungsberichte der Königlich Preussischen Akademie der Wissenschaften zu Berlin (Zweite Halbband)*, (41):929–941, 1900.

573. H. Rubens and F. Kurlbaum. Anwendung der Methode der Reststrahlen zur Prüfung des Strahlungsgesetzes. *Annalen der Physik* [Vierte Folge], 4(4):649–666, 1901.

574. H. Rubens and F. Kurlbaum. On the heat-radiation of long wave-length emitted by black bodies at different temperatures. *Astrophysical Journal*, 14:335–348, 1901.

575. H. Rubens and G. Michel. Beitrag zur Prüfung der Planckschen Strahlungsformel. *Sitzungsberichte der Preussischen Akademie der Wissenschaften (Zweite Halbband)*, (38):590–610, 1921.

576. H. Rubens and G. Michel. Prüfung der Planckschen Strahlungsformel. *Physikalische Zeitschrift*, 22:569–577, 1921.

577. H. Rubin. Walter M. Elsasser 1904–1991: a biographical memoir. In *Biographical Memoirs of the National Academy of Science*, volume 68, pages 103–165. National Academies Press, Washington, DC, 1995.

578. C. Runge. Über die Dimensionen physikalischer Größen. *Physikalische Zeitschrift*, 17:202–212, 1916.

579. G. A. W. Rutgers. Temperature radiation of solids. In S. Flügge, editor, *Encyclopedia of Physics*, volume 26, pages 129–170. Springer-Verlag, Berlin, 1958.

580. M. N. Saha and B. N. Srivastava. *A Treatise on Heat (including kinectic theory of gases, thermodynamics and recent advances in statistical thermodynamics)*. The Indian Press, Ltd., Allahabad, third edition, 1950.

581. A. Sala. *Radiant Properties of Materials: Tables of Radiant Values for Black Body and Real Materials*. Elsevier/PWN–Polish Scientific Publishers, Amsterdam/Warsaw, 1986.

582. J. Salpeter. Über ein Seitenstück zum Wienschen Verschiebungsgesetz. *Physikalische Zeitschrift*, 15(16):764–765, 1914.

583. H. E. Salzer and R. Zucker. Table of the zeros and weight factors of the first fifteen Laguerre polynomials. *Bulletin of the American Mathematical Society*, 55(10):1004–1012, 1949.

584. J. A. Sanderson. Emission, Transmission, and Detection of the Infrared. In A. S. Locke, editor, *Guidance*, pages 126–175. D. Van Nostrand Company, Inc., Princeton, New Jersey, 1955.

585. R. A. Sapozhnikov. Spectral distribution of radiant energy. *Soviet Physics Uspekhi*, 3(1):172–174, 1960.

586. S. L. Sargent. *A Compact Table of Blackbody Radiation Fractions*. Center for Climatic Research, University of Wisconsin, Madison, Wisconsin, 1972.

587. S. L. Sargent. A Compact Table of Blackbody Radiation Fractions. *Bulletin of the American Meteorological Society*, 53(4):360, 1972.

588. J. Satterly. Stefan's radiation law. *Nature*, 157(3996):737, 1946.

589. K. Schaum. *Photochemie und Photographie*. Verlag von Johann Ambrosius Barth, Leipzig, 1908.

590. A. Schleiermacher. Ueber die Abhängigkeit der Wärmestrahlung von der Temperatur und das Stefan'sche Gesetz. *Annalen der Physik und Chemie* [Neue Folge], 26(10):287–308, 1885.

591. R. Schulze. Zur graphischen Auswertung der Planckschen Strahlungsformel. *Licht-Technik*, 1(2):45–50, 1949.

592. S. E. Scrupski. Figure it out for yourself. *Electronics*, 49:118, 1976.

593. F. H. Seares and M. C. Joyner. Effective wave lengths of standard magnitudes; color temperatures and spectral type. *Astrophysical Journal*, 98(2):302–330, 1943.

594. SENSIAC. Infrared Radiance Calculator. http://www.sensiac.org, 2010.

595. Sensors, Inc. *Sensors, Inc. Infrared Radiation Calculator example problem guide*. Sensors, Inc., Ann Arbor, Michigan, 1975.

596. A. V. Sheklein. Some peculiarities of representation of the spectral distribution of solar radiation. *Applied Solar Energy*, 2(1):34–38, 1966.

597. A. Shipway and S. Shipway. Black Body Emission Max. http://www.calctool.org/CALC/phys/p_thermo/wien, 2015.

598. W. Shockley and H. J. Queisser. Detailed balance limit of efficiency of p-n junction solar cells. *Journal of Applied Physics*, 32(3):510–519, 1961.

599. W. A. Shurcliff. A logarithm-of-wavelength scale for use in absorption spectrophotometry. *Jouranl of the Optical Society of America*, 32(4):229–233, 1942.

600. R. Siegel and J. R. Howell. *Thermal Radiation Heat Transfer. Volume 1. The Blackbody, Electromagnetic Theory, and Material Properties*. National Aeronautics and Space Administration, Washington, DC, 1968.

601. R. Siegel and J. R. Howell. *Thermal Radiation Heat Transfer*. McGraw-Hill Book Company, New York, 1972.

602. R. Siegel and J. R. Howell. *Thermal Radiation Heat Transfer*. Hemisphere Publishing Corporation, Washington, second edition, 1981.

603. R. Siegel and J. R. Howell. *Thermal Radiation Heat Transfer*. Taylor & Francis, New York, fourth edition, 2002.

604. C. E. Siewert. An exact expression for the Wien displacement constant. *Journal of Quantitative Spectroscopy and Radiative Transfer*, 26(5):467, 1981.

605. J. F. Skogland. Tables of spectral energy distribution and luminosity for use in computing light transmissions and relative brightness from spectrophotometric data. *National Bureau of Standards. Miscellaneous Publications*, 86, 1929.

606. G. Slussareff. A reply to Max Reiss. *Journal of the Optical Society of America*, 36(12):707, 1946.

607. C. J. Smith. Some notes and suggestions on the teaching of physics. *The London, Edinburgh, and Dublin Philosophical Magazine and Journal of Science* [Seventh Series], 33(226):775–815, 1942.

608. P. A. S. Smith. Review of 'Radiant Properties of Materials: Tables of Radiant Values for Black Body and Real Materials.' By Aleksander Sala. Elsevier Scientific Publishers: Amsterdam and New York. 1984. xvi+479 pp. $125.00. ISBN 0-444-99599-4. *Journal of the American Chemical Society*, 109(23):7244, 1987.

609. R. Smith-Hughes. The F5100 black body radiation slide rule. *Slide Rule Gazette*, 10(2):75–80, 2009.

610. J. F. Snell. Radiometry and Photometry. In W. G. Driscoll, editor, *Handbook of Optics*, pages 1.1–1.30. McGraw-Hill Book Company, New York, 1978.

611. B. H. Soffer and D. K. Lynch. Some paradoxes, errors, and resolutions concerning the spectral optimization of human vision. *American Journal of Physics*, 67(11):946–953, 1999.

612. B. H. Soffer and D. K. Lynch. The spectral optimization of human vision: some paradoxes, errors and resolutions. In T. Asakura, editor, *International Trends in Optics and Photonics*, volume 74 of *Springer Series in Optical Science*, pages 390–405. Springer-Verlag, Berlin, 1999.

613. A. Sommerfeld. *Thermodynamik und Statistik. Vorlesungen über Theoretische Physik, Band 5*. Diederich'sche Verlagsbuchhandlung, Wiesbaden, 1952.

614. E. M. Sparrow and R. D. Cess. *Radiation Heat Transfer*. Hemisphere Publishing Corporation, New York, Augmented edition, 1978.

615. SpectralCal.com. Calculating Blackbody Radiance. http://www.spectralcalc.com, June 2010.

616. I. J. Spiro. Spectral Interval Conversion. *Optical Engineering*, 17(1):SR–3, 1978.

617. W. R. Stahl. Dimensional analysis in mathematical biology I. General discussion. *The Bulletin of Mathematical Biophysics*, 23(4):355–376, 1961.

618. R. Stair and W. O. Smith. A tungsten-in-quartz lamp and its applications in photoelectric radiometry. *Journal of Research of the National Bureau of Standards*, 30(6):449–459, 1943.

619. J. Stefan. Über die Beziehung zwischen der Wärmestrahlung und der Temperatur. *Sitzungsberichte der Mathematisch-Naturwissenschaftlichen Classe der Kaiserlichen Akademie der Wissenschaften* (Wien), 79, II:391–428, 1879.

620. R. Steiner. History and progress on accurate measurements of the Planck constant. *Reports on Progress in Physics*, 76(1):016101, 2013.

621. E. Steinke. Über eine lichtelektrische Methode zur Prüfung des Wien–Planckschen Strahlungsgesetzes im Bereich ultravioletter Strahlung. *Zeitschrift für Physik*, 11(1):215–238, 1922.

622. C. P. Steinmetz. Transformation of electric power into light. *Electric Review*, 49(23):924–926, 1906.

623. C. P. Steinmetz. *Radiation, Light and Illumination: A Series of Engineering Lectures Delivered at Union College*. McGraw-Hill Book Company, New York, 1909.

624. J. Stern. Radiation slide rule. *Science*, 137(3528):439, 1962.

625. G. T. Stevenson. *Black-body Radiation Functions*. US Naval Ordnance Test Station, China Lake, California, September 1961.

626. G. T. Stevenson. *Black-body Radiation Functions*. US Naval Ordnance Test Station, China Lake, California, second printing, May 1963.

627. S. M. Stewart. Terahertzing visible light. *American Journal of Physics*, 79(8):797, 2011.

628. S. M. Stewart. Wien peaks and the Lambert W function. *Revista Brasileira de Ensino de Física*, 33(3):3308, 2011.

629. S. M. Stewart. Blackbody Radiation Functions and Polylogarithms. *Journal of Quantitative Spectroscopy and Radiative Transfer*, 113(3):232–238, 2012.

630. S. M. Stewart. Spectral peaks and Wien's displacement law. *Journal of Thermophysics and Heat Transfer*, 26(4):689–692, 2012.

631. S. M. Stewart. The Aristo System Czerny slide rule for thermal radiation calculations. *Journal of the Oughtred Society*, 22(1):16–24, 2013.

632. S. M. Stewart. A note on the general class of blackbody fractional functions. *Infrared Physics and Technology*, 67:338–340, 2014.

633. S. M. Stewart. Analysis of the relative merits of the 3–5 µm and the 8–12 µm spectral bands using detected thermal contrast. In S.-J. Hsieh and J. N. Zalameda, editors, *Thermosense: Thermal Infrared Applications XXXVII*, volume 9485 of *Proceeding of SPIE*, pages 9485–47, SPIE, Bellingham, WA, 2015.

634. S. M. Stewart and R. B. Johnson. Exact expressions for thermal contrast detected with thermal and quantum detectors. In D. A. Huckridge and R. Ebert, editors, *Electro-Optical and Infrared Systems: Technology and Applications XI*, volume 9249 of *Proceeding of SPIE*, page 92490D, SPIE, Bellingham, WA, 2014.

635. A. M. Stone. Infrared research in Germany. *European Scientific Notes*, 1(1):10–13, 1947.

636. G. J. Stoney. On the advantage of referring the positions of lines in the spectrum to a scale of wave-numbers. In *Report of the Forty-First Meeting of the British Association for the Advancement of Science*, pages 42–43. John Murray, London, 1872.

637. L. Strum. Über eine mögliche Verallgemeinerung der Planckschen Strahlungsformel. *Zeitschrift f'ur Physik*, 51(3-4):287–291, 1928.

638. M. Suzuki, M. Habu, and T. Nagasaka. Tables of blackbody radiation func-

tions for the ultraviolet, visible and infrared regions. *Researches of the Electrotechnical Laboratory* (Tokyo), 715, 1970.

639. T. Szirtes. *Applied Dimensional Analysis and Modeling.* Butterworth-Heinemann, Burlington, MA, 2011.

640. C. B. Tanner and S. M. Robinson. *Black-body function σT^4.* Soils Bulletin 1, Department of Soils, College of Agriculture, University of Wisconsin, Madison 6, Wisconsin, January 1959.

641. B. N. Taylor, W. H. Parker, and D. N. Langenberg. Determination of e/h, using macroscopic quantum phase coherence in superconductors: implications for quantum electrodynamics and the fundamental physical constants. *Reviews of Modern Physics*, 41(3):375–496, 1969.

642. M. Taylor, A. I. Díaz, L. A. Jódar Sánchez, and R. J. Villanueva Micó. A matrix generalisation of dimensional analysis: new similarity transforms to address the problem of uniqueness. *Advanced Studies in Theoretical Physics*, 2(20):979–995, 2008.

643. K. Terada. Tables and graphs of the black body radiation for meteorological use. *The Geophysical Magazine* (Tokyo), 13:137–143, 1939.

644. M. Thiesen. Über das Gesetz der schwarzen Strahlung. *Verhandlungen der Deutschen Physikalischen Gesellschaft*, 2(5):65–70, 1900.

645. L. H. Thomas. M. W. Makowski, Slide rule for radiation calculations, *Rev. Sci. Instruments*, v. 20, 1949, p. 876–884. *Mathematical Tables and Other Aids to Computation*, 4(31):176, 1950.

646. E. C. Titchmarsh. *The Theory of Functions.* Oxford University Press, London, 1932.

647. D. C. Todd. *Blackbody Radiation, Photon Emission, and the Calculation of Debye Functions.* Air Force Systems Command, Arnold Air Force Station, Tennessee, December 1968.

648. L. Tonelli. Sull'integrazione per parti. *Atti della Accademia Nazionale dei Lincei (5)*, 18(2):246–253, 1909.

649. L. T. Troland. Report of committee on colorimetry for 1920–21. *Journal of the Optical Society of America*, 6(6):527–596, 1922.

650. S. A. Twomey. Table of the Planck function for terrestial temperatures. *Infrared Physics*, 3(1):9–26, 1963.

651. J. Tyndall. Ueber leuchtende und dunkle Strahlung. *Annalen der Physik und Chemie* [Zweite Folge], 124(1):36–53, 1865.

652. S. R. Valluri, D. J. Jeffrey, and R. M. Corless. Some applications of the Lambert W function to physics. *Canadian Journal of Physics*, 78(9):823–831, 2000.

653. W. van der Bijl. The maximum of a distribution- or spectrum-function. *Nature*, 178(4535):691, 1956.

654. E. Vandekerkhove. Tables et graphiques concernant les fonctions de Planck, Stefan et Wien. *Communications de l'Observatoire Royal de Belgique*, 52, 1953.

655. A. Vaschy. Sur les lois de similitude en physique. *Annales Télégraphiques* (3e série), 19(1):25–28, 1892.

656. A. Vial. Fall with linear drag and Wien's displacement law: approximate solution and Lambert function. *European Journal of Physics*, 33(4):751–755, 2012.

657. J. D. Vincent. Radiometric integrals in closed form. *Applied Optics*, 10(11):2546–2547, 1971.

658. M. von Ardenne. *Tabellen der Elektronenphysik, Ionenphysik und Übermikroskopie*. VEB Deutscher Verlag der Wissenschaften, Berlin, 1956.

659. C. F. von Gauss. *Dioptrische Untersuchungen*. Druck und Verlag der Dieterichschen Buchhandlung, Göttingen, 1841.

660. F. Wachendorf. The condition of equal irradiance and the distribution of light in images formed by optical systems without artifical vignetting. *Journal of the Optical Society of America*, 43(12):1205–1208, 1953.

661. G. Wald. Frequency or wavelength? *Science*, 150(3701):1239–1240, 1965.

662. G. Waldman and J. Wootton. *Electro-optical Systems Performance Modeling*. Artech House, Boston, 1993.

663. G. W. Walker. A suggestion as to the origin of black body radiation. *Proceedings of the Royal Society of London. Series A, Containing Papers of a Mathematical and Physical Character*, 89(612):393–398, 1914.

664. G. W. Walker. On the formula for black body radiation. *Proceedings of the Royal Society of London. Series A, Containing Papers of a Mathematical and Physical Character*, 90(615):46–49, 1914.

665. R. G. Walker. *Tables of the Blackbody Radiation Functions for Wavenumber Calculations*. Air Force Cambridge Research Laboratories, L. G. Hanscom Field, Massachusetts, September 1962.

666. J. Walsh. ONR London: two decades of scientific quid pro quo. *Science*, 154(3749):623–625, 1966.

667. J. W. T. Walsh. *Photometry*. Constable & Company Ltd, London, 1926.

668. J. W. T. Walsh. *Photometry*. Constable & Company Ltd, London, third edition, 1958.

669. A. Walther and A.-W. Kron. Nomographie und Rechenschieber. In A. Walther, editor, *Applied Mathematics. Part 1* (FIAT Review of German Science 1939–1946), pages 119–127. Dieterich'sche Verlagbuchhandlung, Wiesbaden, 1948.

670. Y. F. Wang and J. X. Mao. Fast and easy integration of Planck function with MATLAB. *Infrared* (Monthly), 29(4):12–14, 2008.

671. H. Wanner. Photometrische Messungen der Strahlung schwarzer körper. *Annalen der Physik* [Vierte Folge], 2(5):141–157, 1900.

672. E. Warburg. Vérification expérimentale de la formule de Planck pour le rayonnement du corps noir. In P. Langevin and M. de Broglie, editors, *La théorie du rayonnement et les quanta*, pages 78–86. Gauthier-Villars, Paris, 1912.

673. E. Warburg, G. Leithäuser, E. Hupka, and C. Müller. Über die Konstante c des Wien–Planckschen Strahlungsgesetzes. *Annalen der Physik* [Vierte Folge], 40(4):609–634, 1913.

674. E. Warburg and C. Müller. Über die Konstante c des Wien–Planckschen Strahlungsgesetzes. *Annalen der Physik* [Fourth Series], 48(19):410–432, 1915.

675. M. E. Warga. European Scientific Notes. *Journal of the Optical Society of America*, 65(4):476, 1975.

676. J. Watson. Review of 'Electro-optical Systems Performance Modeling' G. Waldman and J. Wootton. Artech House, 1993, ISBN 0890-065411, pp 241 + xiv, $85. *Optics and Laser Technology*, 25(4):273, 1993.

677. H. F. Weber. Untersuchungen über die Strahlung fester Körper. *Sitzungs-berichte der Königlich-Preußischen Akademie der Wissenschaften zu Berlin* (Zweiter Halbband), (37):933–957, 1888.

678. E. W. Weisstein. *CRC Encyclopedia of Mathematics.* CRC Press, Boca Raton, Florida, third edition, 2009.

679. H. T. Wensel. International temperature scale and some related physical constants. *Journal of Research of the National Bureau of Standards*, 22(4):375–395, 1939.

680. J. R. White. private communication, 2012.

681. W. K. Widger, Jr. Some further aspects concerning the Wien displacement law. *Bulletin of the American Meteorological Society*, 49(12):1142–1143, 1968.

682. W. K. Widger, Jr. and M. P. Woodall. Integration of the Planck blackbody radiation function. *Bulletin of the American Meteorological Society*, 57(10):1217–1219, 1976.

683. J. A. Wiebelt. *Engineering Radiation Heat Transfer.* Holt, Rinehart and Winston, New York, 1966.

684. W. Wien. Eine neue Beziehung der Strahlung schwarzer Körper zum zweiten Hauptsatz der Wärmetheorie. *Sitzungsbericht der Königlich Preussischen Akademie der Wissenschaften zu Berlin*, pages 55–62, 1893.

685. W. Wien. Die obere Grenze der Wellenlängen, welche in der Wärmestrahlung fester Körper vorkommen können; Folgerungen aus dem zweiten Hauptsatz der Wärmetheorie. *Annalen der Physik und Chemie* [Neue Folge], 49(8):633–641, 1893.

686. W. Wien. Temperatur und Entropie der Strahlung. *Annalen der Physik und Chemie* [Neue Folge], 52(5):132–165, 1894.

687. W. Wien. Über die Energieverteilung im Emissionsspektrum eines schwarzen Körpers. *Annalen der Physik und Chemie* [Neue Folge], 58(8):662–669, 1896.

688. W. Wien. On the division of energy in the emission-spectrum of a black body. *The London, Edinburgh, and Dublin Philosophical Magazine and Journal of Science* [Fifth Series], 43(262):214–220, 1897.

689. W. Wien. Theorie der Strahlung. In A. Sommerfeld, editor, *Encyklopädie der Mathematischen Wissenschaften mit Einschluss ihrer Anwendungen, V: Physik,* pages 282–357. Vieweg+Teubner Verlag, Leipzig, 1909.

690. W. Wien and O. Lummer. Methode zur Prüfung des Strahlungsgesetzes absolut schwarzer Körper. *Annalen der Physik und Chemie* [Neue Folge], 56(11):451–456, 1895.

691. H. M. Wiesbaden. Marianus Czerny. In K. Bethge and H. Klein, editors, *Physiker und Astronomen an der Johann Wolfgang Goethe-Universität am Main,* pages 144–169. Hermann Luchterhand Verlag, Frankfurt, 1989.

692. A. O. Williams. Review of: 'Miscellaneous Physical Tables. Planck's Radiation Functions and Electronic Functions.' New York, Work Projects Administration, 1941. 7+61 pp. $1.50. *Bulletin of the American Mathematical Society*, 51(5):358–359, 1945.

693. B. W. Williams. A specific mathematical form for Wien's displacement law as $\nu_{max}/T = $ contant. *Journal of Chemical Education*, 91(5):623, 2014.

694. D. Wilson. Solar system science: 1961 literature survey, part III book notes: a selected list. *Icarus*, 1(1):286–295, 1962.

695. W. E. Wilson. Radiation from a perfect radiator. *Astrophysical Journal*, 10:80–86, 1899.

696. W. L. Wolfe. Radiation Theory. In W. L. Wolfe, editor, *Handbook of Military Infrared Technology*, pages 3–30. Office of Naval Research, Department of the Navy, Washington, DC, 1965.

697. W. L. Wolfe. Radiation Theory. In W. L. Wolfe and G. J. Zissis, editors, *The Infrared Handbook*, pages 1-27–1-28. Office of Naval Research, Department of the Navy, Washington, DC, 1978.

698. W. L. Wolfe. Radiometry. In R. R. Shannon and J. C. Wyany, editors, *Applied Optics and Optical Engineering*, volume 8, pages 117–170. Academic Press, New York, 1980.

699. W. L. Wolfe. Radiation Theory. In W. L. Wolfe and G. J. Zissis, editors, *The Infrared Handbook – Revised Edition*, pages 1–17. Office of Naval Research, Department of the Navy, Washington, DC, 1989 (third printing).

700. W. L. Wolfe. Radiation Theory. In G. J. Zissis, editor, *The Infrared and Electro-optical Systems Handbook. Volume 1. Sources of radiation*, pages 1–48. SPIE Optical Engineering Press, Bellingham, Washington, 1993.

701. W. L. Wolfe. *Introduction to Infrared System Design*. SPIE Optical Engineering Press, Bellingham, Washington, 1996.

702. W. L. Wolfe. *Introduction to Radiometry*. SPIE Press, Bellingham, 1998.

703. W. L. Wolfe. private communication, 2012.

704. W. L. Wolfe. Famous last words. In M. G. Turner, editor, *Tribute to William Wolfe*, volume 8483 of *Proceeding of SPIE*, page 84830F, SPIE, Bellingham, WA, 2012.

705. W. L. Wolfe and P. W. Kruse. *Thermal Detectors*, volume II of *Handbook of Optics*, chapter 28. McGraw Hill, New York, third edition, 2010.

706. D. Wood. *The Computation of Polylogarithms*. Technical Report No. 15-92*, University of Kent, Computing Laboratory, 1992.

707. R. W. Wood. *Physical Optics*. The Macmillan Company, New York, 1911.

708. A. G. Worthing. Radiation laws describing the emission of photons by black bodies. *Journal of the Optical Society of America*, 28(5):176, 1938.

709. A. G. Worthing. Radiation laws describing the emission of photons by black bodies. *Journal of the Optical Society of America*, 29(2):97–100, 1939.

710. A. G. Worthing. New λT relations for black body radiation. *Journal of the Optical Society of America*, 29(2):101–102, 1939.

711. E. M. Wright. A non-linear difference-differential equation. *Journal für die reine und angewandte Mathematik*, 194:66–87, 1955.

712. E. M. Wright. Solution of the equation $ze^z = a$. *Proceedings of the Royal Society of Edinburgh. Section A. Mathematical and Physical Sciences*, 65(2):193–203, 1959.

713. Wright Instruments, Inc. *A survey for NOL of high altitude atmospheric temperature sensors and associated problems. Performed under Navy contract N-60921-6136 for the Naval Ordnance Laboratory (code LE) White Oak, Silver Springs, Maryland. Final report*. Vestal, New York, 1961.

714. K. Wurm. *Die Planetarischen Nebel*. Akademie-Verlag, Berlin, 1951.

715. C. L. Wyatt. *Radiometric Calibration: Theory and Methods*. Academic Press, New York, 1978.

716. C. L. Wyatt. *Radiometric System Design*. MacMillan, New York, 1987.

717. G. Wyszecki and W. S. Stiles. *Color Science. Concepts and methods, quantitative data and formulas*. John Wiley & Sons, Inc., New York, 1967.

718. X. Xu, X. Wang, Z. Mi, W. Zhong, and X. Dai. Some new expressions of Planck integrals in blackbody radiation. *Chinese Journal of Infrared Research*, 7A(1):7–13, 1988.

719. Z. Yamauti and M. Okamatu. Concise table of Planck's radiation function. *Journal of the Illuminating Engineering Institute of Japan*, 20(7):5–8, 1936.

720. Z. Yamauti and M. Okamatu. Tables of Planck's formula of radiation (1). *Researches of the Electrotechnical Laboratory* (Tokyo), 395, 1936.

721. Z. Yamauti and M. Okamatu. Tables of Planck's formula of radiation (II). Tables for specified wave-length. *Researches of the Electrotechnical Laboratory* (Tokyo), 402, 1937.

722. Z. Yamauti and M. Okamatu. Tables of Planck's formula of radiation for specified wavelength. *Journal of the Illuminating Engineering Institute of Japan*, 21(2):17–24, 1937.

723. N. O. Young. A new spectrometric resolution chart. *Journal of the Optical Society of America*, 51(2):vi, 1961.

724. N. O. Young. A new radiation calculator – and a note on the visual color sensitivity peak. *Journal of the Optical Society of America*, 52(1):viii, 1962.

725. F. Yu. Review of 'Table of Values for Blackbody Radiation' by H. Zhu, X. Liu, Q. Zheng, C. Yang and F. Yu, Science Press, Beijing, 1984. *Chinese Journal of Infrared Research*, 3(3):240, 1984.

726. D. Zagier. The remarkable dilogarithm. *Journal of Mathematical and Physical Sciences*, 22(1):131–145, 1988.

727. E. F. Zalewski. *Radiometry and Photometry*, volume II of *Handbook of Optics*, chapter 34. McGraw Hill, New York, third edition, 2010.

728. A. Zanker. Monochromatic emissive power of black body found quickly by nomograph. *Optik*, 49(4):409–412, 1978.

729. H. Zanstra. An application of the quantum theory to the luminosity of diffuse nebulae. *Astrophysical Journal*, 65(1):50–70, 1927.

730. H. Zanstra. Luminosity of planetary nebulae and stellar temperatures. *Publications of the Dominion Astrophysical Observatory* (Victoria, BC), 4(15):205–260, 1931.

731. H. Zanstra. Untersuchungen über planetarische Nebel. Erster Teil. Der Leuchtprozeß planetarischer Nebel und die Temperatur der Zentralsterne. *Zeitschrift für Astrophysik*, 2(1):1–29, 1931.

732. D. J. Zhang. Some new constants in infrared physics and new function table of $f(\lambda T)$, $F(\lambda T)$ of blackbody. *Journal of Harbin University of Science and Technology*, 10(2):115–124, 1986.

733. D. J. Zhang, X. Chen, and M. Li. Some new constants in infrared physics and new function table of $f(\lambda T)$, $F(\lambda T)$ of blackbody (II). *Infrared Technology*, 8(6):263–269, 1986.

734. Z. M. Zhang and X. J. Wang. Unified Wien's displacement law in terms of logarithmic frequency or wavelength scale. *Journal of Thermophysics and Heat Transfer*, 24(1):222–224, 2010.

735. H. Zhu, X. Liu, Q. Zheng, C. Yang, and F. Yu. *Table of Values for Blackbody Radiation*. Science Press, Beijing, 1984.

736. G. J. Zissis. Fundamentals of infrared – A review. In I. J. Spiro, editor, *Modern Utilization of Infrared Technology I*, volume 62 of *Proceedings of SPIE*, pages 67–94, 1975.

737. C. J. Willers. *Electro-Optical System Analysis and Design: A Radiometry Perspective*. SPIE Press, Bellingham, WA, 2013.

Reference Author Index

Numbers that appear in square brackets refer to the citation number in the list of References.

Charlier, C. V. L. [132], 29
Chen, J. J. [133], 75
Chernoch, J. P. [134], 164
Chiu, W.-C. [135], 56–58
Chiu, W.-C. [136], 56
Christy, J. W. [137], 132
Clark, B. A. [138], 70, 94
Clark, B. A. [139], 70, 115
Cleveland, L. F. [140], 260, 299
Coblentz, W. W. [141], 250
Cohen, E. R. [142], 265, 278
Cohen, E. R. [143], 295
Corless, R. M. [144], 41
Corless, R. M. [145], 40, 41, 93
Corless, R. M. [146], 40
Corrsin, S. [147], 18
Cox, A. N. [148], 272
Crooker, A. M. [149], 221
Crova, A. [150], 178
Császár, E. [151], 11, 16
Császár, E. [152], 29
Császár, E. [153], 29
Curtis, W. D. [154], 18
Cussen, A. J. [155], 230
Cussen, A. J. [156], 231
Cvijović, D. [157], 37
Czerny, M. [158], 184
Czerny, M. [159], 172
Czerny, M. [160], 69, 73, 285
Czerny, M. [161], 75, 285
Czerny, M. [162], 285
Czerny, M. [163], 286
Czerny, M. [164], 69, 75, 284

Dai, J. [165], 295
Dai, X. [166], 111, 115, 295
D'Ans, J. [167], 159, 263
D'Ans, J. [168], 159, 263
Darwin, C. G. [169], 52
Das, R. [170], 46
Davies, C. N. [171], 28
Davis, R. [172], 251
Day, A. L. [173], 4, 27
de A. Martins, R. [174], 30
de Beauclair, W. [175], 284
de Boissoudy, J. [176], 29
de Boissoudy, J. [177], 29
de Boissoudy, J. [178], 29
de Bruijn, N. G. [179], 41
de Groot, W. [180], 254
de Vos, J. C. [181], 269
de Wiveleslie Abney, W. [182], 48
DeBell, A. G. [183], 281
Debye, P. [184], 115
Dereniak, E. L. [185], 131
Dey, S. [186], 127

Doane, D. P. [187], 85
Dorfman, J. G. [188], 28
Dougal, R. C. [189], 28
Draper, J. W. [190], 34
Drawin, H.-W. [191], 297
Drew, L. A. [192], 273
Driver, B. K. [193], 41
Drude, P. [194], 127
Drysdale, C. V. [195], 94, 118, 123
DuMond, J. W. M. [196], 268
DuMond, J. W. M. [197], 197, 263
Dunkle, R. V. [198], 263
Dunklee, R. V. [199], 216

Eaton, J. W. [200], 128
Eaton, J. W. [201], 128
Eckweiler, H. J. [202], 215
Edwards, D. K. [203], 69, 75, 285
Ehrenfest, P. [205], 18
Ehrenfest, P. [206], 18
Ehrenfest, P. [207], 100
Einstein, A. [208], 23
Eisner, L. [209], 197, 225
Elsasser, W. M. [211], 264
Elsasser, W. M. [212], 264
Emmons, R. B. [213], 111, 115, 116
Epstein, P. S. [214], 19
Erminy, D. E. [215], 100, 103
Estey, R. S. [216], 250
Euler, J. [217], 297
Euler, L. [218], 36
Euler, L. [219], 41
Evans, J. H. [220], 18

Fabry, C. [221], 252
Fastie, W. G. [222], 194
Federman, A. K. [223], 29
Feibelman, W. A. [224], 178
Ferriso, C. C. [225], 278
Flynn, D. S. [226], 304
Foerst, W. [227], 189
Foitzik, L. [228], 52
Foote, P. D. [229], 77, 78
Foote, P. D. [230], 77, 78
Forsythe, W. E. [231], 250
Forsythe, W. E. [232], 268
Fourier, J. B. J. [233], 29
Fowle, F. E. [234], 248
Fowle, F. E. [235], 248
Fowle, F. E. [236], 249
Fowle, F. E. [237], 249
Fowle, F. E. [238], 249
Franklin, A. [239], 12, 29
Frehafer, M. K. [240], 251
Frehafer, M. K. [241], 175, 251
Fritsch, F. N. [242], 41

Moon, P. [459], 259
Moon, P. [460], 266
Moon, P. [461], 101, 104

Nagel, M. R. [462], 286
Nichols, E. L. [463], 94
Nichols, E. L. [464], 94
Nicodemus, F. E. [465], 136, 154
Nyswander, R. E. [467], 94

Olver, F. W. J. [468], 37, 40, 116
Omoto, Y. [469], 170, 255
Overduin, J. M. [470], 56
Özişik, M. N. [471], 69, 285

Páez, G. [472], 127
Páez, G. [473], 127
Páez, G. [474], 127
Palacios, J. [475], 19, 29
Palmer, J. M. [476], 131, 199
Partington, J. R. [477], 28
Partington, J. R. [478], 127
Paschen, F. [479], 8, 14
Paschen, F. [480], 8, 14
Paschen, F. [481], 8, 9, 14, 50
Paschen, F. [482], 14, 51
Paschen, F. [483], 8, 14
Paschen, F. [484], 8, 14
Paschen, F. [485], 14
Paschen, F. [486], 14
Paschen, F. [487], 8, 16
Paschen, F. [488], 16
Paschen, F. [489], 14
Paschen, F. [490], 14
Paterson, C. C. [491], 27
Pavelle, R. [492], 108, 238
Pavelle, R. [493], 108, 238
Penner, S. S. [494], 164
Pepperhoff, W. [495], 170, 178, 297
Peters, S. [496], 305
Peters, S. [497], 305
Picolet, L. E. [498], 178
Pinson, L. J. [499], 199, 220, 298
Pisa, E. J. [500], 277
Pivovonsky, M. [501], 111, 115, 231, 286
Pivovonsky, M. [502], 288
Planck, M. [503], 9, 14
Planck, M. [504], 9, 14
Planck, M. [505], 9, 14
Planck, M. [506], 9, 14
Planck, M. [507], 9, 14, 52, 98
Planck, M. [508], 9, 14, 247
Planck, M. [509], 9, 14
Planck, M. [510], 11
Planck, M. [511], 11, 16, 41
Planck, M. [512], 11, 41

Planck, M. [513], 11
Planck, M. [514], 11
Planck, M. [515], 11
Planck, M. [516], 52
Planck, M. [517], 52
Planck, M. [518], 11, 41
Planck, M. [519], 11, 41
Planck, M. [520], 29
Plendl, J. N. [521], 78
Pobedrya, B. E. [522], 18
Pocklington, H. C. [523], 24
Pólya, G. [524], 41
Pompea, S. M. [525], 133
Pontryagin, L. S. [526], 41
Pontryagin, L. S. [527], 41
Preston, J. S. [528], 288
Preston, T. [529], 127
Pricst, I. G. [530], 29
Priest, I. G. [531], 29
Priest, I. G. [532], 29
Pringsheim, E. [533], 16
Pringsheim, E. [534], 16
Pringsheim, E. [535], 26

Queisser, H. J. [536], 210
Queisser, H. J. [537], 210
Queisser, H. J. [538], 210
Quinn, G. C. [539], 197

Rabinowitz, P. [540], 122
Rathore, M. M. [541], 298
Rayleigh, Lord [542], 29
Rayleigh, Lord [543], 29
Rayleigh, Lord [544], 35, 57
Rayleigh, Lord [545], 5
Rayleigh, Lord [546], 29
Rayleigh, Lord [547], 8
Rayleigh, Lord [548], 11, 14, 98, 99
Rayleigh, Lord [549], 99
Rayleigh, Lord [550], 99
Rayleigh, Lord [551], 29
Reiche, F. [552], 19
Reid, C. D. [553], 221
Reiss, M. [554], 146
Reiss, M. [555], 150
Reynolds, M. M. [556], 275
Riabouchinsky, D. [557], 29
Ribando, R. J. [558], 299
Richardson, O. W. [559], 52, 127
Riedl, M. J. [560], 227
Riemann, B. [561], 64
Riley, K. F. [562], 86
Risch, R. H. [563], 108
Roberts, J. [564], 283, 296
Robinson, S. M. [565], 273
Roessler, D. M. [566], 296

Index

Printed and bound by CPI Group (UK) Ltd, Croydon, CR0 4YY

01/11/2024

01782619-0013